Entropy Randomization in Machine Learning

This book presents a new approach to machine learning—entropy randomization—to obtain optimal solutions under uncertainty (uncertain data and models of the objects under study). Randomized machine-learning procedures involve models with random parameters and maximum entropy estimates of the probability density functions of the model parameters under balance conditions with measured data. Optimality conditions are derived in the form of nonlinear equations with integral components. A new numerical random search method is developed for solving these equations in a probabilistic sense. Along with the theoretical foundations of randomized machine learning, **Entropy Randomization in Machine Learning** considers several applications to binary classification, modelling the dynamics of the Earth's population, predicting seasonal electric load fluctuations of power supply systems, and forecasting the thermokarst lakes area in Western Siberia.

Features

- A systematic presentation of the randomized machine-learning problem: from data processing, through structuring randomized models and algorithmic procedure, to the solution of application-relevant problems in different fields

- Provides new numerical methods for random global optimization and computation of multidimensional integrals

- A universal algorithm for randomized machine learning

This book will appeal to undergraduates and postgraduates specializing in artificial intelligence and machine learning, researchers and engineers involved in the development of applied machine learning systems, and researchers of forecasting problems in various fields.

Chapman & Hall/CRC Machine Learning & Pattern Recognition

For more information on this series please visit: https://www.routledge.com/ Chapman--HallCRC-Machine-Learning-Pattern-Recognition/book-series/ CRCMACLEAPAT

Entropy Randomization in Machine Learning

Yuri S. Popkov
Alexey Yu. Popkov
Yuri A. Dubnov

(Translated into English by Alexander Yu. Mazurov)

CRC Press
Taylor & Francis Group
Boca Raton London New York

CRC Press is an imprint of the
Taylor & Francis Group, an **informa** business

A CHAPMAN & HALL BOOK

First edition published 2023
by CRC Press
6000 Broken Sound Parkway NW, Suite 300, Boca Raton, FL 33487-2742

and by CRC Press
4 Park Square, Milton Park, Abingdon, Oxon, OX14 4RN

CRC Press is an imprint of Taylor & Francis Group, LLC

LCCN no. 2022007067

ISBN: 978-1-032-30628-5 (hbk)
ISBN: 978-1-032-30774-9 (pbk)
ISBN: 978-1-003-30656-6 (ebk)

DOI: 10.1201/9781003306566

Typeset in LM Roman
by KnowledgeWorks Global Ltd.

Publisher's note: This book has been prepared from camera-ready copy provided by the authors.

Contents

Preface

A fundamentally important attribute of civilization's development consists in accumulation of knowledge, with further transformation into solutions to extract new knowledge. This transformation is implemented in different ways, including knowledge extraction from accumulated data. A *model* with quantitatively measured characteristics serves as a basic tool to accumulate knowledge. However, besides quantitative (digital) data, there may exist non-quantitative ones, e.g., audio, video, graphical, text, and other data. In the final analysis, the problem of knowledge extraction and accumulation is reduced to the knowledge about the model that is formalized by *estimating* its characteristics. This process can be interpreted as *learning* of the model using data. Since model learning is performed by data processing using computing machines, the conceptual difference between the model and its computer implementation vanishes accordingly, causing a natural belief that the object of learning is computing machines. Such an approach explains the appearance of *Machine Learning (ML)*, a term and even a science that proceeds from the theoretical grounds of mathematical statistics, optimization, numerical methods and information technology.

As a rule, the general problem of machine learning is broken down into three classes of problems as follows: classification (the allocation of objects to appropriate warehouses) clustering (the grouping of more or less homogeneous objects) and dynamic regression (the restoration of relations between cause data and effect data). A crucial stage of machine learning is the *prediction* of possible changes in the solutions of these problems. According to modern understanding, machine learning yields "learned" deterministic models that are equipped with empirical probabilistic reliability estimates. As a result, the matter concerns empirically probabilistic machine learning.

This book develops a new branch of machine learning called *Randomized Machine Learning (RML)*,[1] which aims at generating the *ensemble* of entropy-learned randomized models. We should bear in mind that machine learning procedures are applied to the problems with a rather high level of uncertainty (imperfectly reliable data, incomplete knowledge about modeled processes, etc.). Hence, transition to the entropy-randomized concept of machine learning can be a fruitful and efficient tool for solving applied problems.

Practical solution of these problems requires a tool kit with integrated mathematical and algorithmic methods (on the one hand) and software utilities oriented to high-performance computing systems (on the other hand). The resources and capabilities of computer machines are doubled annually. As a result, huge amounts of data are accumulated and stored, both in natural and digital formats, for example, the currently accumulated data amount to billions petaflops. There are large data centers intended to accumulate and store these data over very long periods of time. What is to be done with accumulated data besides storage? The scientific community is endeavoring to outline some ways to process and use these data.

One possible way was formulated in 1991 by G. Piatetsky-Shapiro as the concept of *Data Mining (DM)*; see [55, 191]. Within this concept it was suggested to integrate different methods of mathematical statistics and probability theory in a user-friendly information technology environment, whereas the statement of the fundamental problems associated with qualitative and, perhaps, quantitative descriptions of knowledge and adequate data arrays was left to a researcher. The pioneering data mining experiments discovered a whole bunch of negative features of data (errors, reliability, adequacy, use of different scales, temporal irregularity and others) that casted doubt on the resulting knowledge. These features become more essential for large amounts of heterogeneous data. The verification of such data in the wide sense turns out to be a serious challenge.

The data problem magnetized a part of the scientific community so much that the first volume of *Journal of Data Science* appeared in 2002, announcing a new branch of science called *Data Science (DS)*. This branch was intended to absorb *all* sciences related to data usage and processing.

In 2008, *Nature* editor C. Lynch prepared a special issue [28] devoted to the problem of large amounts of data, formulating the concept of

---

[1] Throughout the book, the term "randomization" means an *artificial* assignment of desired stochastic properties to a certain object.

Big Data (BD). It was expected to develop an information technology environment with processing methods for the large arrays of numerical, audio, video and text data.

The physical core of all concepts involving data is a computing machine implementing computational algorithms for data processing and knowledge extraction. The necessary tool kit forms the subject of the ML concept, which is based on the methods and algorithms of probability theory and statistics, optimization, automata theory and discrete mathematics, theory of dynamic systems, and more. Even this short list of related disciplines shows that the ML concept is to create a synthetic tool kit for solving data processing problems.

ML has a history of over 60 years and a rich experience in solution of many problems. The first publication on the subject dates back to 1957 when F. Rosenblatt created the MARK I perceptron; see [156]. The notion of empirical risk, which is the basis for the *ML* procedures, was introduced by Ya. Tsypkin in the monograph [176]. A fundamental tool to form empirical risk is the so-called *Vapnik–Chervonenkis dimension* (*VC dimension*) suggested by the authors in [179]. The method of potential functions for the classification and pattern recognition problems was published in 1970; see [12]. We also refer to the monographs [21, 57] popular in the ML community.

The modern *ML* concept employs the *deterministic* parametrization of models and estimates using data arrays with hypothetical properties. The quality of estimation is characterized by empirical risk functions, and minimization yields the corresponding optimal estimates in terms of the chosen empirical risk; the details can be found in [180].

Generally, real problems solved by *ML* procedures are immersed in an *uncertain environment*. If the matter concerns data, they are obtained with errors, omissions and dubious validity. Model design and parametrization is a subjective non-formalizable process that depends on the individual knowledge of a researcher. Therefore, a mass application of *ML* procedures causes a rather high uncertainty. Empirical risk minimization gives parameter estimates for an existing data array within an adopted parameterized model; in other words, these estimates are conditional. There is no clarity about their behavior with other data arrays or within another parametrization model.

Due to the above-mentioned circumstances, it is necessary to simulate the uncertainty somehow. An approach involves *randomized* uncertainty simulation, which stems from a randomized representation of model parameters and data measurement errors. This means that model characteristics are assumed to be random whereas data are assumed to contain random errors. In contrast to the existing ML procedures, such randomized machine learning (RML) procedures give best estimates for the probability density functions (PDFs) of the random parameters and measurement noises under maximum uncertainty (entropy).

This book suggests using **information entropy** to characterize the quality of learning under uncertain conditions. In a series of works [80, 81, 157], E. Jaynes demonstrated two principal properties of entropy that were reflected by the RML procedures. First, by the first law of thermodynamics, entropy is a natural functional describing the processes of universal evolution. Second, by the second law of thermodynamics, entropy maximization yields the worst state of the evolutionary process under the worst external influences on it. Since randomization is a model of uncertainty, the RML procedures with entropy as Shannon's measure of uncertainty [165] allow us to obtain solutions under maximum uncertainty.

This book includes ten chapters covering the general issues of data generation and qualitative specifics, as well as the theoretical foundations of randomized machine learning and its applications to the problems of classification and dynamic regression.

Chapter 1 is of methodological character, discussing machine learning problems in general as a tool for knowledge extraction and further transformation into solutions. We consider the structure of a machine learning procedure that represents a man-machine system with feedback. The basic elements of this system are data blocks, parameterized models and algorithms. We introduce the main notions related to these blocks in a rigorous way. Data arrays are described by a fixed-dimension matrix consisting of the vectors of observable input variables or by a matrix containing a "sliding" set of vectors. A parameterized model is defined by a parameterized input–output mapping with a formal description in terms of a vector functional whose properties have a close relation to the qualitative peculiarities of a corresponding applied problem. In addition, we introduce the notions of deterministic and probabilistic ML algorithms that are used for model parameter estimation. In the former case, these are empirical risk minimization algorithms and in the latter case, likelihood maximization algorithms (or relative likelihood minimization algorithms). Section 1.4 presents the main material of this chapter, devoted to the concept of a randomized machine learning procedure based on maximization of an information entropy functional on a set configured by an existing data array. The information entropy functional is defined on the class of probability density functions for the characteristics of the parameterized model and measurement noises that simulate data errors. Finally, we discuss the relation between the maximization principle of the information entropy functional and the minimization principle of the log-likelihood ratio. Several examples illustrate the cognitive capabilities of randomized models.

Chapter 2 deals with data, to be more precise, data sources and primary data processing. We consider analog and digital data sources. Analog sources generate deterministic and also random functions. Some background on the classes of deterministic and random functions is given, with focus on analog-to-digital conversions. Such conversions (amplitude and time quantization) cause errors. According to standard assumptions, they are random with definite probability density functions. We also estimate quantization errors. Certain types of digital data are studied, namely, audio, graphical and text data as well as official government statistics. The technology used to generate data depends on their nature. In particular, audio data conversion is often performed by a pulse-code modulator with subsequent spectral representation using the discrete Fourier transform. There exist two methods to convert images into digital form—bit image graphics (arrays or matrices of pixels) and vector graphics (mathematical models of primitives). A special class of data is texts, as their digital representation contains word-usage frequency characteristics and also semantic features. Among important primary data processing methods, we mention restoration methods for missing data and reduction methods for data arrays. In the former group of methods, our analysis covers the methods of interpolation, design of auxiliary models and entropy decomposition.

An important stage to prepare data for model learning is to reduce their dimension. This stage is considered in Chapter 3. More specifically, common dimension reduction methods, such as the singular decomposition of data matrices, PCA and random projection, are overviewed therein. Particular attention in Chapter 3 is paid to the method of direct and inverse projections and its implementation in the form of sequential and parallel procedures. The projector matrices are optimized by minimizing the Kullback–Leibler divergence, taking into account the information capacity constraints imposed on the data matrix. A key issue in dimension reduction is to estimate its efficiency when solving RML problems. An efficiency estimation technique for linear models used for forecasting and binary classification is considered. The final paragraph of Chapter 3 is devoted to the entropy random projection method, in which the projector matrices are assumed to be random. The concept of a compactness indicator of a data matrix as the average distance between its rows is introduced. The problem of optimizing the probability density functions of the projector matrix is formulated as the problem of maximizing the corresponding entropy functional subject to constraints imposed on the compactness indicator of the projector matrix. A modification of the method with known projector matrices chosen randomly from a given set is considered.

Chapter 4 is devoted to randomized parametric models in the RML procedure. A distinguishing feature of randomized models that separates them into an independent class is their output in the form of an ensemble of random vectors or random trajectories. We consider two subclasses of randomized parametric models (RPMs), namely, the Single Input–Ensemble Output model and the (Single Input, Feedback)–Ensemble Output model. The RPMs of both subclasses are described using parameterized vector functionals whose properties (continuity, differentiability, boundedness and others) agree with the properties of a modeled object. We study their linear and nonlinear versions (power and polynomial nonlinearities) with interval-type parameters. Since the output of an RPM is an ensemble of random events, we introduce empirical probabilistic characteristics (empirical means, variances, PDFs, medians, α-quantiles, confidence intervals and probabilities) as well as k-moment and k-mean characteristics for quantitative assessment.

After perusal of Chapters 1–4, the reader moves on to Chapter 5 with a detailed treatment of the entropy-robust estimation problem for the PDFs of the model parameters and measurement noises. This problem is the core of the RML procedure. RPMs may have the so-called original parameters belonging to n-dimensional parallelepipeds and also relative nonnegative parameters belonging to the n-dimensional nonnegative unit cube. The same classification applies to the measurement errors. The parameters and noises are characterized by the corresponding PDFs. Output data arrays are in balance with the numerical characteristics of the output ensemble of an RPM, i.e., its k-means. The entropy-robust estimation algorithm is formulated as a functional entropy-linear programming problem. We demonstrate that this problem belongs to the class of the Lyapunov-type optimization problems; hence, it can be solved using the Gateaux derivatives and Lagrange multipliers. We establish optimality conditions for the corresponding estimation problems, which yield a structural representation of the optimal PDFs parameterized by the Lagrange multipliers. The latter are defined by the solution of the balance equations written in universal forms. A peculiarity of these equations consists in the so-called integral terms, that is, the parameterized Riemann integrals determining the k-mean characteristics of the output ensemble of an RPM.

Chapter 6 considers a special case of RPMs—the models with non-randomized parameters. Recall that the uncertain model parameters and measurement noises lie within given intervals. Therefore, they can be transformed into nonnegative auxiliary variables belonging to the non-negative unit cube. These variables are interpreted as the probabilities that the original parameters and noises belong to the corresponding original intervals. In other words, the auxiliary variables (the probabilities of belonging) are nonrandomized, and hence a parameterized model outputs a vector or trajectory defined by the values of these probabilities. The ML algorithm to estimate the probabilities of belonging is stated as a constrained maximization problem of the information entropy on a set defined by the empirical balance equations. The properties of the ML algorithm depend on the class of parameterized models under consideration. If the model is linear, then the ML algorithm is reduced to an entropy-linear programming problem. In the general case, it is reduced to an entropy-nonlinear programming problem.

Chapter 7 is focused on the computational methods of randomized machine learning that are intended for solving the functional entropy-linear programming problems of RPMs and also the entropy-nonlinear programming problems in order to estimate the probabilities of belonging of the nonrandomized model parameters to corresponding intervals. The optimality conditions of these problems are formulated as the nonlinear empirical balance equations. For the functional entropy-linear programming problems, these equations contain integral terms. Our idea is to solve the equations by minimizing a residual function, i.e., the solution is reduced to a minimization problem on a compact set. This problem has some specifics: the information about the objective residual function is available in the form of its values calculated algorithmically. A considerable part of calculations consists in the numerical integration of nonlinear exponential functions using the Monte Carlo method. We estimate the size of the random sequence required for achieving a given accuracy. The residual minimization problem is treated as a special case of the global minimization problem on a compact set defined by a system of inequalities with continuous functions. The solution procedure of this system also employs the Monte Carlo method, more precisely, the packet Monte Carlo iterations. This procedure generates packets of random vectors with the uniform distribution on the nonnegative unit cube, performs filtering of the resulting vectors to construct a packet of the vectors belonging to the admissible compact set, and selects the records for each packet. By analyzing the sequence of records, we estimate some numerical characteristics of the residual function (the Hölder constants). The described algorithm has almost sure convergence (with probability 1). Under a finite number of iterations, we calculate an r-neighborhood of the exact global minimum and the probability that the iterative process reaches this neighborhood. The entropy-nonlinear programming problems are solved using multiplicative algorithms with p-active variables. Finally, their convergence is studied (including the rate of convergence) and a series of examples are given.

The RML algorithms yield the entropy-optimal estimates for the PDFs of the model parameters and measurement errors. The next stage of the RML procedure is to generate the randomized parameters and measurement noises with the optimal PDFs. Generally speaking, this problem is traditional for the Monte Carlo method. There exists vast literature on the generation of random sequences with classical PDFs (Gaussian, uniform, etc.). Difficulties arise when it is necessary to generate random sequences and random vectors with a given PDF of arbitrary form. These difficulties are even more pronounced for the random vectors of high dimension.

Chapter 8 is devoted to the generation problems of random vectors with given PDFs of arbitrary form. In the beginning, this chapter surveys the existing generation methods of random sequences with some typical PDFs. Major attention is paid to the Metropolis–Hastings algorithm as a most popular method to generate random vectors with a given PDF. We suggest a new generation method based on the stepwise approximation of a given PDF and the subsequent formation of uniformly distributed random vectors using random mixing. The approximation stage includes two steps as follows. The first step is to choose an optimal size of the grid that covers the PDF support. Here optimization is implemented by calculating the Lipschitz constant of the PDF. The second step deals with the stepwise approximation problem, which is to choose the constant replacing the real values of the PDF on the grid nodes. We consider two options of this choice, namely, the constant minimizing the integral quadratic error functional and the average constant over the corner values of the grid nodes. Concluding Chapter 7, we demonstrate the experimental simulation results of the suggested method, comparing it with the Metropolis–Hastings algorithm on three lite PDF tests—the

Gaussian distribution, the Laguerre distribution and the pseudo gamma distribution.

The RML problems are reduced to computationally intensive procedures that require considerable resources. Chapter 9 describes a general architecture of an information technology and software tools to develop a computational environment that implements the RML procedure. The suggested architecture is heterogeneous and guarantees an efficient utilization of computational devices of different types. We consider the information technology that implements the RML procedure and its software modules for the solution of global optimization problems, the generation of random vectors with a given PDF and the statistical processing of the randomized prediction ensembles.

Chapter 10 demonstrates how to use the RML procedure for the entropy classification. More specifically, we study the classification problems of heterogeneous objects into two classes in which classification is performed with the entropy-optimal probability for an existing data array. In some sense, soft classification represents an alternative to hard classification where the belonging to a certain class is uniquely defined. Under uncertain conditions (which are common for machine learning problems), hard classification often causes considerable errors. The randomized soft classification procedures use a neural network with random parameters as an RPM embedded in the additive model of measurements with errors described by a set of independent random vectors. The RML algorithm is formulated as a functional entropy-linear programming problem on the class of continuously differentiable PDFs for the neural network parameters and measurement errors. We suggest a procedure to find the probabilities of belonging of objects to appropriate classes based on the Monte Carlo method implementing the above entropy-optimal PDFs. Finally, several numerical examples are provided to illustrate the new method of soft classification.

Chapter 11 considers the application of the RML procedure to dynamic regression and randomized prediction. The restoration of the characteristics of dynamic relations between groups of variables is a widespread class of problems, both in context of theoretical studies and applications. A pragmatic goal of this research is to predict based on the dynamic relations between variables rather than to find the relations themselves. We study examples of dynamic regression problems arising in certain fields of demographic and macroeconomic research. Particular attention is given to randomized dynamic models of world population and forecasting of electrical load in the energy system.

The material for this book was accumulated over many years, following the development of information entropy maximization methods and their usage in different fields of science and engineering. The understanding that our studies are actually machine learning (more specifically, a new branch called randomized machine learning) has been formed recently. As information entropy is a natural functional that characterizes the PDFs of random parameters, it can be used as a functional for the variational principle in the following sense: the best PDF corresponds to the maximum information entropy. On the other hand, information entropy is interpreted as a measure of uncertainty. An integration of these two fundamental properties of information entropy has yielded the concept of randomized (entropy-optimal) solutions.

The authors are grateful to Professors B.T. Polyak and B.S. Darkhovsky for fruitful discussions that facilitated a system comprehension of the concept of randomized machine learning by the authors and (which is equally important) delivered the authors from potential delusions and mistakes. The book could not have appeared without support of Yugorsky Research Institute of Information Technologies and its director, Dr. S., Professor A.B. Melnikov. Special thanks to A.S. Aliev, Cand. Sci. (Eng.), for the experimental studies.

Finally, the authors express gratitude to A.Yu. Mazurov, Cand. Sci. (Phys.-Math.), for his careful translation of the Russian text, permanent feedback and negotiations with the publisher.

On behalf of the authors,
Yuri S. Popkov
Moscow, 2022

CHAPTER 1

GENERAL CONCEPT OF MACHINE LEARNING

Machine learning procedures are a backbone element of the toolkit to generate problem-oriented information for the support of different decisions. An important aspect in the formation of these procedures is their structuring under a considerable uncertainty caused by imperfect knowledge about the operation of real objects or phenomena and various data defects (errors, incompleteness, omissions, and so on). In this chapter, we classify three methodological approaches to the machine learning toolkit, each being based on fundamentally different principles—deterministic, probabilistic and randomized. The first approach involves deterministic models, measured data arrays and nonrandom parameter estimation. The second approach proceeds from the hypotheses about the random origin of the model parameters and measurement noises, and their probabilistic characteristics, which are used to construct nonrandom parameter estimates. Within both approaches, the learned models remain deterministic with the parameters whose values are the optimal estimates yielded by learning. Finally, the third approach stems from an artificial randomization of the model parameters and measurement noises and an estimation of their PDFs.

DOI: 10.1201/9781003306566-1

1

1.1   TRANSFORMATION OF KNOWLEDGE INTO DECISIONS

A necessary, useful and, sometimes, favorite activity of an individual is to make short-, medium- and long-term decisions. At the same time, we want to know guidelines of an individual, i.e., his motivational background for a certain decision. Unfortunately, there exist no universal or scientifically grounded approaches to discover this background. Perhaps, such approaches are impossible in principle, as individual decisions are personalized and their motives are implicit. Even questioning procedures or opinion polling techniques do not guarantee that an individual is frank in his answers. To all appearance, one may consider some collective (average) decisions and their motives.

The decisions of an individual affect certain objects of the material, natural and spiritual worlds, thereby changing their spatiotemporal state. The motivational mechanisms of an individual evolve in parallel to these processes but in another time scale. As a result, new decisions occur, often being caused by other motives.

In other words, motivations and decisions based on them interact within a cyclic procedure $M \to D$ (*Motivations–Decisions*) with a feedback loop FL as a crucial element of its "structure". The functional stages of the $M \to D$ procedure and their sequence are illustrated in Fig. 1.1.

Individual motivations (IM) form a basis for the decision-making process (DM) focused on the evolution of a studied object (O) or problem. This leads to the appearance of information in the object, particularly in the form of new data (D), which supplements the existing knowledge (K) about it. Let us accept a "daring" hypothesis that motivations rest on the knowledge accumulated by an individual. In the sequel, knowledge will be interpreted as an abstract intellectual basis that includes knowledge in the utilitarian understanding, intuition as a product of the subcortical processes of the brain, personal and public experience, data about the object, and so on. However, knowledge is a passive substance that does not activate motivations itself. Using knowledge, it is necessary to form a picture of the world, i.e., a model of a current situation for decision-making. A model arises from a combination of existing data and the beliefs of an individual about the character of this situation. Representing a causing factor of motivations, the situation is a set of certain measurable characteristics (*parameters*) of the model and relations among them.

For assigning a quantitative description to the model based on the available data, the model parameters have to be *estimated* numerically.

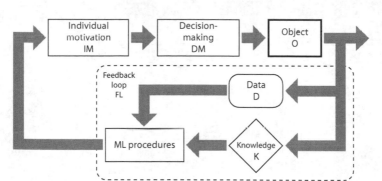

Figure 1.1

In this case, it is possible to solve different problems and predict their future trends. The feedback loop of the $M \to D$ cycle is implemented by a machine learning (ML) procedure with a necessary toolkit to design quantitative models and organize informational support of motivated decision-making.

Therefore, the ML procedure finds a methodological, algorithmic and information technology-based content in the $M \to D$ transformation, as shown in Fig. 1.1.

From the very beginning, note that this logical chain is merely a version of the motivational mechanism that is fruitful for this monograph and does not claim to be of universal value.

As an example, it seems appropriate to consider the $M \to D$ procedure intended for the transport behavior modeling of individuals [125]. Numerous acts implementing this behavior actually form traffic flows in an urban transport network.

Here, two aspects of the transport behavior will be analyzed, namely, the individual and collective behavior. The former is based on the motivations of an individual regarding transport vehicles, the goals of trip and routes. These motivations can be related to the time and cost of trip, the level of comfort and other characteristics. There exist numerous factors that affect the motivations of an individual. They can be specific for each individual. Consequently, the environment of individual motivations becomes essentially uncertain.

On the other hand, there are very many users of transport vehicles (including private and public transport), and their motivations and transport decisions are independent. And so, the stochastic framework

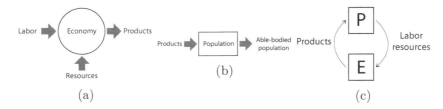

Figure 1.2

is suitable to model the corresponding uncertainty. Really, we cannot follow motivations and decisions of each individual, but their collective motivations can be characterized using some indicators of individual motivations. Due to the stochastic character of the uncertainty model, these characteristics are probabilistic, more specifically, the probability distribution functions of transport events. A widespread indicator of a transport event is trip time. In terms of this indicator, collective motivations are characterized by the so-called resettlement function that describes distributions of individuals by trip time. After normalization, this function can be treated as the probability distribution function of trip-time utilization.

The above-mentioned indicators reflect the collective motivations intrinsic to some average individual. But they have no relation to the real assessment of a current transport situation and its forecasting. However, the former and latter are the integral components of transport decision-making. It is impossible to consider the procedural part of these components for each individual. We can study collective decisions only, e.g., the choice of routes. If there exists a finite set of admissible routes from a certain location, then the probability of route choice can be an aggregated result of estimation and forecasting.

Another example of the $M \to D$ procedure is the interaction process of population and an economy that runs in a demoeconomic system (Figs. 1.2a–c). Due to intelligence, population is a tool of self-development, which results in generation of different types of human activity. Economic activity represents one of such types. By assumption, an economically motivated individual is the main tool of self-development.

Being involved in economic activity, population depends on its results. A key role in this process is played by the assessment of economic situation and the forecasting of possible consequences for decisions. The internal status of population also changes accordingly.

There exists a some kind of closed loop containing population and economic activity, with assessment, forecasting and feedback. Note an obvious analogy with the interaction procedure of motivations and decisions; see the structural diagram in Fig. 1.1.

An economy as a macroobject transforms resources into a product (see Fig. 1.2a). Labor, i.e., the labor activity of people in the economic sphere, is a resource. A certain product manufactured by an economy and consumed by population (in some form, see Fig. 1.2b) modifies the structure of population—its internal clustering and the quantitative form, and status of clusters (groups of population). As a result, other groups of different sizes are involved in economic processes at a new spiral of development.

Therefore, one can speak about an integral population-economy (PE) system consisting of the population (P) subsystem and the economy (E) subsystem that form a closed loop, as shown in Fig. 1.2c. An important field of analysis of this integral system deals with the procedures of situation assessment (state estimation) and forecasting, i.e., the evolution with the course of time. For each subsystem, these problems are studied by corresponding sciences. For example, the structural dynamics of population is the subject of macrodemography. A whole bunch of economic disciplines (political economy, macro- and microeconomics, mathematical economics, etc.) consider the formation and evolution of economic structures. Note that the economic factors affecting different processes in the Population subsystem are treated in macrodemography as the external factors related to the impact of a metasystem. Actually, a similar situation occurs in economic disciplines. The behavior of the Economy subsystem is studied under the condition that the Population subsystem reproduces labor resources of a definite volume and structure.

In particular, a popular line of demographic research is the forecasting of population state indicators under a given economic scenario described in quantitative terms [40]. Moreover, such scenarios reflect certain economic goals that (generally) have no relation to the potential changes of population within a forecasting period. But these changes considerably affect the indicators of economic development to be achieved. The above-mentioned feedback relation objectively exists, and an attempt to ignore it often causes unrealizable forecasts.

Also, note another peculiarity of this object, which is estimated and forecasted using ML procedures. Each of its parts has internal behavior (without the influence of a metasystem) and also disturbed behavior (under certain impacts applied by the metasystem). In this context, we mean states (equilibria, cycles, attractors, etc.), their stability and other properties. However, when the Population and Economy subsystems are united in an integral system, the latter may acquire new properties differing from the initial properties of these parts. Such a system effect particularly reveals itself in the case of nonlinear relations within each subsystem, and also in the case of nonlinearities in the channels used by one subsystem to influence the other. As a result, the integral system may have other equilibria and also other nonequilibrium processes. For such a system, state estimation and forecasting are actually not a trivial problem. Here additional difficulties arise due to a high level of uncertainty of the environment, for the separate subsystems and also for the entire system. The source of this uncertainty is an individual, the main component of both subsystems, with his motivations, social and professional characteristics, and behavioral responses.

Consequently, it seems that the situation here resembles the previous example. The Population–Economy system cannot be replicated, i.e., the internal processes running in this system never recur even for the same operation conditions. This system is self-developing, which eliminates the real repetitiveness of its processes. Therefore, the uncertainty that accompanies the operation of the Population–Economy system is reproduced by a simulation model with parameter estimation using available information.

1.2   STRUCTURE OF MACHINE LEARNING PROCEDURE

Methods and algorithms embedded in the ML procedures are quite diverse. However, they all have the same methodological orientation and structure. The latter is understood as a sequence of functional stages to transform knowledge into informational decision support. The structure of this procedure is shown in Fig. 1.3. The procedure includes four stages as follows.

The first stage (PM) lies in the development of a mathematical model that describes the operation mechanisms of a studied object and a measurement model of its input and output. The mathematical model of the object transforms knowledge about it into a structure with a list of definite characteristics. This stage cannot be formalized in such a general statement. However, often there exists prior information about the class of objects and problems that have to be modeled using a mathematical framework. Here "class" means an aggregate of objects or problems that have a set of common properties. If this set is known, then it is possible

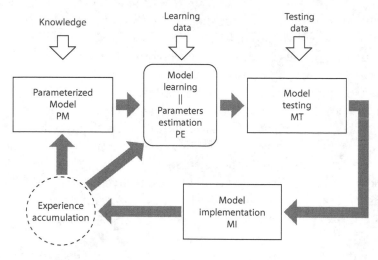

Figure 1.3

to construct a conceptual model for a corresponding class [68]. Such a model consists of a list of interconnected concepts that describe a given class, together with the properties, characteristics and classification of these concepts. An important component of conceptual modeling is to reflect the regularities of internal processes running within the objects of a given class [16]. In other words, a conceptual model systematizes and structures the existing knowledge about a class of objects and problems under study. The formation of such classes is a nontrivial problem whose complexity essentially depends on the degree of abstraction in their description. For example, the general theory of systems operates abstract sets and their relations (hierarchical systems, feedback, etc.) [119]. The classes produced by such a descriptive approach possess a rather high power (as they incorporate a large number of systems) but their conceptual models are often unconstructive. On the contrary, any attempt to add specific properties to an abstract class reduces the power of the new class, albeit yielding a better conceptual model in the sense of constructiveness. A good example is the class of controlled technical systems [91].

In most cases, a conceptual model represents a graph where nodes correspond to its functional elements and arcs to the mutual relations among them. As a matter of fact, this comprehension of a conceptual model was first introduced in [54]. Further complication and deeper specification of the properties intrinsic to different classes of systems outlined a gradual transition to a more efficient description, namely, to specially structured lists called ontologies [169]. Still, the graphical and ontological descriptions of classes retained the level of functional elements. These elements are described using various mathematical tools, depending on the regularities embedded in their operation mechanisms and also on their characteristics. In the case of *parametric characteristics*, the mathematical models of the elements are functional relations (for the description of static elements) or the systems of differential equations of different types (for the description of dynamic elements) [50, 161]. A rather efficient parameterization approach to static relations involves neural networks [96]. A relative simplicity of the basic elements and parameter tuning algorithms of these models allows to use neural networks for high-dimensional spaces.

Sometimes, for convenience the dynamics of an object are described using mathematical models with *functional characteristics*. For example, this class includes the models defined by integral equations or functional series [111, 137, 146, 152]. The role of functional characteristics in such models is played by the kernels of integral equations or functional series.

For a finite-state object, adequate models are automata [72, 107], nonlinear sequential machines [18, 177], and dynamic intelligent systems that operate in accordance with certain rules [126].

An important element of this stage is to model measurements. The real measurements of the object's input and output incorporate errors, and hence it is necessary to simulate them somehow. By a standard hypothesis, an independent random process is a rather adequate simulation model for the source of measurement errors. There exist two structural representations of the influence exerted by the errors on the observable output of the object's model, namely, the additive and multiplicative ones. The choice of an appropriate representation depends on the design properties of the device used to measure the object's output and input.

The second stage (PE) of the ML procedure (see Fig. 1.3) serves to estimate the characteristics of the mathematical model using real data that are generated as a *learning* sequence. As a matter of fact, this stage can be implemented with numerous methods. If the model characteristics are uncertain parameters, the corresponding estimation methods form the subject of mathematical statistics [11, 34, 90]. The estimation problems of increasing complexity arise when the mathematical model has functional characteristics, e.g., the kernel of an integral equation or the impulse-response characteristics of a functional series. Of course, in this case a finite-dimensional approximation of the estimation problems can be performed. But this approach often leads to a high dimension of the resulting approximation, thereby requiring greater volumes of data about the object's operation.

ML procedures consider the estimation problem as the minimization of the empirical risk functional with respect to the corresponding characteristics of the model, i.e., parameters or functions. Here the term "empirical" means that the functional is designed for a learning sequence. A prominent part is assigned to data packets for the object's input and output. Under fixed characteristics of the model, the input data are used to generate its output, which is then compared with the real output in terms of the empirical risk functional. This functional depends on the model characteristics so that their variation also affects the value of the functional. Empirical risk minimization yields the estimates of the model characteristics. The model together with its characteristics calculated in this way is *optimal* for the learning sequence and functional under consideration.

In the third stage (MT) of the ML procedure, the optimized ("learned") model is tested on other packets of the retrospective input and output data. Testing lies in comparing the real output of the object with the output generated by the optimized model, in terms of an accepted criterion. If the optimized model demonstrates a satisfactory level of quality on the test data packets, the model can be used in the normal operation mode.

The last (fourth) stage of the ML procedure (the MI stage) is intended to accumulate experience, i.e., new knowledge for a structural correction of the parameterized model at the PM stage and also for a refinement of the estimation algorithms at the PE stage.

Note that this procedure runs under uncertainty caused by limited knowledge about a studied object and its mathematical model, a finite size of learning sequences that incorporate information about the object's operation, measurement errors, and so on. Therefore, each stage of the ML procedure proceeds from certain hypotheses on the character of uncertainty.

1.3   MAIN CONCEPTS OF MACHINE LEARNING PROCEDURE

As illustrated by the block diagram in Fig. 1.3, the ML procedures consist of four stages, namely, the choice and parameterization of the model (the PM stage); the formation of input (cause) and output (effect) data arrays and model parameter estimation using these arrays (the

PE stage); the formation of data arrays for optimized model testing (the MT stage); and the generation of the optimized model (the MI stage).

As mentioned earlier, the first stage—*the choice and parametrization of the model*—is nonformalizable and involves only the available knowledge about a studied object, phenomenon or problem. This choice gives the elements of the n-dimensional input vectors $\vec{x}(t) \in \mathfrak{X}$, where $\mathfrak{X}$ denotes the space of input functions, and also the elements of the m-dimensional output vectors $\vec{y}(t) \in \mathfrak{Y}$, where $\mathfrak{Y}$ denotes the space of output functions; t is a scalar parameter.

The elements of these vectors can be measured in continuous time $t \in \mathcal{T} = [t_0, T]$ and in discrete time with an arbitrary step, $\{t_0, t_1, \ldots, t_s\} \in \tilde{\mathcal{T}}$, or with a fixed step h, $\{t_0, t_0 + h, \ldots, t_0 + sh\} \in \tilde{\mathcal{T}}_h$. For the time being, consider the last-mentioned measurement scheme, which yields the two-dimensional arrays of the input and output data of sizes $(n \times s)$ and $(m \times s)$, respectively, i.e.,

$$\mathbb{X} = \{\vec{x}^{(0)}, \ldots, \vec{x}^{(s)}\}, \qquad \mathbb{Y} = \{\vec{y}^{(0)}, \ldots, \vec{y}^{(s)}\}. \tag{1.1}$$

In addition, define the following input array that contains a current set of ϱ measured vectors:

$$\mathbb{X}_\varrho^{(j)} = \{\vec{x}^{(j-\varrho)}, \vec{x}^{(j-\varrho+1)}, \ldots, \vec{x}^{(j)}\}, \qquad \varrho \leq j \leq s. \tag{1.2}$$

In the first stage, also choose a model with parameters $\vec{a} \in R^p$ using a rather general description in terms of a vector functional $\hat{\Omega}$, that maps the input space $\mathfrak{X}$ into the output space $\hat{\mathfrak{Y}}$ as follows:

$$\hat{\vec{y}}^{(j)}(\mathbb{X}_\varrho^{(j)}, \vec{a}) = \hat{\Omega}\left(\mathbb{X}_\varrho^{(j)}, \vec{a}\right). \tag{1.3}$$

In this expression, the variable ϱ is the number of the input vectors that affect the output vector. If $\varrho > 0$, then the functional under consideration characterizes the dynamic properties of a studied object, phenomenon or problem, i.e., their "memory" in some sense. In the case $\varrho = 0$, the output of the model in the jth observation (measurement) depends only on the input vector in the jth observation (measurement). Such relations are typical for static objects, phenomena or problems.

The parameters $\vec{a}$ of the model (1.3) are assumed to be *deterministic* but with unknown values. These values are estimated using ML algorithms on the basis of available data and certain hypotheses about their prior properties. ML algorithms perform minimization of some functionals with qualitative interpretations that proceed from the accepted hypotheses. Note that ML algorithms yield estimates for the deterministic characteristics of the model, regardless of the functional selected.

One of such hypotheses treats the data and parameters as *deterministic* sets of real or integer numbers, without making additional assumptions on their properties. Then the optimal estimation of the model parameters is performed in terms of a *vector loss function* of the form

$$\vec{e}^{(j)}(\vec{a} \,|\, X^{(j)}, \vec{y}^{(j)}) = \vec{y}^{(j)} - \hat{\vec{y}}^{(j)}(X^{(j)}, \vec{a}), \quad j = \overline{1, s}, \qquad (1.4)$$

and the closeness of the PM output to the output data of the object is characterized by its norm known as *the empirical risk*, i.e.,

$$\mathcal{Q}_{X,Y}(\vec{a}) = \frac{1}{s} \sum_{j=1}^{s} \|\vec{e}^{(j)}(\vec{a} \,|\, X^{(j)}, \vec{y}^{(j)})\|. \qquad (1.5)$$

The empirical risk minimization methods are defined by the family of ML algorithms [53, 106, 188] in which

$$\hat{\vec{a}}(X, \vec{y}) = \arg\min \mathcal{Q}_{X,Y}(\vec{a}). \qquad (1.6)$$

An additional hypothesis about the properties of the data and parameters often lies in the assumption that the data $Z = (X, Y)$ and the parameters $\vec{a}$ are *random objects* with a joint probability density function $P(Z, \vec{a})$. Here Z denotes the universe of data and $\vec{a}$ is a random vector. In this case, the observable data arrays represent samples $Z^{(j)}$ from the universe Z. Note that each of them is random and independent with the PDF $P(Z^{(j)}, \vec{a})$.

The joint probability density function

$$L(Z^{(1)}, \ldots, Z^{(s)}; \vec{a}) = \prod_{j=1}^{s} P(Z^{(j)}, \vec{a}) \qquad (1.7)$$

is called the *likelihood* function for a sample $\{Z^{(1)}, \ldots, Z^{(s)}\}$ [41, 51]. Under a fixed sample, the likelihood function depends on the parameter $\vec{a}$. A quite natural estimate of this parameter is the vector that maximizes the likelihood function. Instead of the likelihood function, a convenient approach is to use its logarithm $\tilde{L} = \log L$, i.e.,

$$\hat{\vec{a}}(Z^{(1)}, \ldots, Z^{(s)}) = \arg\max_{\vec{a}} \tilde{L}(Z^{(1)}, \ldots, Z^{(s)}; \vec{a}). \qquad (1.8)$$

The hypothesis about the random origin of the pair $(Z, \vec{a})$ also has another interpretation related to the accumulation of information about this pair during experiments. Here, we mean its characteristics expected *before experiments* and the real characteristics obtained *after experiments*. To describe this evolution, Bayes suggested the concept of conditional probability, distinguishing between the *prior* and *posterior* probabilistic characteristics of the data and parameters. As a result, the posterior conditional probability density function (or simply posterior conditional density) $p(\mathbb{Z} \,|\, \vec{a})$ of data under fixed parameters, the prior conditional probability density function $w(\vec{a} \,|\, \mathbb{Z})$ of parameters under fixed data, and the prior probability density function $g(\mathbb{Z})$ were introduced accordingly. They have a clear relation defined by Bayes' formula

$$p(\mathbb{Z} \,|\, \vec{a}) = \frac{w(\vec{a} \,|\, \mathbb{Z})g(\mathbb{Z})}{\int\limits_{\mathbb{Z}} w(\vec{a} \,|\, \mathbb{Z})g(\mathbb{Z})d\mathbb{Z}}. \tag{1.9}$$

The posterior conditional density maximization methods are defined by the family of Bayesian ML algorithms

$$\hat{\vec{a}}(\mathbb{Z}) = \max_{\vec{a}} p(\mathbb{Z} \,|\, \vec{a}). \tag{1.10}$$

Both of the families (1.8, 1.10) employ the joint probability density functions of the data and parameters under the accepted hypotheses about the probabilistic properties of random objects.

1.4 PRINCIPLES OF RANDOMIZED MACHINE LEARNING PROCEDURE

The RML concept rests on two principles as follows. The first principle is the computer simulation principle for the *ensembles* of randomized decisions, vectors or trajectories caused by uncertainty. In other words, the existing uncertainty is compensated through the generation of an *ensemble* of decisions, vectors or trajectories. However, it is desirable to generate *optimal* ensembles in a certain sense, not any ensembles. This leads to the second principle known as the optimization principle for the generated ensembles, which *maximizes information entropy*.

The following examples illustrate the new properties arising in the randomized models.

Example 1 Consider a dynamic object (system) described by the first-order differential equation

$$\frac{dx}{dt} = ax, \qquad x(0) = x_0.$$

If the parameter a is nonrandom and the initial state is fixed, then the unique trajectory of this system has the form

$$x(t) = x_0 \exp(at).$$

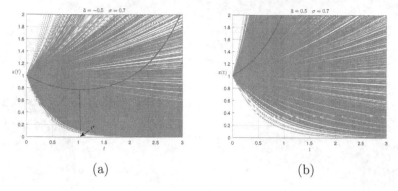

(a) (b)

Figure 1.4

Depending on the sign of a, the trajectory is either decreasing or increasing as $t \to \infty$.
Now, let a be a random parameter with the PDF

$$p(a) = \frac{1}{\sqrt{2\pi}\sigma} \exp\left(-\frac{(a - \bar{a})^2}{2\sigma^2}\right),$$

where $\bar{a}$ and σ^2 denote the mean and the variance of the parameter a, respectively.

In this case, an ensemble of solutions of the first-order equation is obtained. Figures 1.4a and 1.4b show the trajectories of this ensemble for $\bar{a} = -0.5$, $\sigma = 0.7$ and $\bar{a} = 0.5$, $\sigma = 0.7$, respectively. Find the mean trajectory of this ensemble; see Fig. 1.4a. By the definition of a mean, for each t,

$$\bar{x}(t) = x_0 \int\limits_{-\infty}^{\infty} \exp(at)p(a)da = x_0 \exp\left(\bar{a}t + \frac{\sigma^2 t^2}{2}\right).$$

This equality implies that the mean trajectory is given by the mean $\bar{a}$ and also by the variance σ^2. In addition, there exists a time

$$t^* = \begin{cases} 0, & \text{if } \bar{a} \geq 0, \\ -\frac{\bar{a}}{\sigma^2} & \text{if } \bar{a} < 0, \end{cases}$$

such that the mean trajectory decreases for $t < t^*$ and infinitely increases for $t > t^*$. Of course, this property takes no place in the deterministic model.

A similar effect occurs under other PDFs, e.g., of the form

$$p(a) = \begin{cases} \frac{1}{a^+ - a^-} & \text{for } a \in [a^-, a^+], \\ 0 & \text{for } a < a^-, a > a^+. \end{cases}$$

For $0 > a^+ > a^-$ (and hence $\bar{a} < 0$), the resulting ensemble of the trajectories of the first-order system is shown in Fig. 1.5a. For $a^- < 0$ and $a^+ > 0$ (and $\bar{a} < 0$), the resulting ensemble can be seen in Fig. 1.5b. The mean trajectory achieves the minimum value at the point t^* that is the solution of the equation

$$-\left(\frac{1}{t} - a^+\right) + \exp\left((\bar{a} - a^+)t\right)\left(\frac{1}{t} - a^-\right) = 0.$$

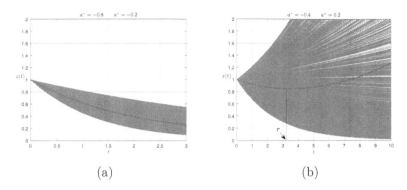

(a)   (b)

Figure 1.5

Example 2. Consider a dynamic system described by the second-order differential equation

$$\frac{dy_1}{dt} = y_2,$$

$$\frac{dy_2}{dt} = -a\,y_2 - b\,y_1.$$

The parameters a, b are random with the PDF

$$p(a) = \begin{cases} \frac{1}{a^+ - a^-} & \text{for } a \in [a^-, a^+], \\ 0 & \text{for } a < a^- \text{ and } a > a^+, \end{cases}$$

where $a^- < 0, a^+ > 0$.

$$w(b) = \begin{cases} \frac{1}{b^+ - b^-}, & \text{for } b \in [b^-, b^+], \\ 0, & \text{for } b < b^-, b > b^+, \end{cases}$$

where $b^+ \geq b^- \geq 0$.

Denote by

$$\bar{a} = \frac{a^+ - a^-}{2}$$

the mean of the parameter a and by

$$\bar{b} = \frac{b^+ - b^-}{2}$$

the mean of the parameter b.

For this example, choose the following limits of the parameter ranges: $a^- = -1$, $a^+ = 2$, $b^- = 4$ and $b^+ = 9$. The corresponding means of the parameters are $\bar{a} = 0.5$ and $\bar{b} = 6.5$. The initial point is $y_1(0) = -2, y_2 = 2$. The resulting ensemble of the system trajectories on the phase plane can be seen in Figs. 1.6a and 1.6b (here the black curve indicates the mean trajectory whereas the grey curve the mean values trajectory).

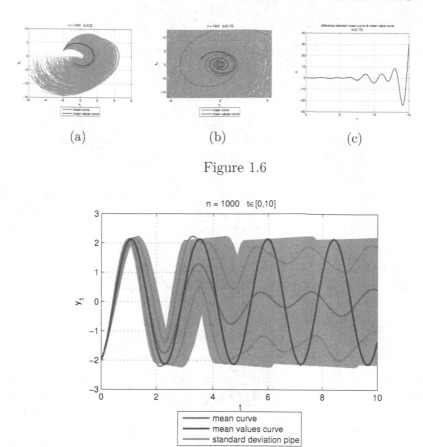

Figure 1.6

Figure 1.7

Since the range of the parameter a also contains negative values, some of the trajectories in the ensemble are divergent. Therefore, for a small time interval (e.g., $t \in [0, 2]$), there is a slight difference between the mean trajectory and the mean values trajectory (see Fig. 1.6a). For a larger time interval, the difference becomes considerable, as illustrated in Fig. 1.6b. The mean values trajectory converges to the equilibrium $(0, 0)$ whereas the mean trajectory diverges on the time interval $t \in [0, 10]$. The difference between the mean and mean values trajectories is shown in Fig. 1.6c.

Figure 1.7 presents the ensembles of the trajectories of the second-order system on the phase plane for $a^- = a^+ = 0$ and $t \in [0, 10]$. The ensemble consists of the periodic curves with different periods. As before, the grey curve (cycle) corresponds to the mean $\bar{b} = 6.5$ and the black curve to the mean trajectory.

Consequently, this random-parameter system has new modes differing from the ones observed in its deterministic counterpart.

Summarizing the aforesaid, one of the implementation principles of the RML concept lies in the generation of the ensembles of random events (decisions, vector or trajectories) with desired probabilistic characteristics. They are generated using the model (1.3) with the randomized parameters **a** obeying a desired PDF. In other words, it is necessary to generate an ensemble of the random vectors **a** with given PDFs.

The second fundamental principle of the RML concept is the optimization principle of the randomized model adopted to calculate the optimal PDFs of the model parameters. Here an important stage consists on the choice of an appropriate quality functional. In our case, this (often debatable) question has a well-justified answer. As emphasize earlier, the RML concept is a tool to obtain best decisions under maximum uncertainty. Recall that the uncertain knowledge about an object under study is described by randomized model parameters, uncertain data by additional noises and uncertain estimates by probability density functions.

In the sequel, **the information entropy functional** will be used to characterize the quality of a randomized model. The maximum of this entropy corresponds to the highest uncertainty of our knowledge about the object and noises intrinsic to measured data. An interpretation of information entropy as a measure of uncertainty (with the minus sign, as a measure of information) was first suggested by C. Shannon in his pioneering paper [165], which motivated intensive research in mathematics, statistical thermodynamics, communication theory and other fields of science. A next surge of interest in information entropy is related to the name of E. Jaynes [80], who proved that the entropy maximization principle yields best solutions under maximum uncertainty. In particular, this principle allows to find best solutions under the worst-case probabilistic characteristics of data measurement noises [66,88,109,141,153,157]. This feature of the information entropy maximization principle is treated as the "entropy"-*robust* property of the resulting solutions.[1]

---

[1] As a matter of fact, such an interpretation differs from the definition of robustness given in [74].

CHAPTER 2

DATA SOURCES AND MODELS

For machine learning procedures, data is important, like their sources, degree of adequacy (measurement errors), volume, properties, and features. We comprehend data as sets of numbers that are associated with certain times and organized in arrays with appropriate properties. If a data source is not of natural origin (e.g., an analog oscillogram), then the data have to be transformed into a structured numerical array.

Although ML procedures involve numerical data, in many cases it is useful to consider information about the general properties of functions that describe data samples. The character of data has a close relation to the source objects whose states are measured in these data. There exist analog and digital data sources.

An analog source is characterized by the functions of continuous argument. They are measured using analog sensors and mostly represent the functions of time and/or space for spatially distributed objects. The classes of such functions and their common properties generate some properties of samples drawn from them.

A digital source contains a digital sensor that outputs all measured data with amplitude (level) and time quantization. A particular case is a binary digital source in which a digital quantizer has only two levels and the resulting data represent a sequence of zeros and ones associated with discrete times.

Depending on further usage, data can be transformed from one class to another. For example, analog data are transformed into digital data using analog-to-digital converters (ADCs) for further computer processing. Sometimes, it is necessary to transform digital data into analog data, which is performed by digital-to-analog converters (DACs). Such converters find wide application in the field of audio, video and graphical data processing.

A crucial problem consists in the adequacy and argument completeness of data. If argument is time, then complete data contain numbers for all times on a fixed time interval. Data adequacy relates to measurement errors, which are often described by random errors with appropriate properties.

2.1 ANALOG SOURCE OF DATA

An analog source is a measuring system (an analog sensor), i.e., a transducer that performs transformation of a certain indicator of a real process into a scalar function $x(t)$ or a vector function $\vec{x}(t)$ defined on a time interval $\mathcal{T} = [t_0, T]$; times t within this interval are real numbers. Some examples of analog sensors include tachometers measuring speed, accelerometers measuring acceleration, spectrometers measuring the spectral composition of a substance, etc.

2.1.1 Deterministic functions

First, consider in brief the general properties of deterministic functions.[1] Further presentation will employ the fundamental notions of mathematical analysis, namely, limit and convergence. Both are related to the notions of a sequence and an element of a sequence. For a time interval $\mathcal{T}$, an element can be a value $t_n \in \mathcal{T}$ and a sequence then takes the form $\{t_0, t_1, \dots\}$ (in most cases, a sequence is ordered, i.e., $t_0 < t_1 < \cdots$). For introducing the notion of convergence and limit, define a "distance" $\rho_{\mathcal{T}}(t_i, t_j)$ between elements t_i and t_j, which is a nonnegative real-valued function satisfying the three axioms below.

1. The identity axiom: $\rho_{\mathcal{T}}(t_i, t_j) = 0$ only for $t_i = t_j$.

2. The symmetry axiom: $\rho_{\mathcal{T}}(t_i, t_j) = \rho_{\mathcal{T}}(t_j, t_i)$.

3. The triangle axiom: $\rho_{\mathcal{T}}(t_i, t_j) + \rho_{\mathcal{T}}(t_j, t_k) \geq \rho_{\mathcal{T}}(t_i, t_k)$, where $t_i < t_j < t_k$.

Such a function can be always defined on a time interval $\mathcal{T}$. Therefore,

---

[1]For a detailed treatment, the reader is referred to the monographs on theory of functions and functional analysis, e.g., [95, 98].

$\mathcal{T}$ forms a metric space. This metric space (of real numbers) is of course very simple. However, the metric spaces of scalar and vector functions can be constructed by analogy. An element t^* is the limit of a sequence $t_0, t_1, \ldots$ if $\rho(t_n, t^*) \to 0$ as $n \to \infty$.

Consider two metric spaces, $\mathcal{Y}$ and $\mathcal{T}$, and a function $y = x(t)$ taking values in $\mathcal{Y}$. A function $x(t)$ is called *continuous at a point* $t^* \in \mathcal{T}$ if for any $\varepsilon > 0$ there exists a $\delta > 0$ such that $\rho_{\mathcal{Y}}(x(t), x(t^*)) < \varepsilon$ for any other point $t \in \mathcal{T}$ satisfying the inequality $\rho_{\mathcal{T}}(t, t^*) < \delta$. By the definition of a continuous function $x(t)$, the convergence $t_n \to t^*$ implies the convergence $x(t_n) \to x(t^*)$.

If this property applies to all the points $t \in \mathcal{T}$, then a corresponding function $x(t)$ *is uniformly continuous on* $\mathcal{T}$. Denote by $C^{(0)}$ the class of all these functions.

Continuity (and uniform continuity) can be extended to the derivatives of a function $x(t)$, provided that they exist. And the classes of *smooth functions* with a definite degree of smoothness arise accordingly. The degree of smoothness is the maximum order of derivatives that exist for the functions of a given class. For example, the class $C^{(1)}$ contains the continuously differentiable functions (first-order derivatives) whereas the class $C^{(r)}$ the derivatives of order r. A substantial subclass of smooth functions denoted by $C^{(\infty)}$ consists of *analytical functions*, which are differentiable infinitely many times on their domains of definition. In particular, an analytical function $x(t)$ defined on a time interval $\mathcal{T}$ has infinitely many derivatives $x^{(k)}(t)$, $k = 1, 2, \ldots$, on this interval. At any point t_0 of the interval $\mathcal{T}$, it can be expanded into the uniformly convergent Taylor series:

$$x(t) = x(t_0) + \sum_{i=1}^{\infty} \left. \frac{d^i x(t)}{dt^i} \right|_{t_0} (t_0 - t)^i, \quad t_0 \in \mathcal{T}. \qquad (2.1)$$

A certain quantitative characteristic can be assigned to continuity as a qualitative property of functions. More specifically, it is possible to introduce the so-called *module of continuity*[2]

$$\omega(h) = \max_{|t_i - t_j| \leq h} |x(t_i) - x(t_j)|. \qquad (2.2)$$

---

[2]In the general case of a multivariable function $y = f(\vec{u})$, $\vec{u} \in R^n$, the module of continuity is defined by

$$\omega(h) = \max_{\|\vec{v} - \vec{z}\| \leq h} |f(\vec{v}) - f(\vec{z})|.$$

If a function $x(t)$ has a module of continuity

$$\omega(h) \leq Hh^s, \quad (H, s) \geq 0, \qquad (2.3)$$

then it belongs to the continuous Hölder functions, where H and s are the Hölder constants.

If a function $x(t)$ has a module of continuity

$$\omega(h) \leq Lh, \quad L \geq 0, \qquad (2.4)$$

then it belongs to the continuous Lipschitz functions, where L is the Lipschitz constant.

An important characteristic of any function is its variability on a time interval $\mathcal{T}$, more specifically, whether such variability is finite or infinite. The functions with finite variability form the class of functions of bounded variation. To define this property, consider an arbitrary partition of the time interval $\mathcal{T}$ into an ordered sequence of points $t_0, t_1, \ldots, t_n$ and the increments (decrements) of a function $x(t)$ at these points. If

$$\sum_{i=0}^{n} |x(t_{i+1}) - x(t_i)| = X_{\mathcal{T}} < +\infty \qquad (2.5)$$

for any sequence of points on $\mathcal{T}$, then $x(t)$ belongs to the class of functions of *bounded variation*. The value $X_{\mathcal{T}}$ is called the total variation of the function $x(t)$ on the time interval $\mathcal{T}$.

The class of functions of bounded variation includes the functions that have additional properties. For example, the continuous functions and the functions with any degree of smoothness belong to the class of functions of bounded variation. If a function $x(t)$ is monotonic, i.e., the expression

$$\frac{x(t_i) - x(t_j)}{t_i - t_j}$$

is of the same sign or vanishes for any t_i, t_j from $\mathcal{T}$, then it belongs to the class of functions of bounded variation, and its total derivative is given by $X_{\mathcal{T}} = x(T) - x(t_0)$. Note that the class under consideration also contains the functions with jump (simple) discontinuities.

Different approximations are used to study the properties of the above-mentioned classes of functions and to generate the well-grounded sets of discrete data associated with them. The approximation methods operate the fundamental concepts of an approximation quality functional and an approximating function $x_a(t)$ from a given class. The following functionals are widespread:

- the Chebyshev functional

$$C = \max_{\mathcal{T}} |x(t) - x_a(t)|; \qquad (2.6)$$

- the integral quadratic functional

$$J_2 = \int_{\mathcal{T}} |x(t) - x_a(t)|^2 dt; \qquad (2.7)$$

- the normalized integral quadratic functional

$$\Delta_2 = \frac{\int_{\mathcal{T}} |x(t) - x_a(t)|^2 dt}{\int_{\mathcal{T}} x^2(t)dt + \int_{\mathcal{T}} x_a^2(t)dt}. \qquad (2.8)$$

The elementary approximating functions that have a direct relation to discrete data are the piecewise constant (also called step) functions

$$x_c(t) = \begin{cases} x(t_0) & \text{for } t_0 \leq t < t_0 + h, \\ x(t_0 + h) & \text{for } t_0 + h \leq t < t_0 + 2h, \\ \dots & \dots \\ x(t_0 + (n-1)h) & \text{for } t_0 + (n-1)h \leq t < T. \end{cases} \qquad (2.9)$$

As their parameter the step functions incorporate the sampling time h, which can be chosen depending on an acceptable value of the functionals (2.6)–(2.8). For certain classes of functions (e.g., continuous or smooth), the step functions possess a fruitful property of asymptotic behavior,[3] i.e.,

$$\lim_{h \to 0} x_c(t) = x(t). \qquad (2.10)$$

Note that this limit is interpreted depending on the properties of a corresponding function $x(t)$.

2.1.2 Random functions

An analog source may generate functions of random origin. Following [65], two close concepts will be employed, namely, *a random function*

---

[3]This section gives only background information about the functions generated by analog sources. For a comprehensive coverage of the theory of functions, the reader is referred to the monographs on the subject, e.g., [95].

$x(t)$ (*a random vector function* $\vec{x}(t)$) and *a random process* $X(t)$ (*a vector random process* $\vec{X}(t)$). A process represents an ensemble of random functions $x(t)$. Under certain conditions, a random process (r.p.) observable in some sense is called *a sample* (a sample function) or *a realization* of this random function. The scalar functions will be considered below.

For a fixed t, $x(t)$ gives a value of a random variable whereas $X(t)$ an ensemble of values of this random variable. A random function is completely defined by all possible probability-theoretical relations for any finite set of values of random variables:

$$x(t_1), x(t_2), \ldots, x(t_n), \quad t_i \in \mathcal{T}, \ i = \overline{1,n}, \ n = 1, 2, \ldots, \qquad (2.11)$$

i.e., by corresponding probability distributions. In other words, a random function $x(t)$, $t \in \mathcal{T}$, is described by a corresponding family of distributions

$$P_{t_1,\ldots,t_n}(\eta_1, \ldots, \eta_n), \qquad t_i \in \mathcal{T}, \ i = \overline{1,n}, \ n = 1, 2, \ldots, \qquad (2.12)$$

where $\eta_1, \ldots, \eta_n$ denote the values of the random variables.

Each probability distribution function is treated as a joint probability distribution function for a collection of the random variables (2.11). Such a treatment becomes justified under some necessary conditions, the so-called *consistency conditions* for the family of distributions (2.12) given by

$$P_{t_1,\ldots,t_n,t_{(n+1)},\ldots,t_{(n+p)}}(\eta_1, \ldots, \eta_n, +\infty, \ldots, +\infty) = P_{t_1,\ldots,t_n}(\eta_1, \ldots, \eta_n);$$
$$P_{t_1,\ldots,t_n}(\eta_1, \ldots, \eta_n) = P_{t_{i_1},\ldots,t_{i_n}}(\eta_{i_1}, \ldots, \eta_{i_n}), \qquad (2.13)$$

where $i_1, \ldots, i_n$ forms any permutation of the indexes $1, 2, \ldots, n$.

A set of functions $P_{t_1,\ldots,t_n}(\eta_1, \ldots, \eta_n)$ is called *finite-dimensional distributions of a random function* $x(t)$.

Applications often lead to several random functions, which will be considered below as the elements of a random vector $\vec{x}(t) = \{x_1(t), \ldots, x_m(t)\}$. A random vector is completely defined by a family of finite-dimensional probability distributions of the form

$$P_{t_1,\ldots,t_n}(\eta_{11}, \eta_{12}, \ldots, \eta_{nm}) = \mathbf{P}\{x_1(t_1) < \eta_{11}, x_2(t_1) <$$
$$< \eta_{12}, \ldots, x_m(t_n) < \eta_{nm}\}. \qquad (2.14)$$

The finite-dimensional probability distributions of a random function $x(t)$ can be defined by the finite-dimensional probability density functions $w_{t_1,\ldots,t_n}(\eta_1,\ldots,\eta_n)$:

$$P_{t_1,\ldots,t_n}(\eta_1,\ldots,\eta_n) = \int\limits_{-\infty}^{\eta_1} \cdots \int\limits_{-\infty}^{\eta_n} w_{t_1,\ldots,t_n}(y_1,\ldots,y_n)dy_1\cdots dy_n. \quad (2.15)$$

In particular, this implies the formula

$$w_{t_1,\ldots,t_n}(\eta_1,\ldots,\eta_n) = \int\limits_{-\infty}^{\infty} \cdots \int\limits_{-\infty}^{\infty} w_{t_1,\ldots,t_n,t_{(n+1)},\ldots,t_{(n+p)}}(\eta_1,\ldots,$$

$$\eta_n, y_1,\ldots,y_p)dy_1\cdots dy_p, \quad (2.16)$$

which can be viewed as the consistency condition (2.1.2) of the finite-dimensional probability distributions.

A useful tool to analyze finite-dimensional probability density functions (or distributions) is their *characteristic function* $\Phi_{t_1,\ldots,t_n}(u_1,\ldots,u_n)$, which represents the Fourier transform [48] of a probability density function:

$$\Phi_{t_1,\ldots,t_n}(u_1,\ldots,u_n) = \int\limits_{-\infty}^{\infty} \cdots \int\limits_{-\infty}^{\infty} \exp\left(j\sum_{k=1}^{n}\eta_k u_k\right) \times$$

$$\times w_{t_1,\ldots,t_n}(\eta_1,\ldots,\eta_n)\,d\eta_1\cdots d\eta_n, \quad (2.17)$$

$$j = \sqrt{-1}.$$

The variables $u_1,\ldots,u_n$ takes values in space R^n. If a probability density function satisfies some analytical conditions (see [48]), then it can be restored through its characteristic function using the Fourier integral as follows:

$$w_{t_1,\ldots,t_n}(\eta_1,\ldots,\eta_n) = \frac{1}{(2\pi)^n} \int\limits_{-\infty}^{\infty} \cdots \int\limits_{-\infty}^{\infty} \exp\left(-\sum_{k=1}^{n}\eta_k u_k\right) \times$$

$$\times \ \Phi_{t_1,\ldots,t_n}(u_1,\ldots,u_n)\,du_1\cdots du_n. \quad (2.18)$$

These definitions of a random function suffice if we are interested in the values of a random function at n points, i.e., the values of argument $(t_1,\ldots,t_n) \in \mathcal{T}$. However, if the set $\mathcal{T}$ contains all real values of time t, then even arbitrarily many finite-dimensional probability distributions of the form (2.12, 2.1.2) say nothing about the constructive properties of random functions such as continuity, differentiability, and others.

In some applications, the complete description of a random function in the form of finite-dimensional probability distributions becomes redundant. One needs only know its numerical characteristics—the moment functions

$$m_{i_1,\ldots,i_s}(t_1,\ldots,t_s) = \mathcal{M} \prod_{k=1}^{s} [x(t_k)]^{i_k}, \quad i_k \geq 0, \; k = \overline{1,s}. \qquad (2.19)$$

The value $q = \sum_{k=1}^{s} i_k$ is called the order of the moment function. In the class of moment functions, a special place belongs to the moment functions of orders 1 and 2, namely,

- the mean

$$m(t_1) = m_1(t_1) = \mathcal{M}\{x(t_1)\}; \qquad (2.20)$$

- the correlation function

$$R(t_1,t_2) = m_{11}(t_1,t_2) = \mathcal{M}\{[x(t_1) - m(t_1)][x(t_2) - m(t_2)]\}. \qquad (2.21)$$

If $t_1 = t_2 = t$, then $R(t,t) = \sigma^2$ are the variances of a random variable $x(t)$ with a fixed t. If $R(t_1,t_2) = 0$, then the values $x(t_1)$ and $x(t_2)$ of a random function $x(t)$ are uncorrelated. In a special case where the two-dimensional joint probability distribution of random variables $x(t_1)$ and $x(t_2)$ is Gaussian and the uncorrelatedness condition holds, the random variables $x(t_1)$ and $x(t_2)$ are independent.

Continuity is an important property of a function. For a random function, this property has the probabilistic sense formalized in terms of the distribution of all pairs $x(t_1)$, $x(t_2)$. Introduce the distance $\rho(t_1,t_1) = |t_1 - t_2|$ on the set $\mathcal{T}$. A random function $x(t)$, $t \in \mathcal{T}$, is called *stochastically continuous* at a point t^* if, for all $\varepsilon > 0$,

$$\mathbf{P}\{|x(t^*) - x(t)| > \varepsilon\} \to 0 \quad \text{as } \rho(t^*,t) \to 0. \qquad (2.22)$$

If condition (2.22) holds for all values $t \in \mathcal{T}$, then a corresponding function $x(t)$ is stochastically continuous on $\mathcal{T}$.

According to the definition (2.22), stochastic continuity is an "asymptotic" property to be declared only. A more constructive definition would operate some quantitative values. In fact, *uniform stochastic continuity* forms such a property. A random function is uniformly stochastically continuous on $\mathcal{T}$ if, for an arbitrary $\varepsilon > 0$, there exists a $\delta > 0$ such that

$$\mathbf{P}\{|x(t) - x(\tilde{t})| > \varepsilon\} < \varepsilon \qquad (2.23)$$

for all the pairs $(t,\tilde{t})$ satisfying the inequality $\rho(t,\tilde{t}) < \varepsilon$.

Since the set $\mathcal{T}$ is compact, the stochastic continuity of a random function on $\mathcal{T}$ implies its uniform stochastic continuity on $\mathcal{T}$.

Consider some classes of random functions (processes).

1. Gaussian random functions

In many applications, a crucial role is played by the random functions with Gaussian finite-dimensional probability distributions. As the matter concerns the finite-dimensional distributions, a random function can be described by a random n-dimensional vector $\vec{x} = \{x(t_1), \ldots, x(t_n)\}$. A random vector $\vec{x}$ obeys the Gaussian distribution if its characteristic function can be represented in the form

$$\Phi(\vec{u}) = \mathcal{M}\{\exp(j\langle \vec{u}, \vec{x}\rangle)\} = \exp\left(j\langle \vec{u}, \vec{m}\rangle - \frac{1}{2}\langle G\,\vec{u}, \vec{u}\rangle\right). \qquad (2.24)$$

Here $\vec{m}$ denotes the n-dimensional vector of the means for the corresponding elements of a vector $\vec{x}$; $\vec{u}$ is the n-dimensional vector of real variables defined on R^n; G indicates a nonnegative definite real symmetric matrix (the correlation matrix); finally, $\langle \bullet, \bullet \rangle$ stands for the scalar product of appropriate vectors.

The moments of a Gaussian random function can be calculated by expanding its characteristic function. If the central moments ($m_1(t_1) = 0$) suffice, then

$$\begin{aligned}\Phi(\vec{u}) &= \exp\left(-\frac{1}{2}\langle G\,\vec{u}, \vec{u}\rangle\right) = 1 - \frac{1}{2}\langle G\,\vec{u}, \vec{u}\rangle + \frac{1}{2^2 2!}\langle G\,\vec{u}, \vec{u}\rangle^2 - \ldots \\ &+ (-1)^n \frac{1}{2^n n!}\langle G\,\vec{u}, \vec{u}\rangle^n + \ldots. \end{aligned} \qquad (2.25)$$

Hence, the odd-order moment functions are

$$m_{i_1,\ldots,i_s}(t_1,\ldots,t_s) = 0, \qquad \sum_{k=1}^{s} i_k = 2n + 1; \qquad (2.26)$$

the even-order moment functions take the form

$$m_{i_1,\ldots,i_s}(t_1,\ldots,t_s) = \frac{\partial^{2n}}{\partial t_1^{i_1}\cdots\partial_s^{i_s}}\left(\frac{1}{2^n n!}\langle G\,\vec{u}, \vec{u}\rangle\right),$$

$$\sum_{k=1}^{s} i_k = 2n. \qquad (2.27)$$

The Gaussian random functions are of paramount importance in different (theoretical and applications-oriented) studies. Really, under rather nonrestricting conditions, the sum of a large number of independent random functions with small values can be approximated by a Gaussian random function, regardless of the probability-theoretic nature of the terms. This significant result—the normal correlation theorem—is a generalization of the central limit theorem [166].

2. Random processes with independent increments

The research in the field of random processes with independent increments started with Brownian motion and the Wiener process as its appropriate mathematical model. At a later time, the generalized mathematical models with integration in the class of processes with independent increments were developed.

A random process $\vec{x}(t)$, $t \in \mathcal{T}$, taking values in R^n is called a process with independent increments if, for any n, $t_0 \leq t_1 \leq \cdots \leq t_n = T$, the random vectors $\vec{x}(t_0), \vec{x}(t_1) - \vec{x}(t_0), \ldots, \vec{x}(t_n) - \vec{x}(t_{n-1})$ are mutually independent.

The vector $\vec{x}(t_0)$ is called the initial state (value) of the process, and its probability distribution is called the initial distribution of the process. For defining a process with independent increments, it suffices to define an initial distribution $P_0(\mathcal{B}) = \mathbf{P}\{\vec{x}(t_0) \in \mathcal{B}\}$, a collection of distributions $P(t, h, \mathcal{B}) = \mathbf{P}\{\Delta\vec{x}(t, h) \in \mathcal{B}\}$ for the vector $\Delta\vec{x}(t, h) = \vec{x}(t + h) - \vec{x}(t)$, where $t \in \mathcal{T}$, $h > 0$, and $\mathcal{B} \in R^m$ is a set of elementary events.

If a distribution $P(t, h, \mathcal{B})$ does not depend on t, i.e., $P(t, h, \mathcal{B}) = P(h, \mathcal{B})$, then a corresponding process with independent increments is called *homogeneous*.

Consider examples of some processes with independent increments. Let us begin with **Brownian motion**. For this process, the probability distribution $P(t, h, \mathcal{B})$ is Gaussian. The physical model of the process is the motion of a small particle in a liquid, which can be observed using a high-power microscope. Such a particle has constant movements, and its path represents a very complicated polygonal line with chaotically directed segments. This motion is explained by the collisions of the particle with the molecules of the liquid. The particle has relatively large dimensions in comparison with the dimensions of each molecule and suffers from a huge number of collisions with them each second. The visible motion of the particle is called Brownian motion. In a rough approximation, the displacements of the particle caused by collisions can be considered mutually independent, and hence Brownian motion can be treated as a process with independent increments. Due to small displacements, we can hypothesize that their sum satisfies the central limit theorem; in other words, Brownian motion can be described by a Gaussian random process with independent increments.

In particular, a homogeneous Brownian random process is defined by two parameters m and σ:

$$\mathcal{M}\{x(t)\} = mt, \quad \mathcal{M}\{x^2(t)\} = \sigma^2 t. \tag{2.28}$$

For $m = 0$ and $\sigma = 1$, the Brownian motion process is called *the standard Wiener process*.

Now consider *Poisson processes*. A stochastically continuous process with independent increments that takes nonnegative integer values is called a Poisson process if, for any $(s, t) > 0$ and $s < t$, the variable $\vec{x}(t) - \vec{x}(s)$ obeys the Poisson probability distribution.

Let $x(t_0) = 0$. Then the variable $x(t)$ has the probability distribution

$$\mathbf{P}\{x(t) = m\} = \frac{a^m(t)}{m!} \exp(-a(t)), \quad \mathcal{M}\{x(t)\} = a(t). \tag{2.29}$$

Hence, it follows that

$$\mathbf{P}\{x(t) - x(s) = m\} = \frac{[a(t) - a(s)]^m}{m!} \exp\left(-[a(t) - a(s)]\right), \tag{2.30}$$
$$m = 0, 1, 2, \ldots$$

If the process is homogeneous, then

$$a(t) = at,$$
$$\mathbf{P}\{x(t) - x(s) = m\} = \frac{a^m(t-s)^m}{m!} \exp(-a(t-s)), \tag{2.31}$$
$$s < t, \quad m = 0, 1, 2, \ldots$$

3. Markov processes

The concept of a Markov process is based on the hypothesis about nonhereditary systems (also known as systems "without aftereffect"). At each time t, such a system has some state $\vec{x}(t)$, which depends on its state $\vec{x}(s)$ at a preceding time $s < t$ only. All admissible states form a certain set $B \in X$, where X is the phase space of the system. Assume that the system evolves stochastically, i.e., its state at a time t is not uniquely defined by the state at a preceding time s, thereby being random. Denote by $P(s, \vec{x}, t, B)$ the probability of the event $\{\vec{x}(t) \in B (s < t)\}$ under the condition $\vec{x}(s) = \vec{x}$, which is often called *the transition probability* or the conditional probability of the system state's belonging to a set B at a time t given its state $\vec{x}$ at a time s in the phase space X.

Suppose that there exists an intermediate state $\vec{x}(u)$ between the states $\vec{x}(s)$ and $\vec{x}(t)$, where $s < u < t$. Denote by $P(s, \vec{x}, u, \vec{y}, t, B)$ the conditional probability of the event $\{\vec{x}(t) \in B\}$ given $\vec{x}(s) = \vec{x}$ and $\vec{x}(u) = \vec{y}$. Using the general properties of conditional probabilities, write

$$P(s, \vec{x}, t, B) = \int_X P(s, \vec{x}, u, \vec{y}, t, B)\, P(s, \vec{x}, u, \vec{y})\, d\vec{y}. \qquad (2.32)$$

For a nonhereditary system, a natural assumption is

$$P(s, \vec{x}, u, \vec{y}, t, B) = P(u, \vec{y}, t, B). \qquad (2.33)$$

In this case, equality (2.34) takes the form

$$P(s, \vec{x}, t, B) = \int_X P(u, \vec{y}, t, B)\, P(s, \vec{x}, u)\, d\vec{y}. \qquad (2.34)$$

This is *the Chapman–Kolmogorov equation*, which underlies the theory of Markov processes [47, 65, 166].

A Markov process is called *homogeneous* if the transition probabilities depend on the difference of the arguments t and s only, i.e.,

$$P(s, \vec{x}, t, B) = P(t - s, \vec{x}, B) = P_{t-s}(\vec{x}, B). \qquad (2.35)$$

For the homogeneous Markov processes, the Chapman–Kolmogorov equation (2.34) can be written in the simpler form

$$P_{u+v}(\vec{x}, B) = \int_X P_u(\vec{x}) P_v(\vec{y}, B)\, d\vec{y}. \qquad (2.36)$$

The definitions given in this section are wide, thereby embracing the Markov processes that differ in their sources and properties. In the sequel, we will discuss some classification principles and certain classes of Markov processes. The first attribute for classification consists in the phase space of the system in which a Markov process is observed. From this viewpoint the elementary Markov processes are the ones with finite or denumerable sets of admissible states. In the latter case, by imposing some analytical constraints on transition probabilities, we can linearize the Chapman–Kolmogorov equations and obtain their linearized counterparts. For general phase spaces, it is possible to define the classes of Markov processes in a wide sense, enduing transitional probabilities with more adequate properties in terms of real system behavior. This viewpoint allows introducing the following classes of Markov processes:

1. *Jump processes.* They correspond to the idea of the processes that move to a certain point of a phase space, staying at this point during a random period of time, and then jump to another point of the space, and so on.

2. *Processes with discrete interference of events.* These processes are generated by a dynamic system whose trajectories suffer from simple discontinuities with random jumps at random times.

3. *Diffusion processes.* They represent the processes in finite-dimensional dynamic systems that have Brownian motion on small periods of time.

4. *Markov processes with independent increments.* The sources of such processes are the finite-dimensional dynamic systems that behave as the processes with independent increments on small periods of time.

For the above-mentioned classes of Markov processes, it is also possible to perform linearization of the Chapman–Kolmogorov equation, transforming it to a linear partial differential equation or an integro-differential equation.[4]

2.2 DIGITAL SOURCE OF DATA

As a matter of fact, digital data are represented by arrays of decimal or binary numbers. They are generated using digital measuring devices. However, we often deal with an analog measuring device. Then these arrays are the result of a digital transformation of initial analog information about a studied object or phenomenon. In other words, an analog-to-digital converter (ADC) is placed after the studied object or phenomenon, as illustrated in Fig. 2.1. For example, if the studied object operates on a given time interval $\mathcal{T}$, thereby generating a certain continuous process, an ADC transforms this process into digital arrays. An ADC performs the amplitude and time quantization of an initial continuous process. A photograph is another example in which a visual image acts as an analog source. However, a photograph is transformed into a digital array using a slightly different ADC with scanning.

---

[4]For a detailed description of such transformation procedures, the reader is referred, e.g., to the monograph [65].

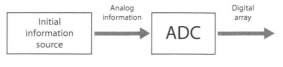

Figure 2.1

2.2.1 Amplitude and time quantization

Amplitude quantization decomposes a certain set of values of a function into a finite number of subsets. If the function under consideration is scalar, then the matter concerns quantization intervals. For a vector function, the concept of quantization has a close relation to the decomposition of the set of values of its components into elementary volumes. The simplest type of quantization is the division of a range $\Delta\phi = \phi_{max} - \phi_{min}$ of an analog function by an integer number, with rounding up or down to the nearest integer number. As a rule, quantization runs in binary code. The range $\Delta\phi$ is divided by 2^n, where n denotes the register length (or the digit count). Each amplitude is coded by a sequence of zeros and ones of length n.

Time quantization, further termed time sampling (in our case, the argument of a function is time), is performed using a pulse generator. Note that sampling rate must be at least twice as large as the maximum frequency of the function spectrum.

Consequently, values of an analog function (information, signal) are represented by a sequence of binary numbers that corresponds to the values of this function at discrete times.

Amplitude quantization is a nonlinear transformation for the values of an analog function [197]. Figure 2.2 shows the characteristic of a nonlinear element that performs the amplitude quantization of an analog function. Amplitude quantization can be uniform and nonuniform. In the former case, the quantization step $h_k = y_{k+1} - y_k$ is a constant h. Under nonuniform amplitude quantization, this step depends on the amplitude. Nonuniform amplitude quantization reduces the relative error of digital data generation. Figures 2.3a–2.3c demonstrate the characteristics of quantizers with different rounding rules: a—rounding up to the nearest integer number, b—rounding down to the nearest integer and c—rounding to the middle of the elementary interval.

Time sampling causes no measurement errors whereas amplitude quantization is accompanied by an error ε due to rounding. For

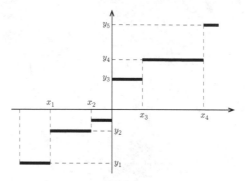

Figure 2.2

estimating this error, a certain model has to be chosen. Consider two models, namely, deterministic and stochastic ones. Within the framework of the first model, the above-mentioned rounding rules lead to the maximum error

$$\varepsilon_{max} = h. \tag{2.37}$$

In other words, it coincides with the quantization step h. Assuming the quantization error to be a random variable, we have to declare its probability density function (PDF). For example, let this variable obey the uniform distribution on the interval $[-\frac{h}{2}, \frac{h}{2}]$, i.e.,

$$p(\varepsilon) = \begin{cases} \frac{1}{h} & \text{for } |\varepsilon| \le \frac{h}{2}, \\ 0 & \text{for } |\varepsilon| > \frac{h}{2}. \end{cases} \tag{2.38}$$

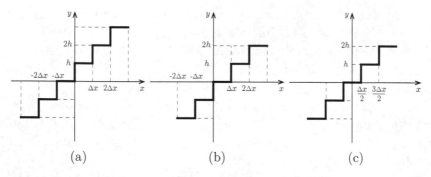

(a)   (b)   (c)

Figure 2.3

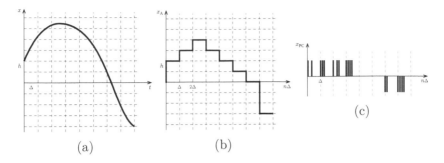

Figure 2.4

Since the PDF has symmetry, the mean of the quantization error is zero whereas its variance is given by

$$\sigma^2 = \frac{1}{h} \int\limits_{-h/2}^{h/2} \varepsilon^2 d\varepsilon = \frac{h^2}{12}. \tag{2.39}$$

Interestingly, the derived estimate is considerably smaller than the one of the deterministic model (2.37). However, recall that this estimate of the quantization error proceeds from the hypotheses of uniform distribution and uncorrelatedness with the analog function.

2.2.2 Audio data

The digital conversion of sound signals is performed by an ADC, which transforms an analog signal into digital audio data. These transformations use different types of digital modulation. Pulse code modulation (PCM) and delta modulation (DM) are most widespread.

PCM transforms a sequence of the instantaneous amplitudes of an analog signal $x(t)$ (Fig. 2.4a) into the quantized values with a fixed step h that correspond to discrete times $n\Delta$ (Fig. 2.4b). Each value $x_{\mathrm{PC}}[n\Delta]$ is transformed into a binary code, as illustrated in Fig. 2.4c. Such a modulator outputs packets of 0's and 1's with time binding.

The accuracy of PCM depends on the step h and the number N of quantization amplitudes. The largest quantized value is $(N-1)h$. The number of binary code combinations makes up $n = [\log_2(N-1)h]$, where $[\cdot]$ denotes the integer part operator.

DM is a special type of PCM that also represents an analog signal in the form of discrete samples of time with amplitude quantization. By assumption, there exists a relation between the samples in an audio signal.

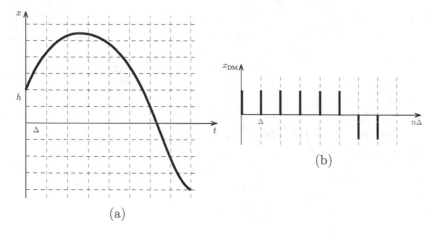

Figure 2.5

Therefore, it is possible to use only one digit for quantization of neighbor samples. Figures 2.5a–2.5b illustrate operation of a delta modulator. As its output this modulator (see Fig. 2.5a) generates a binary sequence of pulses, +1 for positive difference and −1 for negative difference.

The digital representation of sound signals is used in two ways as follows. First, for improving the quality of sound and further conversion back to the analog form. Second, for forming quantitatively measured attributes that characterize sound signals [124]. In this context, above all attributes, note the spectral composition of a signal and its basic parameters.

Consider again an analog signal described by a continuous function $x(t)$. Let this function be defined and square integrable on the infinite interval $(-\infty, +\infty)$. Then the Fourier transform of the function is given by

$$\Phi_1(\nu) = \int_{-\infty}^{+\infty} x(t)\exp(-2\pi\iota\nu t)dt = \Phi(2\pi\nu),$$

$$\Phi_2(\omega) = \frac{1}{\sqrt{2\pi}}\int_{-\infty}^{+\infty} x(t)\exp(-\iota\omega t)dt = \sqrt{2\pi}\Phi(\omega),$$

$$\Phi(\omega) = \int_{-\infty}^{+\infty} x(t)\exp(-\iota\omega t)dt. \tag{2.40}$$

Here $\iota = \sqrt{-1}$, ν denotes linear frequency (in Hz), and $\omega = 2\pi\nu$ is angular frequency (in rad/s). Formulas (2.40) incorporate the complex exponents

$$\exp(-\iota\omega t) = \cos(\omega t) - \iota \sin(\omega t). \tag{2.41}$$

The function $\Phi(\omega)$ is continuous for $\omega \in (-\infty, +\infty)$.

A real analog signal is defined on a finite time interval $\mathcal{T} = [0, T]$. A digital transformation of this signal yields a sequence of quantized values called samples of the form

$$x[n\Delta] = \int\limits_{-\infty}^{+\infty} x(\tau)\,\delta(n\Delta - \tau)d\tau, \tag{2.42}$$

where $\delta(\bullet)$ denotes the delta-function, $\Delta = \frac{T}{N-1}$, and N is the number of samples. Therefore, the digital representation of an analog signal on a finite time interval $\mathcal{T}$ is described by an N-dimensional vector $\mathbf{x}$ with the elements $x[n\Delta]$ (2.42). According to (2.40), the Fourier transform of this signal is given by

$$\Psi(\omega) = \Psi\left(\frac{2\pi}{T}k\right) = \Psi(k) = \sum_{n=0}^{N-1} x[n\Delta]\,\exp\left(-\iota\frac{2\pi}{T}k\,n\right) =$$

$$= \sum_{n=0}^{N-1} x[n\Delta]\left[\cos\left(\frac{2\pi}{T}kn\right) - \sin\left(\frac{2\pi}{T}kn\right)\right], \tag{2.43}$$

$$k = \overline{0, N-1},$$

and called the discrete Fourier transform (DFT). Clearly, the DFT as the discrete spectrum of a digital signal is characterized by a vector $\bar{\Psi}$ with the elements $\Psi(k)$ (2.2.2). Note that the dimension of this vector coincides with that of the vector $\mathbf{x}$ composed of the digital samples (2.42); see Figs. 2.6a and 2.6b.

The vector $\bar{\Psi}$ determines the frequency distribution of the energy of the sound signal under consideration. As a rule, the interval Δ is sufficiently small to preserve the informational characteristics of the analog signal on the time interval $\mathcal{T}$. Subsequently, the vector $\bar{\Psi}$ has a rather high dimension, which complicates the use of the discrete spectrum in applications. On the basis of this vector, it is necessary to obtain a reduced vector $\mathbf{z}$ with an essentially smaller dimension in comparison with the initial vector $\bar{\Psi}$. A simple (albeit, not most objective and informative) solution is to assign a small number of intervals

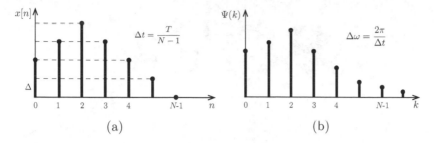

Figure 2.6

$[0, k_1], [k_1+1, k_2], \ldots, [k_s+1, N-1]$ and to construct the so-called energy spectrum, i.e., a vector $\mathbf{z}$ with the elements

$$z_j = \sum_{i=0}^{k_1} \Psi(i) + \sum_{i=k_{(j-1)}+1}^{k_j} \Psi(i), \qquad j = \overline{2, m}, \quad m \ll (N-1). \quad (2.44)$$

Then the total energy takes the form

$$E = \sum_{k=0}^{N-1} \Psi(k). \qquad (2.45)$$

2.2.3 Graphical data

Graphical data are the result of digitizing 2D analog monochrome or color images. The examples of such images include paintings, manuscripts, photo and video images, etc. The images are transformed into graphical data in two ways, using bit image (raster) graphics and vector graphics.

Bit image graphical data represent a matrix of fixed-size *pixels*. The digitization of analog images is performed through spatial discretization. More specifically, an image is split into separate elements (pixels) and each pixel is associated with 2D coordinates and a grey gradation (for monochrome images) or a color palette (for color images). The coordinates and palettes have digital form. Spatial discretization expresses graphical data as a matrix of finite dimensions in which each row contains a definite number of pixels with an appropriate color in a digital scale. At the hardware level, the bit image conversion of an analog image is implemented by a scanner. An example of bit image graphical data is given in Fig. 2.7a.

Vector graphical data involve a mathematical description of the elementary graphical objects called *primitives*. Among such objects, note

(a)

(b)

Figure 2.7

points, straight lines, some curves, splines, circles, polygons, etc. An example of vector graphics primitives is shown in Fig. 2.7b. The major advantage of vector data over bit image data consists in a rather simple scalability without compromise to the original quality of an image.

In most cases, analog images are transformed into digital for further automatic processing using digital devices such as computers, smartphones, video and photo cameras, and so on. Consider the images of standard (fixed-form) documents, e.g., billing and banking documents, fiscal documents, voting ballots and questionnaires. These documents have a definite structure with fields to be filled. In other words, useful graphical data is contained only in corresponding boxes. Hence, there is no need to scan the entire document, just the requisite fields only. Therefore, the bit image representation of such a document has a fixed structure and a bit image representation of its fields.

Note that vector formats are often transformed into bit image ones. At the same time, the converse transformation is computationally intensive.

2.2.4 Text data

Text is a peculiar object for digitization that essentially differs from the above-mentioned ones. Information about audio, video and graphical objects is completely reflected by physical measurements. In contrast, text is the result of integration of some rules and human perceptions of these rules. Subsequently, text has meaning as its distinctive attribute, unlike the objects of the other types. As a matter of fact, the notion of meaning is ambiguous, which causes its implicit description in terms of numerical characteristics.

Therefore, text data are numerical data yielded by some model of text [167]. We will comprehend a model of text as a set of numerical indicators used for its characterization. Besides easily calculated indicators, there exist the so-called syntax trees (or diagrams) and semantic networks with construction procedures of giant computational complexity.

Text modeling involves different types of information about text, i.e., text-relevant information and external information. The following types of text-relevant information are widespread:

- the frequency characteristics of text elements (in the elementary case, the presence or absence of words (terms) from a given list in a document);

- the mutual position of words in a document. Words have a definite order in accordance with the syntax rules of a given language, thereby forming groups and sentences. The mutual position of words contain some useful information;

- formatting of a document. As a rule, before publication a text is appropriately formatted. Separate words, sentences or even paragraphs can be set in different types (bold, italic, etc.) or aligned;

- the logical structure of a document. A document often consists of chapters, sections, subsection and paragraphs. Such a logical structure of a document reflects the meaning of text;

- references to other documents. Documents often refer to other documents (e.g., in the form of hypertext links or bibliographic citations).

External information includes information about the source, author, date of creation or modification of a document, subject classification, user's behavior with respect to a document (e.g., user's access rate, etc.).

The frequency characteristics of a document are most widely used, in the first place, the histogram of homogeneous attributes, e.g., the vector of entries of keywords [160]. In this case, a common glossary of terms has to be managed for all documents analyzed. Each document is represented as a vector whose elements show the occurrence of corresponding terms in the document. This vector has a very large dimension that matches the total number of terms in the glossary.

In the general setting, an element of this vector represents an informational weight of a term in a document. Actually, the informational weight of a term is treated in a wider sense with additional attributes, e.g., the informational weight of a short sequence of letters (e.g., a combination of two letters), syllables, words, or normal-form word combinations [175]. The weight of an element is often defined using some statistical characteristics of its occurrence in text, with certain coefficients that reflect the format and position of this element in a given document.

In the first stage of a numerical characterization of text, the extended term weight allows to separate out a set (bag) of most adequate words for a document. The bag-of-words model was suggested in 1975, still remaining a commonly used model of text [170]. Within the framework of this model, text is represented as a set of words, regardless of their order. A certain weight is assigned for each word in this set. Therefore, the model of a document consists of a finite set of word-weight pairs. Note that weights can be assigned to words (terms) or stems. There exist the following standard methods to assign weights [93]:

- *the binary method (BI)*. This method defines only the presence or absence of certain terms in a document;

- *the number of term entries in a document*. If a term appears many times in text, the document under consideration is likely to have a semantic relation to this term;

- *the term frequency (TF)*. The term frequency is calculated as the number of term entries divided by the total number of words in a document;

- *the logarithmical term frequency (LOGTF)*. This method defines the weight of a term as $1 + \log(\text{TF})$, where TF is the term frequency. Due to the logarithmical scale of informational weight assignment, the model has higher stability against overestimation of documents; *the inverse document frequency (IDF)*. This parameter is the inversion of the term frequency.

A conventional approach is to analyze texts using a collection of documents described by a term-weight matrix.

2.2.5 Government statistical data

There exist many objects that are assessed by government statistics services using problem-oriented registers associated with definite time scales. In the first place, note the measurements and state assessments for the economic system (macroindices, labour resources, GDP, finances, etc.), the demographic system (fertility, mortality and migration rates), and the environmental system (atmosphere, earth, water, soil, and so on). A considerable part of statistical measurements are performed for different regions of a country.

A government statistics system implements the acquisition, analysis and structuring of all measured parameters, as well as their further transformation into data in accordance with an accepted register. Each of these stages, from the acquisition of information from existing sources to data structuring in tabular form, is accompanied by inevitable errors, also causing *errors* in government statistical data. There exist two types of errors, namely, *systematic errors* and *coverage errors*. The former occur due to imperfect methods for transforming initial information into data, whereas the latter due to the mismatch between the target object and the elements actually observed.

To elucidate this point, let us consider an example of data formation to calculate per capita national income. This problem seems trivial: just divide the annual GDP in monetary terms by the amount of able-bodied population. However, the annual GDP in monetary terms is calculated using generalized prices that are defined by certain models. The input variables and characteristics (parameters) of these models are measured and estimated with errors (often interpreted as random variables). Therefore, the annual GDP is the arithmetical mean by month with the estimated variance. A similar situation takes place for able-bodied population, as its size varies unpredictably in the course of a year. In other words, the calculated value of the annual GDP incorporates both types of errors.

Consider the following mathematical model of this situation. Denote by x an index of interest that is associated with a time t. Assume that this index depends on n factors, $y_1, \ldots, y_n$. Such an assumption leads to coverage errors: there is no evidence that the accepted set of factors (i.e., their number and meaning) is complete for defining the index x.

The next stage in formation of the index x is to model a parameterized relation between the factors and x, that is,

$$x = F(\mathbf{y}, \mathbf{a}), \tag{2.46}$$

where $\mathbf{a} = \{a_1, \ldots, a_p\}$ act as parameters. The function F and its parameterization are chosen via expertise. Exactly this stage introduces systematic errors due to limited, incomplete and subjective knowledge of experts.

As a matter of fact, a government statistics system has a distributed structure and involves various primary processing methods for initial information. In combination with the human factor, these features almost rule out the possibility of adequate estimation for the government statistics errors. A remaining alternative is to use expert appraisals, which have a rather large spread from 5% to 20%.

The existing errors essentially affect (a) the estimated characteristics of relations between the input and output data and also (b) the solutions of other problems involving government statistical data.

2.3   RESTORATION METHODS FOR MISSING DATA

Digital sources generate sequences of data that are bound to time scales. A common situation in practice is the existence of different time scales or omissions for a certain time scale. However, for further usage of data sequences, one often needs to perform their synchronization (reduction to the same scale) or to restore missing data. As a matter of fact, data synchronization and restoration are close problems and their solution employs similar methods. Consider some of the methods in detail.

2.3.1   Interpolation

A digital source of data generates a sequence of values of an analog function $x(t)$ for a finite set of values of its argument. As a rule, these values are associated with a certain time scale with a step h. Take another function $y(t)$ with a generated sequence of values in a different time scale with a step $s < h$ such that $\{h/s\}$ is an integer. Then we naturally arrive at the synchronization of these functions, i.e., it is necessary to restore the values of $x(t)$ within the intervals $[ih, (i+1)h]$ with the step s (the so-called data interpolation problem). The classical approach to solve this problem is to construct a rather simple function $z(t)$ so that its values in the h-scale coincide with the values of the function $x(t)$ in the same scale, that is, $z[0] = x[0], \cdots, z[ih] = x[ih]$.

Consider an arbitrary real function $x(t) \in R$ defined on an interval $\mathcal{T} = [0, T]$. In space R choose a finite collection of elementary (simple) functions $\{\varphi_k(t)\}_0^n$ so that they are linearly independent. In most cases, this is a sequence of power functions $1, t, t^2, \ldots$; or trigonometric functions $1, \sin t, \cos t, \sin 2t, \cos 2t, \ldots$; or exponential functions $1, \exp(\alpha_1 t), \exp(\alpha_2 t), \ldots$, where $\{\alpha_i\}$ denotes a number sequence.

Using the first $n+1$ functions $\{\varphi_k(t)\}_0^n$, construct all possible linear combinations

$$a_0\varphi_0(t) + a_1\varphi_1(t) + \cdots + a_n\varphi_n(t) = \psi(t) \qquad (2.47)$$

with real coefficients $a_0, \ldots, a_n$. In interpolation theory, these coefficients are calculated in the following way. Choose $n+1$ values of the variable t, i.e., $t_0, \ldots, t_n$, which are called interpolation nodes. Let the function $x(t)$ and its interpolating function $\psi(t)$ have the same values at these nodes. This leads to the system of linear equations

$$x(t_j) = a_0\varphi_0(t_j) + a_1\varphi_1(t_j) + \cdots + a_n\varphi_n(t_j), \qquad j = \overline{0, n}. \qquad (2.48)$$

The nodes should be chosen so that the determinant

$$\Delta = \begin{vmatrix} \varphi_0(t_0) & \cdots & \varphi_n(t_0) \\ \cdots & \cdots & \cdots \\ \varphi_0(t_n) & \cdots & \varphi_n(t_n) \end{vmatrix} \neq 0. \qquad (2.49)$$

Then the system possesses the unique solution

$$a_i = \frac{\sum_{j=0}^n x(t_j)\Delta_{ij}}{\Delta}, \qquad i = \overline{0, n}, \qquad (2.50)$$

where Δ_{ij} gives the algebraic complement of element (ij). Thus, according to (2.47),

$$\psi(t) = x(t_0)\Phi_0(t) + x(t_1)\Phi_1(t) + \cdots + x(t_n)\Phi_n(t). \qquad (2.51)$$

The functions $\Phi_i(t)$ are linear combinations of the functions $\phi_i(t)$ and do not depend on the function $x(t)$. Since the equalities $x(t_j) = \psi(t_j)$ hold at the interpolation nodes, the functions $\Phi_i(t)$ satisfy the following conditions:

$$\Phi_i(t_j) = \begin{cases} 0 & \text{for } i \neq j, \\ 1 & \text{for } i = j. \end{cases} \qquad (2.52)$$

See, e.g., the monograph [17] for a discussion of refined interpolation procedures and their properties.

As background information consider some classes of interpolating functions that are widely used in applications. First, *the Lagrange interpolating polynomial.* Consider a sequence of power functions $1, t, t^2, \ldots, t^n, \ldots$ as a basic one. This sequence is linearly independent on any interval. For defining the function $\psi(t)$ (2.47), find the functions

$\Phi_i(t)$ that are linear combinations of the functions $\phi_0(t), \ldots, \phi_n(t)$ and satisfy conditions (2.52). In other words, using the elements of the basic sequence, it is required to construct a polynomial that vanishes at the points $t_0, \ldots, t_{i-1}$ and takes value 1 at the point t_i. Such a polynomial has the form

$$\Phi_i(t) = A(t - t_0)(t - t_1) \cdots (t - t_{i-1})(t - t_{i+1}) \cdots (t - t_n). \qquad (2.53)$$

Since $\Phi_i(t_i) = 1$, it follows that

$$1 = A(t_i - t_0)(t_i - t_1) \cdots (t_i - t_{i-1})(t_i - t_{i+1}) \cdots (t_i - t_n). \qquad (2.54)$$

Subsequently,

$$\Phi_i(t) = \frac{(t - t_0)(t - t_1) \cdots (t - t_{i-1})(t - t_{i+1}) \cdots (t - t_n)}{(t_i - t_0)(t_i - t_1) \cdots (t_i - t_{i-1})(t_i - t_{i+1}) \cdots (t_i - t_n)}. \qquad (2.55)$$

And the Lagrange interpolating polynomial is given by

$$
\begin{aligned}
\psi(t) = L_n(t) \quad = \quad & x(t_0) \frac{(t - t_1) \cdots (t - t_n)}{(t_0 - t_1) \cdots (t_0 - t_n)} + \\
+ \quad & x(t_1) \frac{(t - t_0)(t - t_2) \ldots (t - t_n)}{(t_1 - t_0)(t_1 - t_2) \ldots (t_1 - t_n)} + \cdots + \\
+ \quad & x(t_n) \frac{(t - t_0) \ldots (t - t_{n-1})}{(t_n - t_0)(t_n - t_1) \ldots (t_n - t_{n-1})}. \qquad (2.56)
\end{aligned}
$$

Introduce the function

$$\omega_n(t) = \prod_{i=0}^{n} (t - t_i) \qquad (2.57)$$

and write the Lagrange interpolating polynomial as

$$L_n(t) = \sum_{i=0}^{n} x(t_i) \frac{\omega_n(t)}{(t - t_i)\omega_n'(t_i)}. \qquad (2.58)$$

In the sense of applications, the Lagrange interpolating polynomial is somewhat inconvenient, as adding a new node requires forming a new polynomial. This drawback takes no place for the interpolating polynomial in the Newton representation. More specifically, *the Newton interpolating polynomial* involves the notion of divided differences [17] as follows.

Consider $(n + 1)$ nodes and as many values of a function $x(t)$, i.e., $x(t_0), \ldots, x(t_n)$. The first divided difference of the function $x(t)$ at nodes t_{i-1}, t_i is the value

$$\Delta_{i-1,i} = \frac{x(t_i) - x(t_{i-1})}{t_i - t_{i-1}}. \tag{2.59}$$

The higher-order divided differences are calculated recursively. The divided difference of order m has the general formula

$$\Delta_{i-1,i,i+1,\ldots,i+m} = \frac{\Delta_{i,i+1,\ldots,i+m} - \Delta_{i-1,i,i+1,\ldots,i+m-1}}{t_{i+m} - t_i}. \tag{2.60}$$

Express the the Lagrange interpolating polynomial as

$$L_n(t) = L_0(t) + (L_1(t) - L_0(t)) + \cdots + (L_n(t) - L_{n-1}(t)). \tag{2.61}$$

Clearly, this is the identity $L_n(t) = L_n(t)$. It is possible to show [17] that

$$L_i(t) - L_{i-1}(t) = \Delta_{0,\ldots,i}\omega_i(t), \tag{2.62}$$

where $\omega_i(t) = (t - t_0) \cdots (t - t_i)$. Then the Newton representation $N_n(t)$ of the Lagrange interpolating polynomial can be written as

$$L_n(t) = N_n(t) = x(t_0) + \Delta_{0,1}(t - t_1) + \cdots + \\ + \Delta_{0,\ldots,n}(t - t_0)(t - t_1) \cdots (t - t_{n-1}). \tag{2.63}$$

Once again, note that an interested reader may derive deep knowledge of the interpolating polynomials from the monograph [17].

2.3.2 Auxiliary dynamic models

In this subsection, we describe a restoration method for missing data (in a time scale) that is based on mathematical models of a data source. Let $\mathcal{X}_h^N = \{x[0], x[h], \ldots, x[Nh]\}$ be a packet of numerical data bound to a time scale with a step h. Using these data, construct a step function $\tilde{x}(t)$ of the form

$$\tilde{x}(t) = x[ih] \quad \text{for } ih \leq t < (i+1)h, \quad i = \overline{0, N-1}. \tag{2.64}$$

Consider a dynamic system with the above step function as its input and an output $y(t)$ of the form

$$y[nh] = \mathcal{L}(x[ih], \mathbf{a} \mid 0 \leq i < n). \tag{2.65}$$

In this expression, $\mathcal{L}$ denotes a functional that characterizes the dynamic properties of the system with parameters $\mathbf{a} \in R^m$, where m is the dimension of the parameter vector. The parameters $\mathbf{a}$ actually affect the function $y(t)$, i.e., $y(t, \mathbf{a})$. For some values of the elements of the vector $\mathbf{a}$, the system outputs a packet $\mathcal{Y}_h^N = \{y[0, \mathbf{a}], y[h, \mathbf{a}], \ldots, y[(N-1)h, \mathbf{a}]\}$ at discrete times ih.

Define the total residual between packets $\mathcal{X}_h^N, \mathcal{Y}_h^N$ as

$$J(\mathbf{a}) = \sum_{i=0}^{N} (x[ih] - y[ih, \mathbf{a}])^2. \tag{2.66}$$

The minimum of this residual gives the optimal parameters under which the difference between the above-mentioned packets at the nodes of the h-scale is as small as possible (in terms of the functional J (2.66)), i.e.,

$$\mathbf{a}^* = \arg\min J(\mathbf{a}). \tag{2.67}$$

Substituting the optimal parameters into (2.65) yields the packet $\mathcal{Y}_s^{Nr} = \{y^*[0], y^*[s], \ldots, y^*[(l-1)s], y^*[h], y[h+s], \ldots, y^*[h+(l-1)s]; y^*[2h], y^*[2h+s], \ldots, y^*[Nh]\}$, where $s < h$, $h/s = r$ and r is an integer.

This method will be illustrated using an example of a linear dynamic system with a pulse response $w(t, \mathbf{a})$, $t \geq 0$. Let

$$w(t) = a_1 \exp(-a_2 t), \tag{2.68}$$

where a_1, a_2 are interval-type unknown parameters, i.e., the ones belonging to some intervals $0 < a_1 \leq a_1^+$ and $a_2^- \leq a_2 \leq a_2^+$, respectively. Generally speaking, the limits of these intervals are unknown. A possible way to find them consists in the global minimization of J (2.66) in a_1^+, a_2^- and a_2^+ (or calculation of an admissible small value if the global minimum does not exist).

Supply the step function $\tilde{x}(t)$ to the input of this system to generate the output

$$y(t) = a_1 \int_0^t \exp(-a_2(t - \tau)) \tilde{x}(\tau) d\tau. \tag{2.69}$$

The interval $[0, t]$ contains an integer number of subintervals $N = \{t/h\}$ and $t = Nh + \mu$, where $0 \leq \mu < (N+1)h$. Then the expression (2.69) can be written as

$$y(t) = \frac{a_1}{a_2} \exp(-a_2 t) \sum_{i=0}^{N} x[ih] \exp(a_2 ih) (\exp(a_2 h) - 1). \tag{2.70}$$

The system output at the nodes of the h-scale is given by

$$y[ih] = \frac{a_1}{a_2}\left(\exp(a_2 h) - 1\right)\exp(-a_2 ih)\sum_{j=0}^{i} x[jh]\exp(a_2 jh). \qquad (2.71)$$

Consequently, the total residual functional (2.66) takes the form

$$J(a_1, a_2) = \sum_{i=0}^{N}\sum_{j=0}^{i}\left(A_{i,j}(a_1, a_2)x[jh]\right)^2 \Rightarrow \min, \qquad (2.72)$$

where

$$A_{i,j} = \begin{cases} \frac{a_1}{a_2}\left(\exp(a_2 h) - 1\right)\exp(a_2 h(j - i)) & \text{for } j < i, \\ \frac{a_1}{a_2}\left(\exp(a_2 h) - 1\right) - 1 & \text{for } j = i. \end{cases} \qquad (2.73)$$

Note that the functional (2.72) possesses a rather complicated structure for an analytical study of its properties such as unique minimum, convexity (concavity), and so on. Therefore, in applications the missing data should be restored (or synchronized with other data packets) using numerical global minimization methods.

As soon as the parameters a_1 and a_2 are calculated, the missing data with the step s can be restored by

$$\begin{aligned} y[ih + ks] &= \frac{a_1}{a_2}\exp(-a_2(ih + ks))\sum_{i=0}^{N} x[ih]\exp(a_2 ih)\left(\exp(a_2 h) - 1\right), \\ i &= \overline{0, N}, \ k = \overline{0, M}, \ M = \{h/s\}. \end{aligned} \qquad (2.74)$$

2.3.3 Spatial entropy decomposition

As emphasized earlier, the data formed by government statistics services are associated with certain spatial units (regions), each having a set of attributes, i.e., indicators. The dimensions of such regions and information about the values of these indicators are often related to each other. In many cases, the smaller is a region under consideration, the less accessible is information about its indicators. This leads to a rather topical problem of "spreading" the available state indicators of large regions to small districts within them using some indirect information about the latter.

Consider an economic space $\mathbb{E}$ consisting of I regions. Each region i contains M_i districts, as shown in Fig. 2.8. Let the regions i and districts within them (i, j) be characterized by unique homogeneous indicators

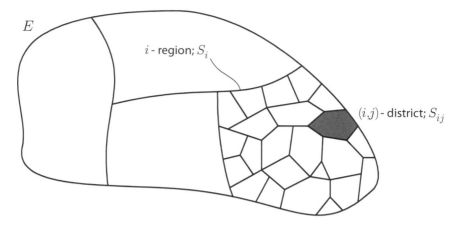

Figure 2.8

S_i and S_{ij}, respectively. For example, the list of possible indicators includes population size, production density, service endowment, and so on. The regional and district indicators, S_i and S_{ij}, satisfy the balance constraints

$$S_i = \sum_{j=1}^{M_i} S_{ij}, \qquad i = \overline{1, I}. \tag{2.75}$$

Each region and hence district produces the goods of r types or disposes of r types of resources. Assume that, for each type k of goods (resources), the production functions of region i depend on the state indicators of all regions, i.e., $\vec{S} = \{S_1, \dots, S_I\}$. Denote these functions by $Q_{ik}(\vec{S})$, $i = \overline{1, I}$, $k = \overline{1, r}$. The same hypothesis also applies to the districts; denote their production functions by $q_{ijk}(\mathbb{S})$, where $\mathbb{S} = \{S_{11}, \dots, S_{1,M_1}; \dots; S_{I,1}, \dots, S_{I,M_I}\}$. Then the resource constraints have the form

$$\sum_{j=1}^{M_i} q_{ijk}(\mathbb{S}) = Q_{ik}(\mathbf{S}), \qquad i = \overline{1, I}, \ k = \overline{1, r}. \tag{2.76}$$

Suppose that the regional indicators $\mathbf{S}$ are known whereas the district ones $\mathbb{S}$ are unknown. The problem is to find the district indicators taking into account the balance and resource constraints given by (2.75) and (2.76). The total number of the district indicators makes up $m = \sum_{i=1}^{I} M_i$, as by assumption each district and region has a single indicator. In addition, the problem includes $(r+1)I$ balance and resource constraints totally, and $m > (r + 1)I$. Therefore, numerous collections of district indicators satisfy equalities (2.75, 2.76).

An approach to obtain a unique collection of district indicators proceeds from the entropy maximization principle. As a matter of fact, entropy is a measure of uncertainty in the distribution of regional indicators over corresponding districts [138], i.e.,

$$H(\mathbb{S}) = - \sum_{i=1,j=1}^{I,M_i} S_{ij} \ln \frac{S_{ij}}{a_{ij}e}, \qquad (2.77)$$

where $\mathbb{S}$ is a matrix with entries S_{ij} that characterizes the distribution of district indicators. The distribution of highest uncertainty, $\mathbf{S}$, answers the maximum value of the entropy H under the balance and resource constraints (2.75, 2.76). Any prior information about the districts within the regions of interest can be incorporated into the parameters a_{ij} of the entropy function H.

Thus, the distribution of the regional indicators over the districts, called *entropy decomposition*, is the solution of the optimization problem

$$H(\mathbb{S}) \Rightarrow \max, \quad \mathbb{S} \in D, \qquad (2.78)$$

where

$$D = \{S_{ij} | \sum_{j=1}^{M_i} S_{ij} = S_i; \ \sum_{j=1}^{M_i} q_{ijk}(\mathbb{S}) = Q_{ik}(S), \quad i = \overline{1,I}, \ k = \overline{1,r}\}. \qquad (2.79)$$

In this general statement, all regions and districts are related to each other through corresponding production functions. Another extreme case is when all regions and districts are isolated, i.e., the regional and district production functions depend only on the state indicators of their own regions or districts:

$$q_{ijk}(\mathbb{S}) = q_{ijk}(S_{ij}), \quad Q_{ik}(\mathbf{S}) = Q_{ik}(S_i); \ i = \overline{1,I} \ k = \overline{1,r}. \qquad (2.80)$$

For the isolated regions, the entropy decomposition problem (2.78) is reduced to I optimization problems of the form

$$H_i(\mathbb{S}_i) \Rightarrow \max, \quad \mathbb{S}_i \in D_i, \qquad (2.81)$$

where

$$H_i(\mathbb{S}_i) = - \sum_{j=1}^{M_i} S_{ij} \ln \frac{S_{ij}}{a_{ij}e}, \qquad (2.82)$$

$$D_i = \{S_{ij}| \sum_{j=1}^{M_i} S_{ij} = S_i, \sum_{j=1}^{M_i} q_{ijk}(S_{ij}) = Q_{ik}(S_i), \quad i = \overline{1,I}, \ k = \overline{1,r}\}.$$

(2.83)

The production functions q_{ijk} of these districts form nonlinear functions of their state indicators. However, if an economy is in a neighborhood of an equilibrium, then the linear representation of the production functions becomes a completely adequate description of these functions, i.e.,

$$q_{ijk}(S_{ij}) = c_{ijk}S_{ij}, \quad i = \overline{1,I}, \ k = \overline{1,r}.$$

(2.84)

Here c_{ijk} indicate the specific outputs of good k in district (i,j). In this case, the admissible set D_i (2.83) is given by the system of linear equalities

$$D_i = \{S_{ij}| \sum_{j=1}^{M_i} S_{ij} = S_i; \sum_{j=1}^{M_i} c_{ijk}S_{ij} = Q_{ik}(S_i), \quad i = \overline{1,I}, \ k = \overline{1,r}\}.$$

(2.85)

Consider the specifics of the entropy decomposition procedure with the linear production functions for one region $(I = 1)$ and one type of resources $(r = 1)$:

$$H_1(\mathbb{S}_1) = -\sum_{j=1}^{M_1} S_{1j} \ln \frac{S_{1j}}{a_{1j}e} \Rightarrow \max,$$

$$\sum_{j=1}^{M_1} S_{1j} = S_1, \quad \sum_{j=1}^{M_1} c_{1j1}S_{1j} = Q_{11}(S_1).$$

(2.86)

This problem is characterized by the following parameters: the prior probabilities a_{1j}; the state indicator S_1 of the region; the specific productivities c_{1j1} and production function Q_{11} of the region. The solution of the problem, i.e., the entropy-optimal distribution of the regional state indicators over their districts depends on these parameters.

Introduce new variables in order to slightly reduce the number of parameters in the problem. Denote

$$w_{1j} = \frac{S_{1j}}{S_1}, \quad h_{1j1} = \frac{c_{1j1}}{Q_{11}(S_1)}, \ j = \overline{1, M_1}.$$

(2.87)

If the regional indicator S_1 is given, then the variables h_{1j1} are also fixed. With these notations, the entropy decomposition problem can be written as

$$H(\mathbf{w}) = -\sum_{j=1}^{M} w_j \ln \frac{w_j}{a_j e} \Rightarrow \max,$$

$$\sum_{j=1}^{M} w_j = 1; \quad \sum_{j=1}^{M} h_j w_j = q, \tag{2.88}$$

where $q = 1/S_1$ and the subscripts $i = k = 1$ are omitted.

Solve the problem using the Lagrange method of multipliers. The Lagrange function has the form

$$L(\mathbf{w}, \mu, \lambda) = H(\mathbf{w}) + \mu(1 - \sum_{j=1}^{M} w_j) + \lambda(q - \sum_{j=1}^{M} h_j w_j). \tag{2.89}$$

The first-order optimality conditions of the Lagrange function yield the equations

$$w_j = a_j \exp(-\mu - \lambda h_j) = a_j\, u\, z^{h_j};$$
$$u = \exp(-\mu), \ z = \exp(-\lambda). \tag{2.90}$$

Taking into account the balance and resource constraints in (2.88), write the following equation for the Lagrange multiplier λ:

$$\sum_{j=1}^{M} a_j\, z^{h_j}\, (h_j - q) = 0, \qquad j = \overline{1, M}. \tag{2.91}$$

Set

$$c_j = a_j(h_j - q), \qquad j = \overline{1, M}, \tag{2.92}$$

and transform equation (2.91) to

$$f(z) = \sum_{j=1}^{M} c_j z^{h_j} = 0, \quad z > 0. \tag{2.93}$$

Order the parameters h_j so that

$$0 < h_1 < h_2 < \cdots < h_l < q < h_{l+1} < \cdots < h_M. \tag{2.94}$$

As a result,

$$c_j < 0 \ \text{ for } j = \overline{1, l} \quad \text{and} \quad c_j > 0 \ \text{ for } j = \overline{l+1, M}. \tag{2.95}$$

Consider equation (2.93) with the coefficients c_j of arbitrary sign. Note that it has the root $z_0 = 0$. Hence, this equation can be written as

$$f(z) = z^{h_1} f_1(z) = 0, \quad z > 0, \tag{2.96}$$

where the function

$$f_1(z) = c_1 + \sum_{j=2}^{M} c_j z^{h_j - h_1} \tag{2.97}$$

has no zero roots.

Theorem 2.1. *Assume that the positive numbers $h_1, \ldots, h_M$ satisfy condition (2.94). Then the equation*

$$f_1(z) = 0, \tag{2.98}$$

where $f_1(z)$ is defined by (2.96), has at most $M - 1$ nonzero solutions.

Proof. To establish this, take advantage of the Gauss principle in the following formulation. Let the equation $\phi(x) = 0$, where ϕ is a differentiable sufficiently many times function and x is a scalar variable, have n roots; then the equation $\phi'(x) = 0$ has $n - 1$ nonzero roots located between the roots of the original equation.

Analyze the equation

$$f_1'(x) = \sum_{j=2}^{M} c_j (h_j - h_1) z^{h_j - h_1 - 1} = 0.$$

Since $\min_j (h_j - h_1 - 1) = h_2 - h_1 - 1 > 0$, this equation has a zero root. Hence, by analogy with (2.96) it can be written as

$$f_1'(x) = z^{h_2 - h_1 - 1} f_2(z) = 0,$$

where the function

$$f_2(z) = c_2 (h_2 - h_1) + \sum_{j=3}^{M} c_j (h_j - h_1) z^{h_j - h_1 - 1}$$

has no zero roots. Continue this procedure to obtain

$$f_{M-1}'(z) = z^{h_M - h_{M-1} - 1} f_M(z) = 0,$$

where the function

$$f_M(z) = c_{M-1}(h_{M-1} - h_1) \cdots (h_{M-1} - h_{M-2}) + c_M (h_M - h_1) \cdots (h_M - h_{M-1}) z^{h_M - h_{M-1}}$$

possesses no zero roots too. Consider the equation

$$f_M(z) = 0.$$

It has the nonzero root

$$y_M^* = z_*^{h_M - h_{M-1}} = -\frac{c_{M-1}(h_{M-1} - h_1) \cdots (h_{M-1} - h_{M-2})}{c_M (h_M - h_1) \cdots (h_M - h_{M-1}) z^{h_M - h_{M-1}}}.$$

The derivative $f'_{M-1}(z)$ possesses one zero root and one nonzero root y^*_M. Due to the Gauss principle, the equation $f_{M-1}(z) = 0$ has at most three roots one of which is zero (i.e., no more than two nonzero roots). On the other hand, $f_{M-1}(z) = f'_{M-2}(z)$. Subsequently, the derivative $f'_{M-2}(z)$ possesses at most two nonzero roots. Using the Gauss principle again, it is easy to find that the function $f_{M-2}(z)$ has at most three nonzero roots. Following this line of reasoning, we finally get the desired result. ■

Now consider equations (2.96, 2.97) in which the coefficients c_j are ordered in accordance with (2.95), i.e.,

$$c_j < 0 \text{ for } j = 1, \dots, l, \qquad c_j > 0 \text{ for } j = l+1, \dots, M. \qquad (2.99)$$

Theorem 2.2. *Under conditions (2.95), equation (2.98) has a unique positive solution.*

Proof. Write equation (2.98) as

$$\phi(z) = \frac{\sum_{i=l+1}^{M} c_i z^{h_i}}{\sum_{j=1}^{l} c_j z^{h_j}} = 1, \ z \geq 0.$$

Study the qualitative properties of the function $\phi(z)$, expressing it as

$$\phi(z) = z^{h_{l+1}-h_1} \frac{\sum_{i=l+2}^{M} c_i z^{h_i} + c_{l+1}}{\sum_{j=2}^{l} c_j z^{h_j} + c_1}.$$

Clearly, $\phi(0) = 0$. Now reduce the function $\phi(z)$ to the form

$$\phi(z) = z^{h_M - h_l} \frac{\sum_{i=l+1}^{M-1} c_i z^{h_i - h_M} + c_M}{\sum_{j=1}^{l-1} c_j z^{h_j - h_l} + c_l}.$$

Due to condition (2.95), $\phi(\infty) = \infty$. Consider the derivative of the function $\phi(z)$, i.e.,

$$\phi'(z) = \frac{\sum_{j=1}^{l} \sum_{i=l+1}^{M} c_j c_i (h_i - h_j) z^{h_j + h_i}}{(\sum_{j=1}^{l} c_j z^{h_j})^2}.$$

Condition (2.95) leads to $h_i - h_j > 0$ for $i = l+1, \dots, M$, $j = 1, \dots, l$, and therefore $\phi'(z) > 0$. ■

Thus, it has been shown that equation (2.98) possesses a unique positive solution under conditions (2.95).

Example 2.1 For problems of transportation infrastructure modeling, analysis and prediction, a typical scenario is the spatial decomposition of a given territory into rather small districts (traffic cells) with state indicators measured only for their groups.

As an example, consider a region of China that contains 26 traffic cells [189]. The region under study occupies the area $P = 47 \text{ km}^2$ accommodating $S = 0.430$ million of people. This population is somehow distributed over the traffic cells. Denote the real distribution by $\tilde{S}_j$, $j = \overline{1, M}$.

The problem is to synthesize a model-based distribution S_j, $j = \overline{1,M}$, using the entropy decomposition method and to assess its adequacy to the real distribution. In this example, the state indicator of a traffic cell is the population located in it.

By assumption, there exists no prior information about the distribution. Therefore, the prior probabilities a_j in the entropy function (2.82) can be treated constant, i.e., $a_j = a$. Then the entropy expression takes a simpler form:

$$H(\mathbf{S}) = -\sum_{j=1}^{M} S_j \ln S_j.$$

The population distribution over the traffic cells must satisfy the constraints (2.83). Recall that the first group of constraints applies to balance, and the second to resources (in our case, there is only one resource). Choose the generalized accommodation cost Q of population in the given region as a resource that characterizes the distribution. Then the parameters c_j of the resource constraints are interpreted as the specific cost, i.e., the accommodation cost per one resident in cell j. Assume that this specific cost is proportional to the spatial weight of cell j, i.e.,

$$c_j = c\frac{p_j}{P}, \qquad j = \overline{1,M}.$$

Here p_j denotes the area of cell j, and c is a normalized coefficient of proportionality measured in cost units. Table 2.1 shows the corresponding values of the relative specific cost $\tilde{c}_j = c_j/c$.

In terms of the relative specific cost, the resource constraints take the form

$$\sum_{j=1}^{M} \tilde{c}_j S_j = q,$$

where $q = Q/c$ is the relative cost as an aggregated characteristic of the population distribution over the traffic cells. Let it be an exogenous parameter of the model. Table 2.1 presents the calculated model-based distributions S_j for the relative cost $q = 0.0097$. For the sake of comparison, the last columns also contain the real population distribution $\tilde{S}_j$ over the traffic cells in this region.

The adequacy of the obtained distributions was assessed using the standard deviation

$$\delta = \frac{\left(\sum_{j=1}^{M}(S_j - \tilde{S}_j)^2\right)^{\frac{1}{2}}}{\left(\sum_{j=1}^{M} S_j^2\right)^{\frac{1}{2}} + \left(\sum_{j=1}^{M} \tilde{S}_j^2\right)^{\frac{1}{2}}}.$$

In fact, the value δ considerably varies depending on the relative cost q, as illustrated by the graph in Fig. 2.9. In this graph, the minimum standard deviation corresponds to $q = 0.0097$.

2.3.4 Randomized restoration method for missing data

1. Structure of randomized restoration procedure for missing data

Consider components and properties of the informational support (infoware) of machine learning problems. A learning collection consists

Table 2.1: Regional state indicators.

no.	$\tilde{c}_j$	S_j	$\tilde{S}_j$
1	0.035	0.0098	0.0144
2	0.018	0.0251	0.0247
3	0.035	0.0098	0.0149
4	0.018	0.0253	0.0245
5	0.028	0.0145	0.0213
6	0.015	0.0296	0.0259
7	0.038	0.0083	0.0142
8	0.029	0.0137	0.0222
9	0.039	0.0079	0.0144
10	0.047	0.0051	0.0102
11	0.035	0.0098	0.0175
12	0.026	0.0161	0.0238
13	0.036	0.0093	0.0168
14	0.028	0.0145	0.0214
15	0.032	0.0116	0.0193
16	0.038	0.0083	0.0126
17	0.045	0.0057	0.0098
18	0.037	0.0088	0.0151
19	0.028	0.0145	0.0185
20	0.038	0.0083	0.0097
21	0.038	0.0104	0.0079
22	0.020	0.0225	0.0292
23	0.024	0.0181	0.0243
24	0.038	0.0084	0.0097
25	0.007	0.0459	0.0433
26	0.028	0.0145	0.0169

of some data on the functioning of an object O (Fig. 2.10) as well as on the measured input $\mathbf{x}(t) \in R^m$ and output $\mathbf{y}(t) \in R^n$ of the object O. The variables $\mathbf{x}(t)$ and $\mathbf{y}(t)$ contain measurement errors $\eta(t) \in R^m$ and $\xi(t) \in R^n$, respectively. As a rule, information about them have a hypothetical character. A common assumption is that they figure in a learning collective additively:

$$\mathbf{x}(t) = \mathbf{x}^0(t) + \eta(t); \qquad \mathbf{y}(t) = \mathbf{y}^0(t) + \xi(t), \qquad (2.100)$$

where $\mathbf{x}^0(t)$ and $\mathbf{y}^0(t)$ are pure variables.

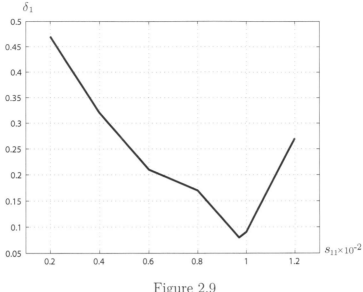

Figure 2.9

Suppose that the input and output are measured in the same time scale $t = kh$ with a step h and on the same time interval $\mathcal{T} = [0, T]$. Hence, a real learning collection consists of a data matrix describing the input and output of the object O.

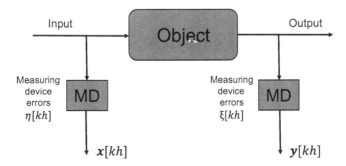

Figure 2.10

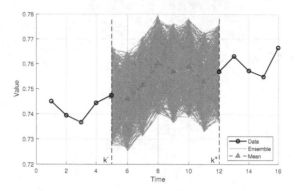

Figure 2.11: Example of missing data ensemble (MDE).

Data arrays for the input and output may have omissions; see the interval $[k^-, k^+]$ in Fig. 2.11. There are three possible cases of such omissions: omissions in output data, omissions in input data and omissions in both types of data (not necessarily at the same time instants).

The main idea of randomized restoration of missing data is to replace the existing omissions by a missing data ensemble (MDE), i.e., the random values of the input and output components (see Fig. 2.11) optimized taking into account the available measurements. This is done using a randomized parametric model (RPM) of the object, whose probabilistic characteristics (the probability density functions (PDFs) of its parameters) are obtained through maximum entropy estimation (MEE). Details will be presented in Appendix A.

The optimized MDE is generated by the sampling of the optimal PDFs [64]. This MDE can be directly used for solving different machine learning problems. Also, it is possible to determine numerical characteristics of the optimized MDE, such as mean, median, variance pipe, interquantile data sets, etc. Figure 2.11 gives an illustration of this approach.

If omissions occur in output data, the procedure employs the input data $\mathbf{x}[kh]$, the RPM-O model and the output data with omissions $\tilde{\mathbf{y}}[kh]$ for calculating the entropy-optimal PDFs (Fig. 2.12a) and also for generating the ensemble $\mathcal{Y}$ of the restored input data.

The existing omissions in input data are restored using an additional RPM-I model with the input represented by an optimized sequence of independent random vectors $\eta[kh]$ and the input data with omissions $\tilde{\mathbf{x}}[kh]$. The PDFs of the parameters and this sequence are obtained

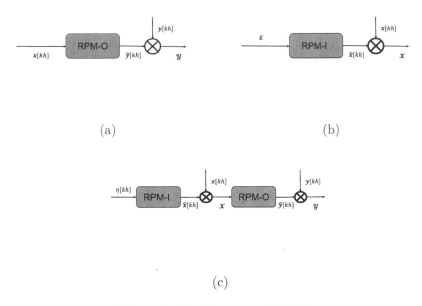

(a)   (b)

(c)

Figure 2.12: Diagrams of RPMs.

through MEE. These functions are then used to generate the ensembles $\mathcal{E}$ of the random vectors $\eta[kh]$ and an ensemble $\mathcal{X}$ of the restored input data; see Fig. 2.12b.

Finally, consider the case of omissions both in the input and output data. First, the ensembles $\mathcal{E}$ of the random vectors $\eta[kh]$ and the ensemble $\mathcal{X}$ of the input data are restored. Second, they are adopted for restoring the ensembles $\mathcal{Y}$ of the output data (Fig. 2.12c).

The set of randomized restoration methods for missing data will be denoted by *RRMD* (Random Restored Missing Data).

2. Optimization algorithms for MDE

a. Omissions in output data

Consider a discrete observation time interval $\tau = [0, N]$ On this interval, there are a set of input data $\mathbf{x} \in R^m$ without omissions, $\mathbf{x}[0], \ldots, \mathbf{x}[N]$, and a set of output data $\mathbf{y} \in R^n$ with omissions, $\mathbf{y}[0], \ldots, \mathbf{y}[k^- - 1], \mathbf{y}[k^+ + 1], \ldots, \mathbf{y}[N]$. The data $\mathbf{y}[k^-], \ldots, \mathbf{y}[k^+]$ are missing.

Construct the matrix

$$X_{(k-\varrho)} = [\mathbf{x}[k - \varrho], \mathbf{x}[k - \varrho + 1], \ldots, \mathbf{x}[k]] \qquad (2.101)$$

from the input data set, where $\rho < N$ is a parameter.

The output data contain additive errors $\xi[k] \in R^n$ of the interval type:

$$\xi[k] \in \Xi_k = [\xi^-, \xi^+]. \tag{2.102}$$

The probabilistic properties of errors are characterized by PDFs $Q_k(\xi[k])$, $k = \overline{0, N}$, which are assumed to be continuously differentiable.

In the general case, the relation between the input and output of the object O is described by a randomized parametric dynamic model RDM-O of the form

$$\hat{\mathbf{y}}[k] = \mathbb{B}\left(\mathbf{a}, X_{(k-\varrho)}\right), \tag{2.103}$$

where $\mathbb{B}$ is a nonlinear functional with random parameters $\mathbf{a} \in R^r$ of the interval type:

$$\mathbf{a} \in \mathcal{A} = [\mathbf{a}^-, \mathbf{a}^+]. \tag{2.104}$$

The probabilistic properties of these parameters are characterized by a PDF $P(\mathbf{a})$, which is assumed to be continuously differentiable.

The observed output of RDM-O is given by

$$\mathbf{v}[k] = \hat{\mathbf{y}}[k] + \xi[k], \qquad k = \overline{0, N}. \tag{2.105}$$

For determining the optimal PDFs, apply the randomized machine learning (RML) algorithm suggested in [148]. In the case under consideration, it has the form

$$\mathcal{H}[P(\mathbf{a}), Q(\xi)] = -\int_{\mathcal{A}} P(\mathbf{a}) \ln P(\mathbf{a}) d\mathbf{a} -$$

$$-\sum_{k=0}^{T} \int_{\Xi_k} Q_k(\xi[k]) \ln Q_k(\xi[k]) d\xi[k] \Rightarrow \max, \tag{2.106}$$

$$\int_{\mathcal{A}} P(\mathbf{a}) d\mathbf{a} = 1; \quad \int_{\Xi_k} Q_k(\xi[k]) d\xi[k] = 1, \ k = \overline{0, N},$$

$$\int_{\mathcal{A}} P(\mathbf{a}) \mathbb{B}\left(\mathbf{a}, X_{(k-\varrho)}\right) d\mathbf{a} + \int_{\Xi_k} Q_k(\xi[k]) \xi[k] d\xi[k] = \mathbf{y}[k],$$

$$k \neq k^-, \ldots, k^+; k = \overline{0, N}.$$

The PDFs are continuously differentiable, and the problem (2.106) belongs to the class of Lyapunov-type functional optimization problems [77]. Hence, it is possible to construct an analytical solution

parameterized by the Lagrange multipliers $\{\theta^{(0)}, \ldots, \theta^{k^--1}, \theta^{k^++1}, \ldots,$
$\theta^{(N)}\}$:

$$P^*(\mathbf{a}, \theta) = \frac{\exp\left[-\sum_{k=0}^{N}\langle\theta^{(k)}, \mathbb{B}\left(\mathbf{a}, X_{(k-\varrho)}\rangle\right)\right]}{\mathcal{P}(\theta)},$$

$$Q_k^*(\xi[k], \theta^{(k)}) = \frac{\exp\left(-\langle\theta^{(k)}, \xi[k]\rangle\right)}{\mathcal{Q}_k(\theta^{(k)})}, \qquad (2.107)$$

$$k \neq k^-, \ldots, k^+; \quad k = \overline{0, N}.$$

Here

$$\mathcal{P}(\theta) = \int_{\mathcal{A}} \exp\left[-\sum_{k=0}^{N}\langle\theta^{(k)}, \mathbb{B}\left(\mathbf{a}, X_{(k-\varrho)}\rangle\right)\right] d\mathbf{a},$$

$$\mathcal{Q}_k(\theta^{(k)}) = \int_{\Xi_k} \exp\left(-\langle\theta^{(k)}, \xi[k]\rangle\right) d\xi[k], \qquad (2.108)$$

$$k \neq k^-, \ldots, k^+; \quad k = \overline{0, N}.$$

The Lagrange multipliers are determined from the system of equations

$$\frac{\mathbf{M}_k(\theta)}{\mathcal{P}(\theta)} + \frac{\mathbf{N}_k(\theta^{(k)})}{\mathcal{Q}_k(\theta^{(k)})} = \mathbf{y}[k], \qquad k \neq k^-, \ldots, k^+; k = \overline{0, N}, \qquad (2.109)$$

where

$$\mathbf{M}_k(\theta) = \int_{\mathcal{A}} \exp\left[-\sum_{k=0}^{T}\langle\theta^{(k)}, \mathbb{B}\left(\mathbf{a}, X_{(k-\varrho)}\rangle\right)\right] \mathbb{B}\left(\mathbf{a}, X_{(k-\varrho)}\right) d\mathbf{a}, \quad (2.110)$$

$$\mathbf{N}_k(\theta^{(k)}) = \int_{\Xi_k} \exp\left(-\langle\theta^{(k)}, \xi[k]\rangle\right) \xi[k] d\xi[k]. \qquad (2.111)$$

b. Omissions in input data

Restoring omissions in input data is more difficult than the ones in output data. Really, the origin mechanism of input data and the source of its excitation are both unknown. In terms of the approach adopted here, this means that the model generating input data and also the process exciting it are unknown. All these factors increase the level of uncertainty in the problem, motivating the choice of randomization.

The randomized approach consists in using a sequence of independent random vectors $\zeta[k] \in R^q$ $(\overline{k0, N})$ with special probabilistic characteristics, and a randomized parametric model, the choice of which is determined by its complexity.[5]

---

[5]It may be necessary to compare and test different models.

Well, consider the observation interval $\tau = [0, N]$ and an input data set $\mathbf{x} \in R^m$ with omissions $\mathbf{x}[0], \ldots, \mathbf{x}[k^- - 1], \mathbf{x}[k^+ + 1], \ldots, \mathbf{x}[T]$. The data $\mathbf{x}[k^-], \ldots, \mathbf{x}[k^+]$ are missing.

To restore the missing input data, assume that they are the result of transforming a sequence of independent random vectors $\zeta[k] \in R^q$, $k \in \tau$. Let the vectors $\zeta[k]$ be of the interval type, i.e.,

$$\zeta[k] \in \mathcal{Z}_k = [\zeta^-[k], \zeta^+[k]], \qquad k = \overline{0, N}, \tag{2.112}$$

Due to the uncertainty and insufficient knowledge about the process and its transformation, choose the randomized parametric (RPM-I) model

$$\hat{\mathbf{x}}[k] = \mathbb{F}\left(\mathbf{c}, Z_{(k-\rho)}\right), \tag{2.113}$$

where $\mathbb{F}$ is a nonlinear vector functional with random parameters $\mathbf{c} \in R^s$ and $Z_{(k-\rho)}$ is the input matrix of dimensions $(q \times \rho)$ composed of the column vectors $\zeta[k] \in R^q$, i.e.,

$$Z_{(k-\rho)} = [\zeta[k - \rho], \zeta[k - \rho + 1], \ldots, \zeta[k]]. \tag{2.114}$$

Introduce the matrix

$$Z = \left[Z_{(-\rho)}, Z_{(1-\rho)}, \ldots, Z_{(N-\rho)}\right], \tag{2.115}$$

which characterizes the input of the RPM-I model (2.113) on the time interval $k = \overline{0, N}$. The matrix Z is random and of the interval type:

$$Z \in \mathcal{Z} = \bigcup_{k=0}^{N} \mathcal{Z}_k = [Z^-, Z^+]. \tag{2.116}$$

The random parameters of this model are

$$\mathbf{c} \in \mathcal{C} = [\mathbf{c}^-, \mathbf{c}^+]. \tag{2.117}$$

The probabilistic properties of the parameters $\mathbf{c}$ and matrix Z (2.115) are characterized by a joint probability density function (PDF) $W(\mathbf{c}, Z)$ with the domain of definition

$$\mathcal{F} = \mathcal{C} \bigcup \mathcal{Z}. \tag{2.118}$$

This function is assumed to be continuously differentiable.

Let the available input data $\mathbf{x}[k]$ contain errors $\eta[k] \in R^m$ of the interval type:

$$\eta[k] \in \mathcal{E}_k = [\eta^-, \eta^+]. \tag{2.119}$$

The probabilistic properties of the errors are characterized by PDFs $G_k(\eta[k]), k = \overline{0, N}$, which are assumed to be continuously differentiable as well.

The observed output of the RPM-I model is given by

$$\mathbf{z}[k] = \hat{\mathbf{x}}[k] + \eta[k], \qquad k = \overline{0, N}. \tag{2.120}$$

Clearly, the missing input data are restored using a model similar to RDM-O, but with the random input $\zeta[k]$.

Apply the RML algorithm [148] for determining the PDFs $W(\mathbf{c}, Z)$ and $G_k(\zeta[k])$, $k = \overline{0, N}$. In the case under consideration, this algorithm has the form

$$\mathcal{H}[W(\mathbf{c}, Z), G(\eta)] = -\int_{\mathcal{F}} W(\mathbf{c}, Z) \ln W(\mathbf{c}, Z) \, d\mathbf{c} \, dZ -$$

$$- \sum_{k \in \mathcal{K}} \int_{\mathcal{E}_k} G_k(\eta[k]) \ln G_k(\eta[k]) d\eta[k] \Rightarrow \max, \tag{2.121}$$

$$\int_{\mathcal{F}} W(\mathbf{c}, Z) \, d\mathbf{c} \, dZ = 1; \quad \int_{\mathcal{E}_k} G_k(\eta[k]) d\eta[k] = 1,$$

$$\int_{\mathcal{F}} W(\mathbf{c}, Z) \mathbb{F}\left(\mathbf{c}, Z_{(k-\rho)}\right) d\mathbf{c} \, dZ + \int_{\mathcal{E}_k} G_k(\eta[k]) \eta[k] d\eta[k] = \mathbf{x}[k],$$

$$k \in \mathcal{K},$$

where the indexing set $\mathcal{K} = \overline{0, N}$ does not include $k = k^-, \ldots, k^+$.

This problem belongs to the same class as (2.106). Its solution is given by

$$W^*(\mathbf{c}, Z, \lambda) = \frac{\exp\left[-\sum_{k \in \mathcal{K}} \langle \lambda^{(k)}, \mathbb{F}\left(\mathbf{c}, Z_{(k-\rho)}\right) \rangle\right]}{\mathcal{W}(\lambda)}, \tag{2.122}$$

$$G_k^*(\eta[k], \lambda^{(k)}) = \frac{\exp\left(-\langle \lambda^{(k)}, \eta[k] \rangle\right)}{\mathcal{G}_k(\lambda^{(k)})},$$

$$k \in \mathcal{K},$$

where

$$\mathcal{W}(\lambda) = \int_{\mathcal{F}} \exp\left[-\sum \langle \lambda^{(k)}, \mathbb{F}\left(\mathbf{c}, Z_{(k-\rho)}\right) \rangle\right] d\mathbf{c}, \quad \mathcal{G}_k(\lambda^{(k)}) =$$

$$= int_{\mathcal{E}_k} \exp\left(-\langle \lambda^{(k)}, \eta[k] \rangle\right) d\eta[k]. \tag{2.123}$$

In the equalities presented above, $\lambda^{(k)}$ is the vector of Lagrange multipliers corresponding to the kth balance constraint. This vector satisfies the system of equations

$$\frac{\tilde{\mathbf{M}}_k(\lambda)}{\mathcal{W}(\lambda)} + \frac{\tilde{\mathbf{S}}_k(\lambda^{(k)})}{\mathcal{G}_k(\lambda^{(k)})} = \mathbf{x}[k], \quad k \neq k^-, \ldots, k^+; k \in \mathcal{K}, \tag{2.124}$$

where

$$\tilde{\mathbf{M}}_k(\lambda) = \int_{\mathcal{F}} \exp\left[-\sum_{k \in \mathcal{K}} \langle \lambda^{(k)}, \mathbb{F}\left(\mathbf{c}, Z_{(k-\rho)}\right)\rangle\right] \mathbb{F}\left(\mathbf{c}, Z_{(k-\rho)}\right) d\mathbf{c}dZ,$$

$$\tilde{\mathbf{S}}_k(\lambda^{(k)}) = \int_{\mathcal{E}_k} \exp\left(-\langle \lambda^{(k)}, \eta[k]\rangle\right) \eta[k]d\eta[k]. \tag{2.125}$$

3. Omissions in input and output data

In this case, the procedures described in the previous paragraphs have to be used sequentially: first, the algorithm (2.121), and then the algorithm (2.106).

The algorithms, applied either separately or sequentially, yield:

- the entropy-optimal PDFs of parameters of the corresponding models, $P^*(\mathbf{a})$ and $W^*(\mathbf{c})$;

- the entropy-optimal PDFs of the measurement noises, $Q_k^*(\xi[k])$ and $G_k^*(\eta[k])$, where $k = \overline{1, N}$;

- the entropy-optimal PDF of the randomized input sequence, $K_k^*(\zeta[k])$, $k = \overline{1, N}$.

3. Sampling of optimal PDFs

Sampling transforms a PDF into an appropriate sequence of random vectors. In the procedure under consideration, this is a way of realizing the ensembles of random vectors on the segments of missing data.

The general method for generating sequences of random vectors with a given PDF was described in [36]. It is intended to sample the sequences of the corresponding random-parameter vectors of the RPM-O and RPM-I models and the sequences of the measurement noise vectors, as well as to generate the ensembles of restored data using the Monte Carlo method [159].

Some useful characteristics of ensembles are as follows:

—the empirical PDFs $\mathcal{P}_k(\hat{\mathbf{v}}_{rst}[k])$ and $\mathcal{W}_k(\hat{\mathbf{z}}_{rst}[k])$ for the restored output and input data, respectively;

—the empirical probability distributions

$$\mathfrak{P}_k(\hat{\mathbf{v}}_{rst}^{(i)}[k]) = \sum_{j=1}^{i} P_k(\hat{\mathbf{v}}_{rst}^{(j)}[k]) \quad \text{(for the output data)}$$

$$\mathfrak{W}_k(\hat{\mathbf{z}}_{rst}^{(i)}[k]) = \sum_{j=1}^{i} \mathcal{W}_k(\hat{\mathbf{v}}_{rst}^{(j)}[k]) \quad \text{(for the input data). (2.126)}$$

The sets of restored data representing the integral characteristics of an ensemble of random trajectories can be used for solving machine learning problems.

1. Data maximizing the PDFs of parameters and noises (max-pn)
—The restored output data:

$$\begin{aligned}
\hat{\mathbf{y}}_{rst}[k] &= \mathbb{B}(\mathbf{a}^*, X_{(k-\rho)}), \quad \mathbf{a}^* = \arg\max P^*(\mathbf{a}), \\
\xi_{rst}[k] &= \arg\max Q_k^*(\xi[k]), \\
\hat{\mathbf{v}}_{rst}[k] &= \hat{\mathbf{y}}_{rst}[k] + \xi_{rst}[k], \quad k = \overline{0, N}.
\end{aligned} \quad (2.127)$$

—The restored input data:

$$\begin{aligned}
\hat{\mathbf{x}}_{rst}[k] &= \mathbb{F}(\mathbf{c}^*, Z_{(k-\rho)}^*), \quad \mathbf{c}^* = \arg\max W^*(\mathbf{a}), \\
Z_{(k-\rho)}^* &= [\zeta^*[k-\rho], \dots, \zeta^*[k]], \quad \zeta^*[k] = \arg\max K_k^*(\zeta[k]), \\
\eta_{rst}[k] &= \arg\max G_k^*(\eta[k]), \\
\hat{\mathbf{z}}_{rst}[k] &= \hat{\mathbf{x}}_{rst}[k] + \eta_{rst}[k], \quad k = \overline{0, N}.
\end{aligned} \quad (2.128)$$

2. Data maximizing the empirical PDFs of observed trajectories for $k = \overline{0, N}$ (max-cPDF)
—The restored output data:

$$\check{\mathbf{v}}_{rst}[k] = \arg\max \mathcal{P}_k(\hat{\mathbf{v}}_{rst}[k]), \quad k = \overline{0, N}. \quad (2.129)$$

—The restored input data:

$$\check{\mathbf{z}}_{rst}[k] = \arg\max \mathcal{W}_k(\hat{\mathbf{z}}_{rst}[k]), \quad k = \overline{0, N}; \quad (2.130)$$

3. Data corresponding to the mean trajectory (mean)
—The restored output data:

$$\bar{\mathbf{v}}_{rst}[k] = \frac{1}{M} \sum_{i=1}^{M} \hat{\mathbf{v}}_{rst}^{(i)}[k], \quad k = \overline{0, N}. \quad (2.131)$$

—The restored input data:

$$\bar{\mathbf{z}}_{rst}[k] = \frac{1}{M} \sum_{i=1}^{M} \hat{\mathbf{z}}_{rst}^{(i)}[k], \qquad k = \overline{0, N}. \tag{2.132}$$

(Here M denotes the number of trajectories in an ensemble.)
4. Data corresponding to the median trajectory (med)
—The restored output data:

$$\hat{\mathbf{v}}_{rst}^{(j^*)}[k] \Rightarrow \sum_{j=1}^{j^*} \mathfrak{P}_k(\hat{\mathbf{v}}_{rst}^{(j)}[k]) = \sum_{j=j^*+1}^{M} \mathfrak{P}_k(\hat{\mathbf{v}}_{rst}^{(j)}[k]). \tag{2.133}$$

—The restored input data:

$$\hat{\mathbf{z}}_{rst}^{(j^*)}[k] \Rightarrow \sum_{j=1}^{j^*} \mathfrak{W}_k(\hat{\mathbf{z}}_{rst}^{(j)}[k]) = \sum_{j=j^*+1}^{M} \mathfrak{W}_k(\hat{\mathbf{z}}_{rst}^{(j)}[k]). \tag{2.134}$$

(Here M denotes the number of trajectories in an ensemble.)
Useful restored data can be of assistance when solving applied problems.

CHAPTER 3

DIMENSION REDUCTION METHODS

3.1 REVIEW OF DIMENSION REDUCTION METHODS

The arrays of problem-oriented data described in the previous sections may contain information of little significance. This hypothesis underlies dimension reduction methods. Although any reduction causes loss of information about a studied object, it is reasonable to consider reduction with "minimum loss." The experience accumulated by solving different machine learning problems indicates that the dimension reduction problems and the problems solved on resulting reduced data are related to each other. Therefore, there exists no universal efficient approach applicable to any problem. Note that dimension reduction becomes a rather topical problem for RML procedures due to the specific properties of the equations yielding the optimal "randomized" estimates. This feature explains the choice of the reduction methods below.

As a matter of fact, the dimension reduction methods employ one of the three basic ideas as follows. The first idea is related to the matrix that describes data of unspecified origin. By assumption, this matrix has a certain core of smaller dimension that concentrates main information about a studied object. In this case, a reduction tool is the singular decomposition of the data matrix and ordering of its singular values in absolute magnitude.

DOI: 10.1201/9781003306566-3

The second idea is to treat existing data as random, sometimes with additional statistical properties. Then, further analysis is focused on covariance matrix using one of two standard approaches. Following the first approach, one should assign some connectivity threshold for variables and eliminate one of two variables whose coefficient of correlation exceeds this threshold. The second approach is based on the singular decomposition of the covariance matrix.

Finally, the third idea for dimension reduction consists in direct and inverse projection. Direct projection is performed onto a space of smaller dimension (on choice). As the name implies, inverse projection is the same operation back onto the original space, yielding the so-called inverse vector. Naturally enough, the inverse vector differs from the original one. Hence, it is necessary to choose an appropriate functional of distance between the original and inverse vectors to be minimized over corresponding projection operators.

3.1.1 Singular decomposition method for data matrix

Consider a rectangular data matrix A of dimensions $(m \times s)$ and recall some definitions from [60]. The left singular vectors $\vec{u}$ of this matrix A are the eigenvectors of the matrix AA' of dimensions $(m \times m)$. The right singular vectors $\vec{u}$ of this matrix A are the eigenvectors of the matrix $A'A$ of dimensions $(s \times s)$. A nonnegative real value β is a singular value of A if there exist two unit-length vectors $\vec{u} \in R^m$ and $\vec{v} \in R^s$ such that

$$A\vec{v} = \beta\vec{u}, \qquad A'\vec{u} = \beta\vec{v}. \tag{3.1}$$

A singular decomposition of the matrix A is given by the representation

$$A = USV', \tag{3.2}$$

where the diagonal matrix S of dimensions $(m \times s)$ contains the singular values of A, whereas the matrices U and V of dimensions $(m \times m)$ and $(s \times s)$, respectively, are unitary[1] and consist of the left and right singular vectors.

By this definition of singular decomposition, the matrices U and V are not unique and hence can be chosen so that the diagonal elements of the matrix S have the form

$$\beta_1 \geq \beta_2 \geq \cdots \geq \beta_r > \beta_{r+1} = \cdots = \beta_m = 0, \tag{3.3}$$

where r denotes the rank of the matrix S.

---

[1]A matrix G is unitary if $GG' = I$.

The chain of inequalities (3.3) plays a key role in principal component analysis. The first principal component is the one with the maximum singular value (in the current notation, β_1). The second successive component corresponds to the singular value β_2, and so on. The last principal component is associated with the last nonzero singular value β_r and its index (subscript) coincides with the rank of the matrix A.

The singular values are calculated by solving the following problem. Consider an approximate linear representation of the matrix A of the form

$$A = USV' + C. \tag{3.4}$$

Minimize

$$\varepsilon^2 = \text{tr}(CC') \tag{3.5}$$

subject to the normalization conditions

$$U'U = VV' = I. \tag{3.6}$$

If the matrix A has a rank $r > \min(m, s)$, then $C \approx 0$. The minimum value r that satisfies equality (3.2) is the rank of the matrix A. To find the left and right singular vectors $\vec{u}_k$, $\vec{v}_k$ for $k = 1, \ldots, r$, introduce auxiliary normalized vectors $\vec{a}_k, \vec{b}_k$ and define vectors $\vec{u}_k$, $\vec{v}_k$ as

$$\vec{u}_k = \frac{\vec{a}_k}{\|\vec{a}_k\|}, \qquad \vec{v}_k = \frac{\vec{b}_k}{\|\vec{b}_k\|}. \tag{3.7}$$

Then the singular value is

$$\lambda_k = \|\vec{a}_k\| \|\vec{b}_k\|. \tag{3.8}$$

The auxiliary vectors are given by the equalities

$$\vec{a}_k = \frac{A\vec{b}_k}{\vec{b}_k \vec{b}_k'}, \quad \vec{b}_{k+1} = \frac{\vec{a}_k'}{\vec{a}_k' \vec{a}_k}, \quad k = 1, 2, \ldots \tag{3.9}$$

At step $k = 1$, the row $\vec{b}_1$ of the matrix A has the maximum norm. At steps $k = 2, 3, \ldots$, this procedure is iteratively applied to the matrix $A_{k+1} = A_k - \vec{u}_k \lambda_k \vec{v}_k$. To construct the matrix $\hat{A}_r = U_r S_r V'$, it suffices to calculate only r columns of the matrix U and r rows of the matrix V'.

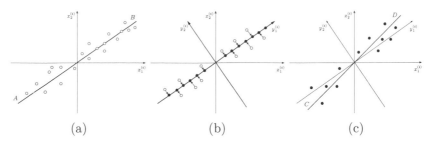

Figure 3.1

3.1.2 Principal component analysis

Apparently, principal component analysis is the most commonly used dimension reduction method for data arrays [10, 84, 127]. There exist at least three basic versions of this method, namely, data approximation by linear smaller-dimension manifolds; searching for smaller-dimension subspaces with the maximum data spread (variance) in the orthogonal projection onto them; searching for smaller-dimension subspaces with the maximum root-mean-square distance between points in the orthogonal projection onto them. The basic versions involve a data matrix $X = [\vec{x}^{(1)}, \ldots, \vec{x}^{(m)}]$ whose column vectors $\vec{x}^{(i)}$ have dimension s.

To clarify the idea of this method, first consider the case $s = 2$. Consider a data matrix composed of two-dimensional columns whose elements are the measured values of some variables (attributes of interest). Interpret them as points $\{x_1^{(i)}, x_2^{(i)}\}$ on the plane; see Fig. 3.1a. Draw a line AB so that the *maximum* variation of data occurs along it. The line AB is the first principal component. Perform the projection of the initial points on this line and use it as a new coordinate axis (Fig. 3.1b). Take the points representing the difference between the initial points and their projections on axis AB and again draw a line CD so that the maximum variation of these differences occurs along it, as shown in Fig. 3.1c. The line CD is the second principal component and so on.

In the general multidimensional case, principal component analysis is performed in accordance with the following procedure:

- identify the center of the existing data set and place the origin (i.e., the zeroth principal component) to this center;

- find the direction of the maximum data variation (the first principal component);

- find the data projections onto the first principal component as well as their variance;

- if the variance is large, find the new principal direction of maximum variations that is orthogonal to the first one, and so on.

This procedure allows to replace the initial data matrix X of dimensions $(s \times m)$ by two matrices T and P of dimensions $(s \times k)$ and $(k \times s)$, respectively, where $k < m$. This representation is obtained using the new variables

$$\vec{t}^{(j)} = \vec{p}^{(j,1)} \vec{x}^{(1)} + \cdots + \vec{p}^{(j,m)} \vec{x}^{(m)}, \quad j = \overline{1, k}, \tag{3.10}$$

where $\vec{t}^{(j)}$ and $\vec{p}^{(j,i)}$, $i = \overline{1, m}$, denote the columns of the matrices T and P, respectively. The new variables (3.10) are called the *principal components*, and the matrix T is called the principal component matrix. An important property of the principal components consists in their orthogonality. Therefore, the matrix T is not reconfigured as we increase the number of principal components.

With these variables, the matrix X can be written as

$$X = \hat{X} + E = \sum_{j=1}^{k} \vec{t}^{(j)} [\vec{p}^{(j)}]' + E, \qquad \hat{X} = T P'. \tag{3.11}$$

Here E stands for the residue matrix of dimensions $(s \times m)$. The residual error is assessed in terms of some numerical characteristic of the residue matrix, e.g., the relative variance

$$D = 1 - \frac{\sum_{j=1}^{s} \sum_{i=1}^{m} e_{ji}^2}{\sum_{j=1}^{s} \sum_{i=1}^{m} x_{ji}^2}, \quad e_{ji} = x_{ji} - \hat{x}_{ji}. \tag{3.12}$$

There exist many recursive algorithms to calculate the principal components, which are oriented towards the specifics of initial information and employ different data processing methods. First, the matrix X is centered, i.e., for each column the average value

$$\bar{x}_j = \frac{1}{s} \sum_{i=1}^{m} x_i^{(j)} \tag{3.13}$$

and the standard deviation

$$\sigma_j = \frac{1}{s} \sqrt{\sum_{i=1}^{m} (x_{ji} - \bar{x}_j)^2}. \tag{3.14}$$

are calculated. The modified data matrix $\tilde{X}$ that is used to calculate the principal components consists of the elements

$$\tilde{x}_{ij} = \frac{x_{ij} - \bar{x}_j}{\sigma_j}. \tag{3.15}$$

The algorithm for searching the principal components includes the following steps.

1. Choose an initial vector $\vec{t}$ (3.10).

2. Calculate $\vec{p}' = \frac{\vec{t}'E_j}{\vec{t}'\vec{t}}$.

3. Calculate $\vec{p} = \frac{\vec{p}}{(\vec{p}'\vec{p})^{1/2}}$.

4. Calculate $\vec{t} = \frac{E_j\vec{p}}{\vec{p}'\vec{p}}$.

5. Find D using (3.12) and get back to Step 2.

The number of the principal components is defined by comparing the relative variance (3.12) with the admissible error.

As mentioned earlier, the singular decomposition (3.2) of a rectangular matrix and principal component analysis are related to one another. This relation has the form

$$T = US, \quad P = V. \tag{3.16}$$

Principal component analysis and its modifications were considered in a vast literature, mostly with applications to the problems of classification and regression [188].

3.1.3 Random projection method

Like principal component analysis, dimension reduction by means of random projections involves a linear transformation $f : \mathbb{R}^n \rightarrow \mathbb{R}^k$ of a data matrix and is applied in problems where objects are represented by points of the multidimensional Euclidean space [20].

The idea of the random projection method is based on the Johnson–Lindenstrauss lemma [83]: for a set X of points in a multidimensional space of a high dimension, there exists a subspace of a smaller, albeit rather high, dimension such that projection onto this subspace will save, on average, the relative distances between all points from the given set. The error of transformation is described by a value $\varepsilon \in (0, 1)$:

$$(1 - \varepsilon) \|x_1 - x_2\|^2 \le \|f(x_1) - f(x_2)\|^2 \le (1 + \varepsilon) \|x_1 - x_2\|^2 \quad (3.17)$$

for all vectors $x_1, x_2 \in X$.

Moreover, the original lemma also gives a lower bound on k, i.e., the dimension of the subspace onto which the objects are projected. For example, for a set of N points, the value k must satisfy the condition

$$k > \frac{8 \ln N}{\varepsilon^2}. \quad (3.18)$$

In other words, the minimum possible dimension of the subspace for projection does not depend on the dimension of the original space, but depends only on the number of projected objects and the error of transformation. In various special cases, the lower bound on the value k was refined in [82, 86].

Consider a linear transformation $f : \mathbb{R}^n \to \mathbb{R}^k$ for projecting objects from a set X. It is defined by a rectangular matrix R of dimensions $[k \times n]$ $(k << n)$ and has the form

$$X_{k \times N}^{RP} = R_{k \times n} X_{n \times N}. \quad (3.19)$$

As it turned out, the elements of the matrix R for such a transformation can consist of random variables, which makes the projection random. In early works devoted to the random projection method [30, 76, 108], it was proposed to use random variables with the standard Gaussian distribution $N(0, 1)$, and to construct the matrix R so that its row vectors are orthogonal and have unit length.

Later in [8], it was demonstrated that the matrix R can be obtained using the uniform distribution, under certain constraints dictated by storing information in databases, due to integer calculations. In this case, it was suggested to compose the matrix R from random elements r_{ij} of the form

$$r_{ij} = \sqrt{3} \times \begin{cases} +1 & \text{with probability } \frac{1}{6}, \\ 0 & \text{with probability } \frac{2}{3}, \\ -1 & \text{with probability } \frac{1}{6}. \end{cases} \quad (3.20)$$

Thus, for calculating the projection X^{RP}, it is necessary to multiply the matrices R and X, and this transformation has the complexity

$\mathcal{O}(knN)$. In principal component analysis, the matrix for projection is square with dimensions $[n \times n]$, and the transformation to projections has the complexity $\mathcal{O}(n^2 N)$. Hence, for dimensionality reduction, the random projection method is preferable to principal component analysis in terns of computational efficiency.

Also, note the existing studies to speed up the transformation to projections. One line of research is focused on sparse matrices: as it was shown in [20], if the data matrix X has at most c nonzero components in each column, then the complexity of transformation is reduced to $\mathcal{O}(ckN)$. The influence of data sparsity on the efficiency of the random projection method was studied in the papers [38, 87]. Another approach to accelerating projection is based on the Fast Johnson–Lindenstrauss Transform (FJLT) [9], a randomized Fourier transform.

The range of possible applications of the random projection method for dimension reduction includes a wide class of problems with multidimensional data. Since the relative distances between different objects of a given set are preserved in the smaller dimension space, the structure of this set is also preserved, which is especially important when dividing objects into groups, e.g., in classification and clustering. Therefore, random projections become an effective dimension reduction solution in metric data analysis methods, such as the k-Nearest Neighbor (kNN) for clustering [30], support vector machine (SVM) for classification [27], and the least squares (LS) method for regression [116].

Random projections turn out to be especially effective when dealing with large collections of multidimensional sparse data, such as images, text, audio and video data, etc. In the field of signal processing, a separate line of research devoted to the use of random projections for data analysis and restoration is called Compressed Sensing; for example, see [29, 46].

Regardless of the field of application, the nature and dimension of the original data represented by a set of points in a multidimensional space, the use of random projections allows reducing the dimension while preserving the data structure. Moreover, the random projection method is easily interpreted and does not require significant computing resources, even in the case of high and ultrahigh-dimensional data, in contrast to alternative dimension reduction methods such as PCA and SVD.

However, note that random projections are intended only for the problems in which analysis and calculations are performed on data of reduced dimension, but the data are not used for visualization or restoration. For example, in the case of images, dimension reduction with random projections can lead to visual degradation of the image, for both the human eye and computer vision systems.

3.1.4 Direct and inverse projection

Consider the space R^{NM} of all real matrices $A_{(N \times M)}$ whose elements take random values.[2] Transform a matrix $A_{N \times M}$ into a matrix $B_{N \times m}$, where $m < M$. This can be done through a two-step procedure as follows. First, project the space R^{NM} onto the space R^{Nm} ($R^{NM} \Rightarrow R^{Nm}$) using a projector matrix $S_{(N \times m)}$, which gives the matrix $D_{(N \times m)} = A_{(N \times M)} S_{(M \times m)}$. This operation will be called *direct projection*.

The transition from the original space R^{NM} to the smaller-dimension one R^{Nm} causes the losses of information contained in the former's elements. For estimating these losses, it is necessary to return to the original space by means of *inverse projection*: $R^{Nm} \Rightarrow R^{NM}$. This operation is performed using a projector matrix $J_{(m \times M)}$, which gives the matrix $X_{(N \times M)} = D_{(N \times m)} J_{(m \times M)} \in R^{NM}$. The matrices $X_{(N \times M)}$ and $A_{(N \times M)}$ are elements of the same space. Hence, the quality of compression of the original matrix can be estimated in terms of an appropriate criterion.

In the sequel, the quality of data compression, which depends on the projector matrices $S_{(N \times m)}$ and $J_{(m \times M)}$, will be estimated in entropy terms. More specifically, introduce the information entropy [139] of a matrix $X_{(N \times M)}$ with respect to a matrix $A_{(N \times M)}$,

$$H(X \mid A) = -\sum_{i=1}^{N} \sum_{j=1}^{M} x_{ij} \ln \frac{x_{ij}}{a_{ij}} = tr\left(X^{\mathsf{T}} Y\right), \qquad (3.21)$$

where tr denotes the matrix trace and Y is a matrix of dimensions $(N \times M)$ consisting of the elements

$$y_{ij} = \ln \frac{x_{ij}}{a_{ij}}. \qquad (3.22)$$

Hereinafter, this entropy will be called *the information cross-entropy* [158]. The direct-inverse projection method described above is the core of the compression method of random data matrices. Consider a nonnegative data matrix U of dimensions $(m \times s)$. It is required to compress the matrix to a matrix Z of dimensions $(n \times r)$, where $r < s$ and $n < m$. This problem will be solved sequentially, in two stages as follows. First, reduce the matrix $U_{(m \times s)}$ to a matrix $Y_{(m \times r)}$, and then

---

[2]Note that matrix *vectorization* [115] is used.

compress the latter to a matrix $Z_{(n \times r)}$. The approach presented below will be called *the entropy dimension reduction (EDR) method.*

Stage 1 Introduce a matrix $Q_{(s \times r)}$ and perform direct projection for obtaining the column-compressed matrix

$$Y_{(m \times r)} = U_{(m \times s)} Q_{(s \times r)}. \tag{3.23}$$

Next, introduce a matrix $T_{(r \times s)}$ and perform inverse projection, which yields the matrix

$$X_{(m \times s)} = U_{(m \times s)} Q_{(s \times r)} T_{(r \times s)}. \tag{3.24}$$

The matrix $X_{(m \times s)}$ has the same dimension as the original matrix $U_{(m \times s)}$ and characterizes the effect (losses) of direct and inverse projections in Stage 1.

The reduction of a data matrix is accompanied by the losses of information contained in it or by an increase of the relative information entropy [139]. For Stage 1 of this method, the information cross-entropy of a matrix $X_{(m \times s)}$ with a fixed original matrix $U_{(m \times s)}$ is given by

$$H_1(X \mid U) = -\sum_{i=1}^{m} \sum_{j=1}^{s} x_{ij}(Q, T \mid U) \ln \frac{x_{ij}(Q, T \mid U)}{u_{ij}} =$$

$$= H_1(Q, T \mid U) \Rightarrow \max_{(Q,T) \geq 0}, \tag{3.25}$$

where, according to (3.24),

$$x_{ij}(Q, T \mid U) = \sum_{\mu=1}^{r} \sum_{\nu=1}^{s} t_{\mu,j} q_{\nu,\mu} u_{i\nu}. \tag{3.26}$$

The maximization of the function (3.1.4) finally yields the entropy-optimal projector matrix $Q^*_{(s \times r)}$ and the compressed first-stage matrix

$$Y^*_{(m \times r)} = U_{(m \times s)} Q^*_{(s \times r)}. \tag{3.27}$$

Stage 2 The second-stage data matrix $Y_{(m \times r)}$ has a reduced number $r < s$ of columns. It is required to compress this matrix to a given number $n < m$ of rows. Introduce matrices $B_{(n \times m)}$ and $A_{(m \times n)}$ for performing the direct and inverse projection operations, respectively, by analogy with Stage 1:

$$Z_{(n \times r)} = B_{(n \times m)} Y_{(m \times r)}, \qquad G_{(m \times r)} = A_{(m \times n)} Z_{(n \times r)}. \tag{3.28}$$

The matrix $G_{(m \times r)}$ has the same dimension as the original second-stage matrix $Y_{(m \times r)}$ and characterizes the effect (losses) of direct and inverse projections in Stage 2. These losses are estimated by the information cross-entropy

$$H_2(G \mid Y^*) = -\sum_{i=1}^{m} \sum_{j=1}^{r} g_{ij}(B, A \mid Y) \ln \frac{g_{ij}(B, A \mid Y^*)}{y_{ij}} =$$

$$H_2(B, A \mid Y^*) \Rightarrow \max_{(B,A) \geq 0}, \tag{3.29}$$

where

$$g_{ij}(B, A \mid Y^*) = \sum_{\beta=1}^{n} \sum_{\alpha=1}^{m} a_{i,\beta} b_{\beta,\alpha} y_{\alpha,j}^* \tag{3.30}$$

and

$$y_{\alpha,j}^* = \sum_{eta=1}^{s} u_{\alpha,\eta} q_{\eta,j}^*. \tag{3.31}$$

3.2 ENTROPY OPTIMIZATION OF SEQUENTIAL PROCEDURE

The first stage of the sequential procedure is to determine the optimal elements of the nonnegative matrices Q and T:

$$(Q^*, T^*) = \arg \max_{(Q,T) \geq 0} H_1(Q, T \mid U). \tag{3.32}$$

This problem will be solved numerically using an iterative algorithm based on projected gradient descent, which represents a combination of the standard gradient descent method and the truncation of all negative elements.

3.2.1 Optimality conditions and algorithm

The gradient of the function $H_1(Q, T \mid U)$ with respect to the elements of the matrices Q and T has the components

$$\frac{\partial H(Q, T \mid U)}{\partial q_{\nu,\mu}} = \sum_{i=1}^{m} \sum_{j=1}^{s} \frac{\partial h_{ij}(Q, T \mid U)}{\partial x_{ij}} \frac{\partial x_{ij}(Q, T \mid U)}{\partial q_{\nu,\mu}}, \tag{3.33}$$

$$\nu = \overline{1, s}, \ \mu = \overline{1, r},$$

$$\frac{\partial H(Q, T \mid U)}{\partial t_{\mu,j}} = \sum_{i=1}^{s} \sum_{j=1}^{m} \frac{\partial h_{ij}(Q, T \mid U)}{\partial x_{ij}} \frac{\partial x_{ij}(Q, T \mid U)}{\partial t_{\mu,j}}, \tag{3.34}$$

$$\mu = \overline{1, r}, \ j = \overline{1, s},$$

where

$$h_{ij}(T, Q \mid X) = x_{ij} \ln \frac{x_{ij}}{u_{ij}},$$

$$\frac{\partial h_{ij}(T, Q \mid X)}{\partial x_{ij}} = \ln \frac{x_{ij}}{u_{ij}} + 1, \qquad i = \overline{1, m}, \ j = \overline{1, s}; \quad (3.35)$$

$$\frac{\partial x_{ij}((Q, T \mid U))}{\partial q_{\nu,\mu}} = t_{\mu,j} \, u_{i,\nu}, \qquad \nu = \overline{1, s}, \mu = \overline{1, r};$$

$$\frac{\partial x_{ij}((Q, T \mid U))}{\partial t_{\mu,j}} = \sum_{\nu=1}^{s} q_{\nu,\mu} u_{i,\nu}, \qquad \mu = \overline{1, r}; \quad (3.36)$$

$$i = \overline{1, m}, \quad j = \overline{1, s}.$$

Perform the *vectorization* of the matrices Q and T (see [115]) to obtain the corresponding vectors $\mathbf{q}$ and $\mathbf{t}$. Denote by $\nabla_Q(\mathbf{q}, \mathbf{t})$ the gradient of the information cross-entropy functional $H(Q, T \mid U)$ with the components (3.2.2), and by $\nabla_T(\mathbf{q}, \mathbf{t})$ the gradient of the information cross-entropy functional $H(Q, T \mid U)$ with the components (3.33–3.36).

The algorithm **max** H_1 has the following structure.

a). Initial step:

$$\mathbf{q}^0 > \mathbf{0}, \quad \mathbf{t}^0 > \mathbf{0}.$$

b). nth iterative step:

$$X^n = U \, Q^n \, T^n,$$

$$H(Q^n, T^n \mid U) = \sum_{i=1}^{m} \sum_{j=1}^{r} x_{ij}^n \ln \frac{x_{ij}^n}{u_{ij}} = H^n;$$

$$\mathbf{q}^{(n+1)} = \begin{cases} \mathbf{q}^n + \gamma_{\mathbf{q}} \nabla_Q(\mathbf{q}^n, \mathbf{t}^n) \text{if } \mathbf{q}^{(n+1)} \geq \mathbf{0}, \\ \mathbf{q}^n \text{if } \mathbf{q}^{(n+1)} < \mathbf{0}; \end{cases}$$

$$\mathbf{q}^{(n+1)} \Rightarrow Q^{(n+1)};$$

$$\mathbf{t}^{(n+1)} = \begin{cases} \mathbf{t}^n + \gamma_{\mathbf{t}} \nabla_T(\mathbf{q}^n, \mathbf{t}^n) \text{if } \mathbf{t} \geq \mathbf{0}, \\ \mathbf{t}^n \text{if } \mathbf{t}^{(n+1)} < \mathbf{0}. \end{cases}$$

$$\mathbf{t}^{(n+1)} \Rightarrow T^{(n+1)};$$

$$X^{(n+1)} = U \, Q^{(n+1)} \, T^{(n+1)};$$

$$H(Q^{(n+1)}, T^{(n+1)} \,|\, U) = \sum_{i=1}^{m} \sum_{j=1}^{r} x_{ij}^{(n+1)} \ln \frac{x_{ij}^{(n+1)}}{u_{ij}} = H^{(n+1)}.$$

c). Stopping condition:

If $H^{(n+1)} - H^{(n)} \leq \delta$, then $STOP$.

3.2.2 Approximation of information cross-entropy functional

In some cases, the elements of the reduced matrix Q can be determined with sufficient accuracy using the logarithmic function approximation in the neighborhood of a point $x_0 = a$, where $a \geq a_{min}$:

$$\ln x < \ln a + \frac{1}{a_{min}} (x - a). \qquad (3.37)$$

For $a = x_{ij}$, this formula gives the following approximate expression for the information cross-entropy functional:

$$H_1(Q, T \,|\, U) \approx \sum_{i=1}^{s} \sum_{j=1}^{m} \left[x_{ij}^2(Q, T \,|\, U) - x_{ij}(Q, T \,|\, U)u_{ij} \right] =$$
$$\tilde{H}(Q, T \,|\, U). \qquad (3.38)$$

This equality can be written in the matrix form

$$\tilde{H}_1(Q, T \,|\, U) = tr(XX^\mathsf{T}) - tr(XU^\mathsf{T}) =$$
$$\lceil X(Q,T), X(Q,T) \rfloor - \lceil X(Q,T), U \rfloor, \qquad (3.39)$$

where $\lceil A, B \rfloor$ denotes the Frobenius inner product [115],

$$tr(AB^\mathsf{T}) = tr(BA^\mathsf{T}) = \lceil A, B \rfloor = \lceil B, A \rfloor. \qquad (3.40)$$

The elements of the matrices Q and T will be found by minimizing the functional $\tilde{H}_1(Q, T \,|\, U)$ on the set of all nonnegative matrices Q and T under a fixed data matrix U, i.e.,

$$(\tilde{Q}^*, \tilde{T}^*) = \arg \min_{(Q,T) \geq 0} \tilde{H}_1(Q, T \,|\, U). \qquad (3.41)$$

This problem is solved by an algorithm involving the components of the gradient of the functional $\tilde{H}_1(Q, T \,|\, U)$ with respect to the matrices Q and T.

Recall some differentiation rules for scalar functions of matrices [115]. Consider the linear space of all matrices A of dimensions $(m \times n)$. In this space, the scalar product and the Frobenius norm are given by

$$\lceil A, B \rfloor = tr(AB^\mathsf{T}) = tr(BA^\mathsf{T})$$

and

$$\|A\|^2 = \lceil A, A \rfloor,$$

respectively. According to these definitions, a matrix is treated as a vector composed of its adjoining rows (the so-called *matrix vectorization*). For a scalar function of a matrix, the gradient is defined as follows. The increment of a scalar function $f(A)$ is written as

$$f(A + \Delta) = f(A) + \lceil H, \Delta \rfloor + o(\Delta).$$

Here $o(\Delta)$ denotes the remainder term of the Taylor series. The gradient of the function $f(A)$ with respect to the matrix A is

$$\triangle_A f(A) = H.$$

Some simple and useful examples of scalar functions of a variable are

$$f(A) = \lceil C, A \rfloor, \qquad \triangle_A f(A) = C$$

and

$$f(A) = \|A\|^2, \qquad \triangle_A f(A) = 2A.$$

Using the matrix differentiation rules described above, derive the following expressions for the gradients of the functional $\tilde{H}_1(Q, T \,|\, U)$ (3.41):

$$\triangle_Q(Q, T) = \frac{\partial \tilde{H}_1(Q, T \,|\, U)}{\partial X} \frac{\partial X}{\partial Q} = 2TPQT - TP, \qquad (3.42)$$

$$\triangle_T(Q, T) = \frac{\partial \tilde{H}_1(Q, T \,|\, U)}{\partial X} \frac{\partial X}{\partial T} = 2Q^\mathsf{T} PQT - Q^\mathsf{T} P, \qquad (3.43)$$

where $P = XX^\mathsf{T}$ and $\triangle_Q(Q, T)$ and $\triangle_T(Q, T)$ mean the gradients with respect to the matrices Q and T, respectively. According to (3.42, 3.43), the gradients of the function $\tilde{H}(Q, T \,|\, U)$ with respect to the matrices Q and T are matrices of dimensions $(s \times r)$.

For maximizing the functional $\tilde{H}_1(Q, T \,|\, U)$, apply gradient descent with the nonnegativity check of the matrices U and T.

The algorithm **max** $\tilde{H}_1$ has the following structure.

a) Initial step:

$$T^0 > 0, \quad Q^0 > 0.$$

b) nth iterative step:

$$X^n = U Q^n T^n,$$

$$H(Q^n, T^n \,|\, U) = \sum_{i=1}^{m} \sum_{j=1}^{r} x_{ij}^n \ln \frac{x_{ij}^n}{u_{ij}} = H^n;$$

$$Q^{(n+1)} = \begin{cases} Q^n + \gamma_Q \triangle_Q (\tilde{H}(Q^n, T^n \,|\, U) \geq 0 & \text{if } Q^{(n+1)} \geq 0, \\ Q^n & \text{if } Q^{(n+1)} < 0; \end{cases}$$

$$T^{(n+1)} = \begin{cases} T^n + \gamma_T \triangle_T (\tilde{H}(Q^n, T^n \,|\, U) \geq 0 & \text{if } T^{(n+1)} \geq 0, \\ T^n & \text{if } T^{(n+1)} < 0; \end{cases}$$

$$X^{(n+1)} = U \, Q^{(n+1)} \, T^{(n+1)};$$

$$H(Q^{(n+1)}, T^{(n+1)} \,|\, U) = \sum_{i=1}^{m} \sum_{j=1}^{r} x_{ij}^{(n+1)} \ln \frac{x_{ij}^{(n+1)}}{u_{ij}} = H^{(n+1)}.$$

c) *Stopping condition:*

If $H^{(n+1)} - H^n \leq \delta$, then *STOP*.

The algorithm **max** H_1 or **max** $\tilde{H}_1$ yields the entropy-optimal projector matrix $Q^*_{(s \times r)}$. Hence, it is possible to determine the matrix $Y^*_{(m \times r)}$ (1.5) for passing to the second stage of the compression procedure of the matrix U, which is characterized by the projector matrices $B_{(n \times m)}$ and $A_{(m \times n)}$.

The optimal elements of the nonnegative matrices B and A are given by

$$(B^*, A^*) = \arg \max_{(B,A) \geq 0} H_2(B, A \,|\, Y^*). \tag{3.44}$$

By analogy with Stage 1, this problem will be solved using the iterative algorithm based on projected gradient descent (a combination of gradient descent and the truncation of all negative elements).

The gradient of the function $H_2(B, A \,|\, Y^*)$ with respect to the elements of the matrices B and A has the components

$$\frac{\partial H_2(B, A \,|\, Y^*)}{\partial a_{i,\beta}} = \sum_{i=1}^{m} \sum_{j=1}^{r} \frac{\partial s_{ij}(B, A \,|\, Y^*)}{\partial g_{ij}} \frac{\partial g_{ij}(B, A \,|\, Y^*)}{\partial a_{i,\beta}}, \tag{3.45}$$

$$i = \overline{1, n}, \ j = \overline{1, r}, \ \beta = \overline{1, n},$$

$$\frac{\partial H_2(B, A \,|\, Y^*)}{\partial b_{\beta,\alpha}} = \sum_{i=1}^{m} \sum_{j=1}^{r} \frac{\partial s_{ij}(B, A \,|\, Y^*)}{\partial g_{ij}} \frac{\partial g_{ij}(B, A \,|\, Y^*)}{\partial b_{\beta,\alpha}}, \tag{3.46}$$

$$i = \overline{1, n}, \ j = \overline{1, r}, \ \beta = \overline{1, n}, \ \alpha = \overline{1, m},$$

where

$$\frac{\partial s_{ij}(B, A \mid Y^*)}{\partial g_{ij}} = \ln \frac{g_{ij}}{y_{ij}^*} + 1, \qquad i = \overline{1, n}, \ j = \overline{1, r}; \tag{3.47}$$

$$\frac{\partial g_{ij}((B, A \mid Y^*))}{\partial a_{i,\beta}} = \sum_{\alpha=1}^{m} b_{\beta,\alpha} y_{\alpha,j}^*, \qquad \beta = \overline{1, n};$$

$$\frac{\partial g_{ij}((B, A \mid Y^*))}{\partial b_{\beta,\alpha}} = a_{i,\beta} y_{\alpha,j}^*, \qquad \beta = \overline{1, n}, \alpha = \overline{1, m}; \tag{3.48}$$

$$i = \overline{1, m}, \quad j = \overline{1, r}.$$

Perform the *vectorization* of the matrices B and A, and introduce the corresponding vectors $\mathbf{b}$ and $\mathbf{a}$. Denote by $\nabla_B(\mathbf{b}, \mathbf{a})$ the gradient of the relative information entropy $H_2(B, A \mid Y^*)$ with the components (3.46–3.48), and by $\nabla_A(\mathbf{q}, \mathbf{t})$ the gradient of the relative entropy functional $H_2(B, A \mid Y^*)$ with the components (3.48).

The algorithm **max** H_2 has the following structure.

a) Initial step:

$$\mathbf{b}^0 > \mathbf{0}, \quad \mathbf{a}^0 > \mathbf{0}.$$

b) nth iterative step:

$$G^n = A^n B^n Y^*,$$

$$H(B^n, A^n \mid Y^*) = \sum_{i=1}^{n} \sum_{j=1}^{r} g_{ij}^n \ln \frac{g_{ij}^n}{y_{ij}^*} = H^n;$$

$$\mathbf{b}^{(n+1)} = \begin{cases} \mathbf{b}^n + \gamma_\mathbf{b} \nabla_B(\mathbf{b}^n, \mathbf{a}^n) \text{if } \mathbf{b}^{(n+1)} \geq \mathbf{0}, \\ \mathbf{b}^n \text{if } \mathbf{b}^{(n+1)} < \mathbf{0}; \end{cases}$$

$$\mathbf{b}^{(n+1)} \Rightarrow Q^{(n+1)};$$

$$\mathbf{a}^{(n+1)} = \begin{cases} \mathbf{a}^n + \gamma_\mathbf{a} \nabla_A(\mathbf{b}^n, \mathbf{a}^n) \text{if } \mathbf{a} \geq \mathbf{0}, \\ \mathbf{a}^n \text{if } \mathbf{a}^{(n+1)} < \mathbf{0}; \end{cases}$$

$$\mathbf{a}^{(n+1)} \Rightarrow A^{(n+1)};$$

$$G^{(n+1)} = A^{(n+1)} B^{(n+1)} Y^*;$$

$$H(B^{(n+1)}, A^{(n+1)} \mid Y^*) = \sum_{i=1}^{n} \sum_{j=1}^{r} g_{ij}^{(n+1)} \ln \frac{g_{ij}^{(n+1)}}{y_{ij}^*} = H^{(n+1)}.$$

c) Stopping condition:

$$\text{If } H^{(n+1)} - H^{(n)} \leq \delta, \text{ then } STOP.$$

3.3   ENTROPY OPTIMIZATION OF PARALLEL PROCEDURE

3.3.1   Definition and structure

The parallel implementation of the direct-inverse projection procedure leads to the matrix equalities

$$U_{(m \times s)} Q_{s \times r} = Y_{(m \times r)}, \quad B_{(n \times m)} Y_{(m \times r)} = Z_{(n \times r)},$$
$$Z_{(n \times r)} W_{(r \times s)} = D_{(n \times s)}, \quad E_{(m \times n)} D_{(n \times s)} = X_{(m \times s)}. \quad (3.49)$$

The first chain in these formulas, related to the direct projection with a matrix $Q_{(s \times r)}$ and the inverse projection with a matrix $B_{(n \times m)}$, yields a reduced matrix $Z_{(n \times r)}$. The second chain, related to the direct projection with a matrix $W_{(r \times s)}$ and the inverse projection with a matrix $E_{(m \times n)}$, yields a matrix $X_{(m \times s)}$.

From equalities (3.49), it follows that

$$X_{(m \times s)} = E_{(m \times n)} \left\{ \left[B_{(n \times m)} \left(U_{(m \times s)} Q_{(s \times r)} \right) \right] W_{(r \times s)} \right\}. \quad (3.50)$$

Here the round, square and curly brackets indicate the sequence of the projection operations. The elements of the matrix $X_{(m \times s)}$ have the form

$$x_{ij} = \sum_{\mu=1}^{n} e_{i,\mu} \sum_{\nu=1}^{r} w_{\nu,j} \sum_{\beta=1}^{m} b_{\mu,\beta} \sum_{\alpha=1}^{s} u_{\beta,\alpha} q_{\alpha,\nu}, \quad i = \overline{1,m}, \, j = \overline{1,s}.$$
$$(3.51)$$

3.3.2   Optimality conditions and algorithm

For measuring the deviation of the transformed matrix $X_{(m \times s)}$ from the original one $U_{(m \times s)}$, choose *the information cross-entropy*

$$\mathcal{H}(X \,|\, U) = - \sum_{i=1}^{m} \sum_{j=1}^{s} s_{ij}(X \,|\, U), \quad (3.52)$$

where

$$s_{ij} = x_{ij} \ln \frac{x_{ij}}{u_{ij}}. \quad (3.53)$$

Due to (3.51), the information cross-entropy (3.52) is a scalar function of the matrices $(Q, B, W, E) \geq 0$, i.e.,

$$\mathcal{H} = \mathcal{H}(Q, B, W, E). \quad (3.54)$$

The optimal elements of the matrices are obtained by maximizing the function $\mathcal{H}$:

$$(Q^*, B^*, W^*, E^*) = \arg \max_{(Q,B,W,E) \geq 0} \mathcal{H}(Q, B, W, E). \tag{3.55}$$

This problem will be solved using the coordinate-wise scheme of projected gradient descent.

In the parallel procedure, there are a pair of matrices (Q, B) for reducing the original matrix U with respect to the first variable $(r < s)$ and a pair of matrices (W, E) for compressing the matrix U with respect to the second variable $(n < m)$. Each iterative step in the coordinate-wise scheme of projected gradient descent consists of two sub-steps, related to iterations over the (Q, B)-gradients and over the (W, E)-gradients, respectively.

According to (3.51–3.53), the gradient of the function $\mathcal{H}(Q, B, W, E)$ has the components

$$\frac{\partial \mathcal{H}}{\partial q_{\alpha,\nu}} = \sum_{i=1}^{m} \sum_{j=1}^{s} \frac{\partial s_{ij}}{\partial x_{ij}} \frac{\partial x_{ij}}{\partial q_{\alpha,\nu}},$$

$$\frac{\partial \mathcal{H}}{\partial b_{\mu,\beta}} = \sum_{i=1}^{m} \sum_{j=1}^{s} \frac{\partial s_{ij}}{\partial x_{ij}} \frac{\partial x_{ij}}{\partial b_{\mu,\beta}},$$

$$\frac{\partial \mathcal{H}}{\partial w_{\nu,j}} = \sum_{i=1}^{m} \sum_{j=1}^{s} \frac{\partial s_{ij}}{\partial x_{ij}} \frac{\partial x_{ij}}{\partial w_{\nu,j}},$$

$$\frac{\partial \mathcal{H}}{\partial e_{i,\mu}} = \sum_{i=1}^{m} \sum_{j=1}^{s} \frac{\partial s_{ij}}{\partial x_{ij}} \frac{\partial x_{ij}}{\partial e_{i,\mu}}, \tag{3.56}$$

where

$$\frac{\partial s_{ij}}{\partial x_{ij}} = \ln \frac{x_{ij}}{u_{ij}} + 1,$$

$$\frac{\partial x_{ij}}{\partial q_{\alpha,\nu}} = w_{\nu,j} \sum_{\mu=1}^{n} e_{i,\mu} \sum_{\beta=1}^{m} b_{\mu,\beta} u_{\beta,\alpha}, \qquad \alpha = \overline{1,s}, \ \nu = \overline{1,r};$$

$$\frac{\partial x_{ij}}{\partial b_{\mu,\beta}} = e_{i,\mu} \sum_{\nu=1}^{r} w_{\nu,j} \sum_{\alpha=1}^{s} u_{\beta,\alpha} q_{\alpha,\nu}, \qquad \mu = \overline{1,n}, \ \beta = \overline{1,s};$$

$$\frac{\partial x_{ij}}{\partial w_{\nu,j}} = \sum_{\mu=1}^{n} e_{i,\mu} \sum_{\beta=1}^{m} b_{\mu,\beta} \sum_{\alpha=1}^{s} u_{\beta,\alpha} q_{\alpha,\nu},$$

$$\frac{\partial x_{ij}}{\partial e_{i,\mu}} = \sum_{\nu=1}^{r} w_{\nu,j} \sum_{\beta=1}^{m} b_{\mu,\beta} \sum_{\alpha=1}^{s} u_{\beta,\alpha} q_{\alpha,\nu}, \qquad \mu = \overline{1,n}. \tag{3.57}$$

Applying *matrix vectorization*, pass to the corresponding vectors $\mathbf{q}, \mathbf{b}, \mathbf{w}$ and $\mathbf{e}$. Denote by $\nabla_{\mathbf{q}}, \nabla_{\mathbf{b}}, \nabla_{\mathbf{w}}$ and $\nabla_{\mathbf{e}}$ the components of the gradient with respect to these vectors.

The algorithm **max** $\mathcal{H}$ has the following structure.

a) Initial step:

$$\mathbf{q}^0 > \mathbf{0}, \quad \mathbf{b}^0 > \mathbf{0}, \quad \mathbf{w}^0 > \mathbf{0}, \quad \mathbf{e}^0 > \mathbf{0}.$$

b) nth iterative step:

$$X^n = E^n \left\{ [B^n (UQ^n)] W^n \right\},$$

$$\mathcal{H}^n = \sum_{i=1}^{m} \sum_{j=1}^{s} x_{ij}^n \ln \frac{x_{ij}^n}{u_{ij}};$$

$$\mathbf{q}^{(n+1)} = \begin{cases} \mathbf{q}^n + \gamma_{\mathbf{q}} \nabla_{\mathbf{q}}(\mathbf{q}^n, \mathbf{b}^n, \mathbf{w}^n, \mathbf{e}^n) \text{ if } \mathbf{q} \geq \mathbf{0}, \\ \mathbf{q}^n \text{ if } \mathbf{q}^{(n+1)} < \mathbf{0}; \end{cases}$$

$$\mathbf{q}^{(n+1)} \Rightarrow Q^{(n+1)};$$

$$\mathbf{b}^{(n+1)} = \begin{cases} \mathbf{b}^n + \gamma_{\mathbf{b}} \nabla_{\mathbf{b}}(\mathbf{q}^n, \mathbf{b}^n, \mathbf{w}^n, \mathbf{e}^n) \text{ if } \mathbf{b}^{(n+1)} \geq \mathbf{0}, \\ \mathbf{b}^n \text{ if } \mathbf{b}^{(n+1)} < \mathbf{0}; \end{cases}$$

$$\mathbf{b}^{(n+1)} \Rightarrow B^{(n+1)};$$

$$\mathbf{w}^{(n+1)} = \begin{cases} \mathbf{w}^n + \gamma_{\mathbf{w}} \nabla_{\mathbf{w}}(\mathbf{q}^{(n+1)}, \mathbf{b}^{(n+1)}, \mathbf{w}^n, \mathbf{e}^n) \text{ if } \mathbf{w} \geq \mathbf{0}, \\ \mathbf{w}^n \text{ if } \mathbf{w}^{(n+1)} < \mathbf{0}; \end{cases}$$

$$\mathbf{w}^{(n+1)} \Rightarrow W^{(n+1)};$$

$$\mathbf{e}^{(n+1)} = \begin{cases} \mathbf{e}^n + \gamma_{\mathbf{e}} \nabla_{\mathbf{e}}(\mathbf{q}^{(n+1)}, \mathbf{b}^{(n+1)}, \mathbf{w}^n, \mathbf{e}^n) \text{ if } \mathbf{e}^{(n+1)} \geq \mathbf{0}, \\ \mathbf{e}^n \text{ if } \mathbf{e}^{(n+1)} < \mathbf{0}; \end{cases}$$

$$\mathbf{e}^{(n+1)} \Rightarrow E^{(n+1)};$$

$$X^{(n+1)} = E^{(n+1)} \left\{ \left[B^{(n+1)} \left(UQ^{(n+1)} \right) \right] W^{(n+1)} \right\};$$

$$\mathcal{H}^{(n+1)} = \sum_{i=1}^{m} \sum_{j=1}^{s} x_{ij}^{(n+1)} \ln \frac{x_{ij}^{(n+1)}}{u_{ij}}.$$

c) Stopping condition:

$$\text{If } \mathcal{H}^{(n+1)} - \mathcal{H}^{(n)} \leq \delta, \text{ then } STOP.$$

3.4 ENTROPY REDUCTION UNDER MATRIX NORM AND INFORMATION CAPACITY CONSTRAINTS

When applying the entropy reduction of a data matrix, it is desirable to obtain a reduced matrix with definite properties. In terms of the EDR method, this means introducing some restrictions on the class of projector matrices, which indirectly affect the resulting reduced matrix. Consider some useful characteristics of projector matrices.

1. Frobenius norms of matrices Above, it has been supposed that the direct Q and inverse T projector matrices have nonnegative elements. Such an assumption is due to the use of cross-entropy to characterize the "distance" between them. However, in applications-relevant machine learning problems, it often makes sense to narrow the class of projector matrices. This can be achieved by considering the Frobenius matrix norm:

$$
\begin{aligned}
n_Q(Q) &= \sum_{i=1}^{s}\sum_{j=1}^{r} q_{ij}^2 = \text{tr}\,(Q\,Q^{\mathsf{T}}) = \lceil Q, Q \rceil; \\
n_T(T) &= \sum_{i=1}^{r}\sum_{j=1}^{s} t_{ij}^2 = \text{tr}\,(T\,T^{\mathsf{T}}) = \lceil T, T \rceil.
\end{aligned}
\tag{3.58}
$$

2. Information capacity of data matrix In the course of reduction, some information contained in a data matrix is lost. Therefore, it should be measured somehow. Introduce the concept of *the information capacity $\mathcal{I}(A)$* of a nonnegative matrix $A_{n \times q}$, which is given by

$$
\mathcal{I}(A) = \sum_{i=1}^{n}\sum_{j=1}^{q} a_{ij} \ln a_{ij}.
\tag{3.59}
$$

Clearly, the information capacity corresponds to the entropy taken with the opposite sign.

3. Cross-entropy reduction algorithm The dimension reduction procedure of a data matrix is to minimize the cross-entropy:

$$
H(Q, T \,|\, U) \Rightarrow \min_{Q,T},
\tag{3.60}
$$

where

$$
H(Q, T \,|\, U) = \sum_{i=1}^{m}\sum_{j=1}^{s} x_{ij}(Q, T \,|\, U) \ln \frac{x_{ij}(Q, T \,|\, U)}{u_{ij}},
\tag{3.61}
$$

$$x_{ij}(Q, T \mid U) = \sum_{\mu=1}^{r} \sum_{\nu=1}^{s} t_{\mu,j} q_{\nu,\mu} u_{i,\nu}, \qquad i = \overline{1, m}; \; j = \overline{1, s}, \qquad (3.62)$$

subject to the following constraints:
—the nonnegative values of the projector matrices,

$$Q \geq 0, \qquad T \geq 0; \qquad (3.63)$$

—the upper bounds (a, b) on the norm of the projector matrices,

$$n_Q(Q) = \sum_{i=1}^{s} \sum_{j=1}^{r} q_{ij}^2 \leq a, \quad n_T(T) = \sum_{i=1}^{r} \sum_{j=1}^{s} t_{ij}^2 \leq b; \qquad (3.64)$$

—the δ-admissible decrease of the information capacity,

$$\frac{\mathcal{I}(U) - \mathcal{I}(Y(Q|U))}{\mathcal{I}(U)} \leq \delta. \qquad (3.65)$$

In this inequality,

$$\mathcal{I}(Y(Q \mid U)) = \sum_{i=1}^{m} \sum_{j=1}^{r} y_{ij}(Q \mid U) \ln y_{ij}(Q \mid U), \qquad (3.66)$$

$$y_{ij}(Q \mid U) = \sum_{\nu=1}^{s} u_{i,\nu} q_{\nu,j}. \qquad (3.67)$$

Due to the properties of the objective function (an entropy-bilinear form) and the constraints specified by quadratic entropy inequalities, the cross-entropy reduction problem (3.60)–(3.64) is nonconvex; for details, see [134, 171]. Therefore, there exists a certain (a priori unknown) number of local optimums, and this problem should be treated as a global optimization one.

Consider the problem (3.60)–(3.64) as a local minimization problem on the nonnegative orthant R_+^{2sr}. Here the variables are the elements of the matrices Q and T each containing rs entries. Using *matrix vectorization*, i.e., the transition to a complex vector through the sequential joining of rows, construct the vectors $\mathbf{q}$, $\mathbf{t}$ and $\mathbf{u}$, corresponding to the matrices Q, T and U, respectively. Then the cross-entropy minimization problem can be represented as

$$\mathcal{H}(\mathbf{q}, \mathbf{t} \mid \mathbf{u}) \Rightarrow \min, \qquad (3.68)$$

$$(\mathbf{q}, \mathbf{t}) \in R_+^{(2sr)},$$
$$N_Q(\mathbf{q}) = -\tilde{N}_Q(\mathbf{q}) + a \geq 0, \quad N_T(\mathbf{t}) = -\tilde{N}_T(\mathbf{t}) + b \geq 0, \text{ (3.69)}$$
$$\mathcal{I}_Y(\mathbf{q}) = I_Y(\mathbf{q} \,|\, \mathbf{u}) - \gamma(\delta) \geq 0, \quad \gamma(\delta) = (1 - \delta)I_U.$$

Recall that the problem (3.68) is considered as a local optimization one, for which the Karush–Kuhn–Tucker conditions are necessary. The Lagrange function for this problem has the form

$$L(\mathbf{q}, \mathbf{t}, \alpha, \beta, \gamma) = \mathcal{L}(\mathbf{q}, \mathbf{t}, \gamma) - \langle \alpha, \mathbf{q} \rangle - \langle \beta, \mathbf{t} \rangle, \tag{3.70}$$

where $\alpha = \{\alpha_1, \ldots, \alpha_{rs}\}$, $\beta = \{\beta_1, \ldots, \beta_{rs}\}$, and $\gamma = \{\gamma_1, \gamma_2, \gamma_3\}$ are the Lagrange multipliers related to the corresponding constraints.

The function

$$\mathcal{L}(\mathbf{q}, \mathbf{t}, \gamma) = \mathcal{H}(\mathbf{q}, \mathbf{t}) - \gamma_1 N_Q(\mathbf{q}) - \gamma_2 N_T(\mathbf{t}) - \gamma_3 \mathcal{I}_Y(\mathbf{q}) \tag{3.71}$$

will be called the truncated Lagrange function.

Assume that there exists a local minimum point $(\mathbf{q}^*, \mathbf{t}^*)$ in the problem (3.68, 3.70) and a neighborhood

$$\mathcal{O}_{(\mathbf{q}^*, \mathbf{t}^*)}(\mathbf{q}, \mathbf{t}) \in \|\mathbf{q}^* - \mathbf{q}\|^2 + \|\mathbf{t}^* - \mathbf{t}\|^2 \leq \varrho \tag{3.72}$$

such that the problem (3.68), (3.70) becomes convex on the set (3.72). Then the Kuhn–Tucker conditions [134] are satisfied on the set (3.72) in terms of the truncated Lagrange function (3.71), i.e.,

$$\begin{aligned}
&\nabla_{\mathbf{q}} \mathcal{L}(\mathbf{q}, \mathbf{t}, \gamma) \geq 0, \quad \nabla_{\mathbf{t}} \mathcal{L}(\mathbf{q}, \mathbf{t}, \gamma) \geq 0, \\
&\mathbf{q} \otimes \nabla_{\mathbf{q}} \mathcal{L}(\mathbf{q}, \mathbf{t}, \gamma) \geq 0, \quad \mathbf{t} \otimes \nabla_{\mathbf{t}} \mathcal{L}(\mathbf{q}, \mathbf{t}, \gamma) \geq 0, \\
&\mathbf{q}^* \geq 0, \quad \mathbf{t}^* \geq 0, \\
&\nabla_{\gamma} \mathcal{L}(\mathbf{q}, \mathbf{t}, \gamma) \leq 0, \quad \gamma \otimes \nabla_{\gamma} \mathcal{L}(\mathbf{q}, \mathbf{t}, \gamma) = 0, \\
&\gamma \geq 0.
\end{aligned} \tag{3.73}$$

This system of equations and inequalities can be solved using the multiplicative algorithm [143]. Introduce the following notations:

$$\begin{aligned}
\mathbf{G_q}(\mathbf{q}, \mathbf{t}, \gamma) &= 1 - \theta \nabla_{\mathbf{q}} \mathcal{L}(\mathbf{q}, \mathbf{t}, \gamma), \\
\mathbf{G_t}(\mathbf{q}, \mathbf{t}, \gamma) &= 1 - \theta \nabla_{\mathbf{t}} \mathcal{L}(\mathbf{q}, \mathbf{t}, \gamma), \\
\mathbf{K}_\gamma(\mathbf{q}, \mathbf{t}, \gamma) &= 1 + \theta \nabla_{\gamma} \mathcal{L}(\mathbf{q}, \mathbf{t}, \gamma).
\end{aligned} \tag{3.74}$$

The multiplicative algorithm, further referred to as **Minimum Cross-Entropy with constraints (MCEC)**, has the following structure.

a) *Initial step:*

$$\mathbf{q}^0 > \mathbf{0}, \quad \mathbf{t}^0 > \mathbf{0}.$$

b) *nth iterative step:*

$$\mathbf{q}^{(n+1)} = \mathbf{q}^n \otimes \mathbf{G_q}(\mathbf{q}^n, \mathbf{t}^n, \gamma^n),$$

$$\mathbf{t}^{(n+1)} = \mathbf{t}^n \otimes \mathbf{G_t}(\mathbf{q}^n, \mathbf{t}^n, \gamma^n);$$

$$\mathbf{t}^{(n+1)} \to T^{(n+1)};$$

$$\gamma^{n+1} = \gamma^n \otimes \mathbf{K_\gamma}(\mathbf{q}^n, \mathbf{t}^n, \gamma^n)$$

$$\mathbf{q}^{(n+1)} \to Q^{(n+1)};$$

$$\mathbf{t}^{(n+1)} \to T^{(n+1)};$$

$$X^{(n+1)} = U \, Q^{(n+1)} \, T^{(n+1)};$$

$$H(Q^{(n+1)}, T^{(n+1)} \,|\, U) = \sum_{i=1}^{m} \sum_{j=1}^{r} x_{ij}^{(n+1)} \ln \frac{x_{ij}^{(n+1)}}{u_{ij}} = H^{(n+1)}.$$

c) *Stopping condition:*

$$H^{(n+1)} - H^{(n)} \leq \Delta.$$

This algorithm converges to a local minimum if the initial point is positive [143].

3.5 ESTIMATING EFFICIENCY OF DIMENSION REDUCTION FOR LINEAR MODEL LEARNING

Linear static models are widely used to describe objects in RML problems. Consider this description in detail for the original and reduced input learning collections.

3.5.1 Linear model

A linear static object has a single output z and s inputs $\mathbf{u} \in R^s$. The state of this object is characterized by s attributes with parametric weights $\mathbf{a}$. The object's mathematical model is randomized, i.e., parameters $\mathbf{a}$ are random of the interval type:

$$\mathbf{a} \in \mathcal{A} = \{\mathbf{a}^- \leq \mathbf{a} \leq \mathbf{a}^+\}. \tag{3.75}$$

The probabilistic properties of these parameters are characterized by a probability density function (PDF) $P(\mathbf{a})$. Since the object is linear and static, the output z of its model has a linear relation to the inputs:

$$z = \langle \mathbf{u}, \mathbf{a} \rangle, \tag{3.76}$$

where $\langle , \rangle$ denotes the scalar product of vectors.

Suppose that there is an input learning collection $\mathcal{U}$ of m such objects, whose state is characterized by vectors $\mathbf{u}^{(1)}, \ldots, \mathbf{u}^{(m)}$ from the attribute space R^s and output is characterized by a vector $\mathbf{y} = \{y_1, \ldots, y_m\}$. Due to the object's linearity, the output of the model has the form

$$\mathbf{z} = U\mathbf{a}, \qquad \mathbf{z} = \{z_1, \ldots, z_m\}, \tag{3.77}$$

where $U = U_{(m \times s)}$ is a nonnegative matrix of dimensions $(m \times s)$ and

$$\mathbf{z} \in \mathcal{Z} = [\mathbf{z}^-, \mathbf{z}^+], \qquad \mathbf{z}^- = U\mathbf{a}^-; \ \mathbf{z}^+ = U\mathbf{a}^+. \tag{3.78}$$

A common assumption is that the outputs of the objects contain measurement errors described by a random vector $\bar{\xi} = \{\xi_1, \ldots, \xi_m\}$ of the interval type, i.e.,

$$\bar{\xi} \in \mathcal{K} = \{\bar{\xi}^- \leq \bar{\xi} \leq \bar{\xi}^+\}, \tag{3.79}$$

with a probability density function $L(\bar{\xi})$. In addition, let the PDFs of the model parameters and measurement noises be continuously differentiable.

The observed output of the model is distorted by some noises and has the form

$$\mathbf{v} = \mathbf{z} + \bar{\xi} = U_{(m \times s)} \mathbf{a} + \bar{\xi}. \tag{3.80}$$

The vector $\mathbf{v}$ belongs to the m-dimensional parallelepiped

$$\mathcal{V} = [\mathbf{v}^-, \mathbf{v}^+], \qquad \mathbf{v}^- = \mathbf{z}^- + \bar{\xi}^-; \ \mathbf{v}^+ = \mathbf{z}^+ + \bar{\xi}^+. \tag{3.81}$$

RML procedures produce the estimates $\hat{P}(\mathbf{a})$ and $\hat{L}(\bar{\xi})$ of the corresponding PDFs. This is done using the learning collection $(\mathcal{U}, \mathbf{y})$ and the randomized model (3.80). The resulting estimates serve for implementing this model, i.e., for generating an ensemble $\mathcal{V}$ of the random vectors $\mathbf{v} \in R^m$. The generation of such ensembles is one of the randomized machine-learning problems [145], further referred to as *the s-problem of RML*.

In many cases, the dimension s of the attribute space turns out to be very high for the existing computational capabilities. Therefore, it makes sense to reduce this space to a given dimension $r < s$.

Introduce an input data matrix $W_{(m \times r)}$, which will characterize a collection $\mathcal{W}$ of m objects in the reduced attribute space R^r, and a randomized model with parameters $\mathbf{b} = \{b_1, \ldots, b_r\} \in R^r$:

$$\mathbf{w} = W_{(m \times r)} \mathbf{b}, \qquad \mathbf{b} \in R^r, \ \mathbf{w} \in R^m. \tag{3.82}$$

The parameter vector $\mathbf{b}$ consists of random components of the interval type, i.e.,

$$\mathbf{b} \in \mathcal{B} = [\mathbf{b}^-, \mathbf{b}^+], \tag{3.83}$$

with a PDF $B(\mathbf{b})$. The observed vector has the form

$$\mathbf{t} = \mathbf{w} + \bar{\zeta}, \qquad \bar{\zeta} \in R^m. \tag{3.84}$$

The random vector $\bar{\zeta}$ describes measurement errors and is of the interval type, i.e.,

$$\bar{\zeta} \in \mathcal{J} = [\bar{\zeta}^-, \bar{\zeta}^+], \tag{3.85}$$

with a PDF $Q(\bar{\zeta})$. Finally,

$$\mathbf{t} \in \mathcal{T} = [\mathbf{t}^-, \mathbf{t}^+], \qquad \mathbf{t}^- = W\mathbf{b}^- + \bar{\zeta}^-, \ \mathbf{t}^+ = W\mathbf{b}^+ + \bar{\zeta}^+. \tag{3.86}$$

All the PDFs characterizing the model parameters and measurement noises are assumed to be continuously differentiable functions. The estimation of the PDF $B(\mathbf{b})$ and the generation of a corresponding ensemble $\mathcal{T}$ is another randomized machine-learning problem, called *the r-problem of RML*.

3.5.2 Comparing s- and r-problems of RML

These problems will be compared via a three-stage method as follows. In the first stage, the s- and r-problems solved by the RML procedures using the learning collections $(\mathcal{U}, \mathbf{y})$ and $(\mathcal{W}, \mathbf{y})$, respectively. This stage yields the entropy-optimal PDFs of the model parameters and measurement noises, $P^*(\mathbf{a})$ and $L^*(\bar{\xi})$ for the s-problem as well as $B^*(\mathbf{b})$

and $Q^*(\bar{\zeta})$ for the r-problem, respectively. In the second stage, the linear models (3.80, 3.84) are used for calculating the PDFs $F_s(\mathbf{v})$ and $F_r(\mathbf{t})$ normalized on the set $\mathcal{C} = \mathcal{V} \cap \mathcal{T}$. Finally, in the third stage, the Kullback–Leibler relative entropy [102] is calculated, which characterizes the absolute divergence between the PDFs $F_s(\mathbf{v})$ and $F_r(\mathbf{t})$. This method will be called *the s&r comparison* (SRC).

Consider *the first stage*, and recall the RML procedure for solving the s-problem; see the model described by (3.80).

For obtaining the best estimates under maximum uncertainty (the highest quality of learning), maximize the entropy functional defined on the PDFs $P(\mathbf{a})$ and $L(\bar{\xi})$,

$$\mathcal{H}[P(\mathbf{a}), L(\bar{\xi})] = -\int_{\mathcal{A}} P(\mathbf{a}) \ln P(\mathbf{a}) d\mathbf{a} - \int_{\mathcal{K}} L(\bar{\xi}) \ln L(\bar{\xi}) d\bar{\xi} \Rightarrow \max,$$

(3.87)

subject to the constraints

$$\int_{\mathcal{A}} P(\mathbf{a}) d\mathbf{a} = 1, \qquad \int_{\mathcal{K}} L(\bar{\xi}) d\bar{\xi} = 1,$$

(3.88)

and

$$\mathcal{M}\{\mathbf{z}\} = \int_{\mathcal{A}} U P(\mathbf{a}) d\mathbf{a} + \int_{\mathcal{K}} \bar{\xi} L(\bar{\xi}) d\bar{\xi} = \mathbf{y}.$$

(3.89)

This is an entropy-linear functional programming problem [138]. Its solution has the form

$$P^*(\mathbf{a}) = \pi^{-1}(\bar{\theta}) \exp\langle -\bar{\theta}, U \mathbf{a} \rangle,$$
$$L^*(\bar{\xi}) = \varpi^{-1}(\bar{\theta}) \exp\langle -\bar{\theta}, \bar{\xi} \rangle,$$

(3.90)

where

$$\pi(\bar{\theta}) = \int_{\mathcal{A}} \exp\langle -\bar{\theta}, U \mathbf{a} \rangle d\mathbf{a},$$
$$\varpi(\bar{\theta}) = \int_{\mathcal{K}} \exp\langle -\bar{\theta}, \bar{\xi} \rangle d\bar{\xi}.$$

(3.91)

The Lagrange multipliers $\bar{\theta}$ are found from the balance equations (3.89):

$$\int_{\mathcal{A}} U P^*(\mathbf{a}) d\mathbf{a} + \int_{\mathcal{K}} \bar{\xi} L^*(\bar{\xi}) d\bar{\xi} = \mathbf{y}.$$

(3.92)

Thus, the solution of the s-problem of RML gives the two entropy-optimal PDFs $P^*(\mathbf{a})$ (model parameters) and $L^*(\bar{\xi})$ (measurement noises).

In *the second stage*, the linear model (3.80) is used, which has independent parameters and measurement noises. The probability density function $F(\mathbf{v})$ of the observed vector $\mathbf{v}$ has the form

$$F(\mathbf{v}) = \int_{\mathcal{K}} G(\mathbf{v} - \bar{\xi}) L^*(\bar{\xi}) d\bar{\xi}, \tag{3.93}$$

where $G(\bullet)$ is the density of the vector $\mathbf{z}$ (3.80). According to this equality,

$$\mathbf{a} = (U^\mathsf{T} U)^{-1} U^\mathsf{T} \mathbf{z}. \tag{3.94}$$

Therefore,

$$P^*((U^\mathsf{T} U)^{-1} U^\mathsf{T} \mathbf{z}) = \eta(\mathbf{z}), \tag{3.95}$$

where the vector $\mathbf{z}$ is defined on the set $\mathcal{Z}$ (3.78). Normalizing this function, obtain the PDF

$$G(\mathbf{z}) = \frac{\eta(\mathbf{z})}{\int_{\mathcal{Z}} \eta(\mathbf{z})} d\mathbf{z} \tag{3.96}$$

of the vector $\mathbf{z}$, which is normalized on the set $\mathcal{Z}$ (3.78).

Well, equality (3.93) gives the PDF $F(\mathbf{v})$, $\mathbf{v} \in \mathcal{V}$ (3.81) for the randomized model of the s-problem. Denote it by $F_s(\mathbf{v})$.

Using a similar procedure for the r-problem (3.82–3.84) with the reduced data matrix W (3.75), find the PDF $F_r(\mathbf{t})$ normalized on the set $\mathcal{T}$ (3.86).

For comparing the PDFs $F_s(\mathbf{v})$ and $F_r(\mathbf{t})$, introduce the set

$$\mathcal{C} = \mathcal{V} \bigcap \mathcal{T}, \tag{3.97}$$

and normalize these functions on it. The normalized PDFs have the form

$$\tilde{F}_s(\mathbf{c}) = \frac{F_s(\mathbf{c})}{\int_{\mathcal{C}} F_s(\mathbf{c}) d\mathbf{c}}, \tilde{F}_r(\mathbf{c}) = \frac{F_r(\mathbf{c})}{\int_{\mathcal{C}} F_r(\mathbf{c}) d\mathbf{c}}.$$

In *the third stage*, the PDFs will be compared using the Kullback–Leibler divergence as a measure of *the absolute information error* Δ between the PDFs $\tilde{F}_r(\mathbf{c})$ and $\tilde{F}_s(\mathbf{c})$:

$$KL(\tilde{F}_s, \tilde{F}_r) = \int_{\mathcal{C}} \tilde{F}_r(\mathbf{c}) \ln \frac{\tilde{F}_r(\mathbf{c})}{\tilde{F}_s(\mathbf{c})} = \Delta \geq 0. \tag{3.98}$$

Note that the minimum $\Delta = 0$ is achieved at $\tilde{F}_r(\mathbf{c}) = \tilde{F}_s(\mathbf{c})$.

In addition, introduce *the relative information error* δ in the form

$$\delta = \frac{\Delta}{H_s + H_r}, \tag{3.99}$$

where

$$H_s = \int_{\mathcal{C}} \tilde{F}_s(\mathbf{c}) \ln \tilde{F}_s(\mathbf{c}) d\mathbf{c},$$

$$H_r = \int_{\mathcal{C}} \tilde{F}_r(\mathbf{c}) \ln \tilde{F}_r(\mathbf{c}) d\mathbf{c}. \tag{3.100}$$

These equalities are the information entropies of the PDFs $\tilde{F}_s$ and $\tilde{F}_r$ defined on the support $\mathcal{C}$ (3.97).

3.6 ESTIMATING EFFICIENCY OF EDR FOR BINARY CLASSIFICATION PROBLEMS

The EDR method will be characterized and compared by efficiency with principal component analysis (PCA) and the random projection (RP) method using a computational experiment on real data for the binary classification problem with a linear classifier.

3.6.1 Linear classifier

Recall some necessary results from the theory of linear classification [21, 57, 188]. The classifier's model has the form

$$z(t_k) = \text{sign}(\sum_{i=1}^{s} a_i u_i(t_k)). \tag{3.101}$$

Here the parameters $\mathbf{a} \in R^s$ are unknown deterministic weights of the attributes, and

$$\text{sign}(x) = \begin{cases} +1, & \sum_{i=1}^{s} a_i u_i(t_k) \geq 0, \\ -1, & \sum_{i=1}^{s} a_i u_i(t_k) < 0. \end{cases} \tag{3.102}$$

Let $\mathcal{L}(\mathbf{y}, U_{(m \times s)})$ be a collection of learning data. The components of the vector $\mathbf{y} \in R^m$ are equal to $+1$ if the object with number t_k belongs to the first class and equal to -1 if this object belongs to the second class. The rows of the matrix $U_{m \times s}$ contain the values of the attributes of all objects t_k, $k = \overline{1, m}$.

The classifier (3.101) is learned by minimizing the empirical risk

$$J(\mathbf{a}) = \sum_{i=1}^{m} \|\mathbf{y} - \mathbf{z}(\mathbf{a} \mid U_{(m \times s)})\|^2 \tag{3.103}$$

over the parameters $\mathbf{a}$. The components of the vector $\mathbf{z}$ are generated by the classifier's model (3.101).

The learned classifier is tested using another data collection $\mathcal{T} = (\mathbf{x}, V_{(n \times s)})$, where $\mathbf{x} \in R^n$. The classifier

$$w(t_k) = \text{sign}(\sum_{i=1}^{s} \hat{a}_i v_i(t_k)), \qquad k = \overline{1, n} \tag{3.104}$$

outputs $w(t_k) = +1$ or $w(t_k) = -1$ depending on the rows of the matrix $V_{(n \times s)}$. They are compared with the test vector $\mathbf{x}$, and the number of matches (correct classification) is calculated:

$$I = \sum_{k=1}^{n} \Lambda(w(t_k) - x(t_k)), \qquad (3.105)$$

where

$$\Lambda(y) = \begin{cases} 1 & \text{for } y = 0, \\ 0 & \text{for } y \neq 0. \end{cases} \qquad (3.106)$$

Then the classification accuracy ε can be characterized by the ratio

$$\varepsilon = \frac{I}{n}. \qquad (3.107)$$

3.6.2 Scheme of computational experiment

Following [162, 190], the EDR method was experimentally studied by solving the binary classification problem in the original and reduced attribute spaces, with subsequent comparison of this method with PCA and the RP method in terms of the classification accuracy (3.107). For making the results stable, Monte Carlo simulations and cross-validation were used.

The experiment consisted of several stages:

1. solving the classification problem in the original space;

2. solving the classification problem in the PCA-reduced spaces;

3. solving the classification problem in the RP-reduced spaces;

4. solving the classification problem in the EDR-reduced spaces.

Stages 2–4 were sequentially implemented for the dimension reduced by 1, i.e., $r = s - 1, \ldots, 1$, where r denotes the dimension of the reduced spaces. Stage 4 was implemented for various values of δ (an admissible decrease of information capacity).

The Python program for the experimental study was developed in OS Linux using the numpy 1.14.3, scipy 1.1.0, sklearn 0.19.1 libraries. Dimension reduction by principal component analysis and the random projection method was carried out using the corresponding classes of the scikit-learn library [128], PCA and GaussianRandomProjection. The classification problem was solved using support vector machines [70], i.e., the SVC class with the linear kernel.

The experimental study was based on two data sets as follows. The first one (*simulation*) was a synthetic data set of 100 objects with 10 attributes. The data were generated by the make_classification function [26] with the parameters n_classes=2, n_clusters_per_class=2, n_informative=10, n_redundant=0 and class_sep=1.0. This function generates points in the space of the dimension n_informative that are normally distributed (with zero mean and unit variance) around the edges of a hypercube with sides 2 * class_sep with some noise added.

The second data set (*heart*) consisted of the real data collected during heart disease research [2]. This data set contains 270 objects with 13 attributes.

In all experiments, the data were normalized to the interval $[0, 1]$.

3.6.3 Results of experiment

All experiments involved cross-validation using the bootstrap method [70]. More specifically, 30 objects were randomly selected 2 times from the original data set; 20 of them were used to learn the classifier, whereas the remaining 10 were used for testing; the results for each sample were averaged. Then the experiment was repeated 10 times, the results were also averaged. Dimension reduction at each stage was carried out for the entire learning sample (30 objects).

The problem (3.32)–(3.35) was solved using the constrained optimization method SLSQP [97] on the nonnegative orthant, implemented in the scipy library by the minimize function with the parameters ftol=1e-4 and maxiter=1000. Note that for some values of δ the method yielded no solution for the given optimization parameters. In such cases, the last solution found was used.

1. Dependence of ε on number of objects (sample size m) It is important to study this dependence for the three reduction methods in order to identify the areas of their application. As is well known, entropy maximization methods and their derivatives, in particular, the EDR method, are usually used with a limited amount of data compared to the dimension of the attribute space. In the case of "big data," there exist no fundamental restrictions on their application, but the computational difficulties will significantly increase. As for PCA, for large amounts of data (sufficiently great m) and their heterogeneity, the method becomes unstable [136].

The dependence of the classification accuracy on small sample sizes for the *simulation* data set was examined. The results are presented in

Table 3.1: Classification accuracy for data set *simulation* with PCA-based dimension reduction for different learning sample sizes m.

m / r	10	9	8	7	6	5
10	0.60	0.60	0.60	0.60	0.67	0.67
9	0.48	0.48	0.48	0.48	0.48	0.52
8	0.46		0.46	0.46	0.46	0.46
7	0.48			0.48	0.48	0.48
6	0.44				0.44	0.44
5	0.40					0.40

Table 3.2: Classification accuracy for data set *simulation* with RP-based dimension reduction for different learning sample sizes m.

m / r	10	9	8	7	6	5
10	0.60	0.67	0.57	0.77	0.60	0.57
9	0.48	0.48	0.59	0.41	0.52	0.48
8	0.46	0.46	0.46	0.46	0.46	0.54
7	0.48	0.48	0.52	0.38	0.43	0.52
6	0.44	0.28	0.39	0.39	0.39	0.39
5	0.40	0.40	0.40	0.40	0.40	0.40

Tables 3.1–3.3 for $m = \overline{5,10}$. For each of these sets, the attribute space was reduced to the size $r = \overline{5,10}$.

According to Tables 3.1–3.3, the EDR and RP methods are operable under any ratios of the sample size and dimension of the attribute space. At the same time, PCA does not work for $m < r$.

2. Classification accuracy for $m \gg r$ Tables 3.4 and 3.5, as well as Fig. 3.2, show the classification accuracy for different dimensions of the reduced space. This accuracy was estimated for two data sets, *simulation* and *heart*, in order to analyze its stability with respect to the data sets.

3. Dependence of ε on admissible decrease δ of information capacity (EDR method) Table 3.6 gives the classification accuracy for different values δ and the data set *simulation*, whereas Fig. 3.3 demonstrates the corresponding graphs. In addition, Table 3.7 and Fig. 3.3 present the same characteristics for the data set *heart*.

4. Information losses caused by reduction As has been emphasized above, the reduction of the attribute space leads to information losses.

Table 3.3: Classification accuracy for data set *simulation* with EDR-based dimension reduction for different learning sample sizes m and $\delta = 0.5$.

$m \,/\, r$	10	9	8	7	6	5
10	0.60	0.70	0.70	0.63	0.50	0.53
9	0.48	0.37	0.52	0.59	0.37	0.37
8	0.46	0.46	0.46	0.46	0.46	0.50
7	0.48	0.43	0.43	0.48	0.48	0.57
6	0.44	0.44	0.28	0.39	0.44	0.39
5	0.40	0.53	0.47	0.53	0.60	0.40

Table 3.4: Classification accuracy for data set *simulation*.

r	10	9	8	7	6	5
PCA	0.65	0.64	0.63	0.65	0.65	0.64
RP	0.65	0.58	0.54	0.58	0.62	0.48
EDR	0.65	0.62	0.62	0.47	0.48	0.49

Table 3.5: Classification accuracy for data set *heart*.

r	13	12	11	10	9	8
PCA	0.76	0.76	0.76	0.76	0.76	0.76
RP	0.76	0.79	0.79	0.73	0.76	0.78
EDR	0.76	0.70	0.78	0.74	0.76	0.69

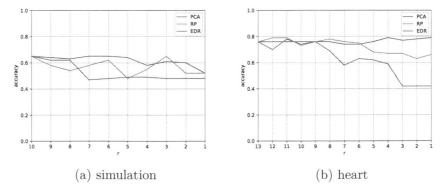

(a) simulation   (b) heart

Figure 3.2: Classification accuracy for data sets *simulation* and *heart*.

Table 3.6: Dependence $\varepsilon(\delta)$ for data set *simulation* (EDR method).

δ / r	9	8	7	6	5
0.1	0.60	0.65	0.50	0.50	0.50
0.2	0.70	0.60	0.55	0.45	0.45
0.3	0.75	0.75	0.40	0.45	0.50
0.4	0.55	0.50	0.45	0.50	0.55
0.5	0.50	0.60	0.45	0.50	0.45

Table 3.7: Dependence $\varepsilon(\delta)$ for data set *heart* (EDR method).

δ / r	12	11	10	9	8
0.1	0.65	0.85	0.80	0.80	0.55
0.2	0.60	0.70	0.70	0.60	0.65
0.3	0.75	0.75	0.75	0.80	0.75
0.4	0.75	0.85	0.80	0.80	0.70
0.5	0.75	0.75	0.65	0.80	0.80

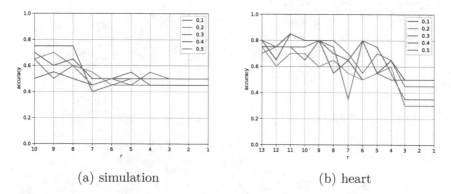

(a) simulation　　　　　(b) heart

Figure 3.3: Dependence $\varepsilon(\delta)$ for data sets *simulation* and *heart*.

Table 3.8: Relative information losses for data set *simulation*.

r	9	8	7	6	5
PCA	0.6691	0.6979	0.7298	0.7616	0.7973
RP	0.6630	0.6818	0.7844	0.7618	0.8777
EDR	0.2688	0.2911	0.3034	0.3501	0.4125

Table 3.9: Relative information losses for data set *heart*.

r	12	11	10	9	8	7
PCA	0.2935	0.3317	0.3857	0.4399	0.4941	0.5671
RP	0.1444	0.3708	0.5190	0.6051	0.6736	0.8167
EDR	0.3083	0.3013	0.3184	0.2772	0.2589	0.2792

The relative value of such losses can be described by

$$I = \frac{\mathcal{I}(U) - \mathcal{I}(Y)}{\mathcal{I}(U)}, \qquad (3.108)$$

where Y is the reduced data matrix. In the course of the experiment, the information losses caused by dimension reduction using PCA, the RP and EDR methods were estimated for the data sets *simulation* and *heart*.

Tables 3.8–3.9 illustrate the relative information losses I (3.108). For the EDR method, the values were averaged over δ.

Figure 3.4 shows the graphs of the relative information losses for the data sets *simulation* and *heart*.

The relative information losses caused by the EDR method for different values δ are combined in Tables 3.10–3.11.

The results of the computational experiment allow drawing the following conclusions:

Table 3.10: Dependence $I(\delta)$ for data set *simulation* (EDR method).

$\delta \ / \ r$	9	8	7	6	5
0.1	0.0959	0.0997	0.1594	0.2795	0.3996
0.2	0.2000	0.1932	0.1982	0.2656	0.3880
0.3	0.2896	0.2896	0.2952	0.2943	0.3805
0.4	0.3615	0.3887	0.3933	0.3981	0.3992
0.5	0.3971	0.4842	0.4707	0.5130	0.4954

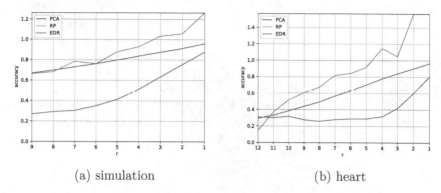

(a) simulation   (b) heart

Figure 3.4: Relative information losses for data sets *simulation* and *heart*.

Table 3.11: Dependence $I(\delta)$ for data set *heart* (EDR method).

δ / r	12	11	10	9	8	7
0.1	0.1071	0.0478	0.0821	0.1244	0.1065	0.1254
0.2	0.5947	0.0898	0.0676	0.0568	0.1435	0.1428
0.3	0.3344	0.7126	0.6659	0.3530	0.2961	0.2992
0.4	0.2921	0.3878	0.3833	0.4066	0.2814	0.3974
0.5	0.2134	0.2683	0.3932	0.4452	0.4670	0.4310

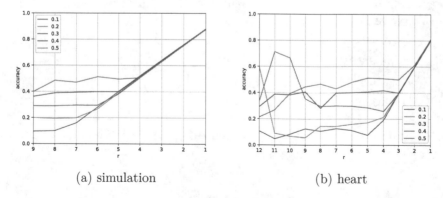

(a) simulation   (b) heart

Figure 3.5: Dependence $I(\delta)$ for data sets *simulation* and *heart* (EDR method).

- The EDR method is operable for any amounts of data in the learning and testing data collections; PCA does not work when the amount of data is smaller than the dimension of the attribute space; the operability of the RP method requires additional research.

- The level of information losses when applying the EDR method is retained up to the two-fold reduction of the attribute space; for PCA and the RP method, the level of information losses grows significantly with decreasing the dimension of the attribute space.

- Thanks to the control parameter δ in the EDR method, the level of relative information losses can be regulated. At the same time, this possibility is absent in PCA and the RP method.

3.7 ENTROPY METHODS FOR RANDOM PROJECTION

3.7.1 Statements of Entropy Randomized Projection Problems

1. Point-set representation of data matrices: indicator of compactness

Consider a data matrix describing some objects and their attributes:

$$U_{(m \times s)} = \begin{pmatrix} \mathbf{u}^{(1)} \\ \cdots \\ \mathbf{u}^{(m)} \end{pmatrix}, \qquad \mathbf{u}^{(i)} = \{u_1^{(i)}, \ldots, u_s^{(i)}\} \in R^s, \; i = \overline{1, m}. \tag{3.109}$$

Here m and s are the numbers of objects and their attributes, respectively; R^s denotes the attribute space. Assume that the elements of this data matrix are standardized, i.e., $u_{ij} \in [0, 1]$, $i = \overline{1, m}$, $j = \overline{1, s}$.

The point-set representation of a matrix $U_{(m \times s)}$ in the attribute space $R_{(m)}^s$ is the set $\mathfrak{U}_m^s$ of the points $\{\mathbf{u}^{(1)}, \ldots, \mathbf{u}^{(m)}\}$.

For example, the set $\mathfrak{U}_{100}^2$ of 100 two-dimensional points represents a matrix $U_{(100 \times 2)}$.

The spread of points in a set $\mathfrak{U}_m^s$ can be characterized by *the indicator of compactness* ρ_U, which will be defined as the arithmetic average of the distances between the points:[3]

$$\rho_U = \frac{2}{m(m-1)} \sum_{(\alpha,\beta)=1,\,\alpha \neq \beta}^{m} \varrho(\mathbf{u}^{(\alpha)}, \mathbf{u}^{(\beta)}), \tag{3.110}$$

---

[3]Other indicators are also possible, e.g., the maximum distance between the points.

where $\varrho(\mathbf{u}^{(\alpha)}, \mathbf{u}^{(\beta)})$ gives the distance between the points $\mathbf{u}^{(\alpha)}$ and $\mathbf{u}^{(\beta)}$. If the elements of the data matrix are standardized, then

$$0 \leq \varrho(\mathbf{u}^{(\alpha)}, \mathbf{u}^{(\beta)}) \leq \sqrt{2} \qquad \text{for all } (\alpha, \beta) = \overline{1, m},$$
$$0 \leq \rho_U \leq 1. \tag{3.111}$$

Introduce the following classes of transformations for data matrices:

- (mr)-*class*: $U_{(m \times s)} \rightarrow Y_{(m \times r)}$, the set $\mathfrak{Y}_m^r$ is in the space $R_{(m)}^r$;

- (ns)-*class*: $U_{(m \times s)} \rightarrow B_{(n \times s)}$, the set $\mathfrak{B}_n^s$ is in the space $R_{(n)}^s$;

- (nr)-*class*: $U_{(m \times s)} \rightarrow Z_{(n \times r)}$, the set $\mathfrak{Z}_n^r$ is in the space $R_{(n)}^r$.

2. Random projector matrices and their probabilistic characteristics

1. Projection onto the space $R_{(m)}^r$ (generation of the set $\mathfrak{Y}_m^r$) This operation will be performed using a projector matrix $Q_{(s \times r)}$:

$$Y_{(m \times r)} = \begin{pmatrix} \mathbf{y}^{(1)} \\ \cdots \\ \mathbf{y}^{(m)} \end{pmatrix} = U_{(m \times s)} Q_{(s \times r)};$$

$$y_j^{(i)} = \sum_{k=1}^{s} u_{ik} q_{kj}, \quad i = \overline{1, m}; \; j = \overline{1, r}. \tag{3.112}$$

The projector matrix $Q_{(s \times r)}$ is random and of the interval type, i.e.,

$$Q_{(s \times r)} \in \mathcal{Q} = [Q^-, Q^+]. \tag{3.113}$$

The probabilistic properties of the projector matrix are characterized by a continuously differentiable probability density function (PDF) $P(Q)$, defined on $\mathcal{Q}$.

The vectors $\mathbf{y}^{(i)}$ form the set $\mathcal{Y}_m^r$ of r-dimensional points, m in total. Define the distance $\varrho(\mathbf{y}^{(\alpha)}, \mathbf{y}^{(\beta)} \,|\, Q_{(s \times r)})$ between them and the indicator of compactness

$$\rho_Y(Q) = \frac{2}{m(m-1)} \sum_{(\alpha, \beta) = 1, \, \alpha \neq \beta}^{m} \varrho(\mathbf{y}^{(\alpha)}, \mathbf{y}^{(\beta)} \,|\, Q_{(s \times r)}). \tag{3.114}$$

Since the elements of the projector matrix have random values, the average distance (3.114) is random as well.

Hence, define the mathematical expectation of the indicator of compactness:

$$\bar{\rho}_Y = \int_{\mathcal{Q}} P(Q)\rho_Y(Q)dQ. \tag{3.115}$$

2. Projection onto the space $R^{(n \times s)}$ (generation of the set $\mathfrak{B}_n^s$) This operation will be performed using a projector matrix $T_{(n \times m)}$ with random elements:

$$B_{(n \times s)} = \begin{pmatrix} \mathbf{b}^{(1)} \\ \cdots \\ \mathbf{b}^{(n)} \end{pmatrix} = T_{(n \times m)}U_{(m \times s)}; \quad b_j^{(i)} = \sum_{k=1}^{m} t_{ik}\, u_{kj}, \tag{3.116}$$

$$i = \overline{1, n}; \; j = \overline{1, s}.$$

The projector matrix $T_{(n \times m)}$ is random and of the interval type, i.e.,

$$T_{(n \times m)} \in \mathcal{T} = [T^-, T^+]. \tag{3.117}$$

Its probabilistic properties are characterized by a PDF $W(T)$ defined on $\mathcal{T}$.

The vectors $\mathbf{b}^{(i)}$ form the set $\mathfrak{B}_n^s$ of s-dimensional points, $n < m$ in total. Define the indicator of compactness

$$\rho_B(T) = \frac{2}{n(n-1)} \sum_{(\alpha,\beta),\, \alpha \neq \beta}^{n} \varrho(\mathbf{b}^{(\alpha)}, \mathbf{b}^{(\beta)} \,|\, T) \tag{3.118}$$

and its mathematical expectation

$$\bar{\rho}_B = \int_{\mathcal{T}} W(T)\rho_B(T)dT. \tag{3.119}$$

3. Projection onto the space $R_{(n)}^r$ (generation of the set $\mathfrak{Z}_n^r$) This operation will be performed using random projector matrices $T_{n \times m}$ and $Q_{(s \times r)}$:

$$Z_{(n \times r)} = \begin{pmatrix} \mathbf{z}^{(1)} \\ \cdots \\ \mathbf{z}^{(n)} \end{pmatrix} = T_{(n \times m)}U_{(m \times s)}Q_{(s \times r)}; \tag{3.120}$$

$$z_j^{(i)} = \sum_{p=1}^{m} t_{ip} \sum_{k=1}^{s} u_{pk}\, q_{kj}, \quad i = \overline{1, n}; \; j = \overline{1, r}.$$

These projector matrices are of the interval type:

$$T_{(n \times m)} \in \mathcal{T} = [T^-, T^+],$$
$$Q_{(s \times r)} \in \mathcal{Q} = [Q^-, Q^+]. \tag{3.121}$$

The probabilistic properties of the projector matrices are characterized by a joint PDF $F(T, Q)$ defined on the set

$$\mathcal{F} = \mathcal{T} \bigotimes \mathcal{Q}. \tag{3.122}$$

The vectors $\mathbf{z}^{(i)}$ form the set $\mathfrak{Z}_n^r$ of r-dimensional points. The indicator of compactness for this set is given by

$$\rho_Z(T, Q) = \frac{2}{n(n-1)} \sum_{(\alpha,\beta),\, \alpha\neq\beta}^{n} \varrho(\mathbf{z}^\alpha, \mathbf{z}^\beta \,|\, T, Q). \tag{3.123}$$

The mathematical expectation of the indicator of compactness has the form

$$\bar{\rho}_Z = \int_{\mathcal{F}} F(T, Q)\rho_Z(T, Q)dTdQ. \tag{3.124}$$

Depending on the relations between the dimensions (m, s) of the original data matrix and the dimensions $(m, r), (n, s)$ and (n, r) of the projector matrices, the former is compressed (if $r < s$ and $n < m$) or expanded (if $r > s$ and $n > m$).

3.7.2 Algorithms for Entropy Randomized Projection

Consider projection onto the space $R_{(n)}^r$, which transforms a data matrix simultaneously by both parameters.

The entropy randomized projection (ERP) algorithm consists in the maximum entropy estimation (MEE) of the PDF $F(T, Q)$ with the additional condition that the data matrix $U_{(m\times s)}$ and the projection $Z_{(n\times r)}$ have equal values of their indices of compactness.

The MEE-estimate of the function $F(B, Q)$ (see Chapter A) is calculated as

$$\mathcal{H}[F(T, Q)] = -\int_{\mathcal{F}} F(T, Q) \ln F(T, Q) \, dTdQ \Rightarrow \max, \tag{3.125}$$

$$\int_{\mathcal{F}} F(T, Q)dTdQ = 1,$$

$$\int_{\mathcal{F}} F(T, Q)\rho_Z(T, Q)dTdQ = \rho_U.$$

The problem (3.125) is a functional constrained optimization problem of the Lyapunov type: all components of such problems are integral functionals [77].

Due to the integral form of this problem, the optimality conditions involve the following Lagrange functional with scalar Lagrange multipliers:

$$\mathcal{L}[F(T,Q)] = \mathcal{H}[F(T,Q)] + \mu\left(1 - \int_{\mathcal{F}} F(T,Q) dT dQ\right) +$$
$$+ \lambda\left(\rho_U - \int_{\mathcal{F}} F(T,Q)\rho_Z(T,Q) dT dQ\right). \quad (3.126)$$

The first-order optimality conditions for this functional in terms of the Gâteaux derivative [77] ($\mathfrak{G}_F$) have the form

$$\mathfrak{G}_F \mathcal{L}[F(T,Q)] = 0. \quad (3.127)$$

The entropy-optimal PDF of the projector matrices is given by

$$F^*(T,Q\,|\,\lambda) = \frac{\exp\left(-\lambda\,\rho_Z(T,Q)\right)}{\mathbb{F}(\lambda)}, \quad (3.128)$$

$$\mathbb{F}(\lambda) = \int_{\mathcal{F}} \exp\left(-\lambda\,\rho_Z(T,Q)\right) dT dQ.$$

According to (3.128), the structure of the PDF depends on the distance (3.123) and is independent of the parameter λ. This parameter satisfies the balance equation for the indicators of compactness:

$$\Phi(\lambda) = \int_{\mathcal{F}} \exp\left(-\lambda\rho_Z(T,Q)\right)\left(\rho_Z(T,Q) - \rho_U\right) dT dQ = 0. \quad (3.129)$$

Since λ is a scale parameter, it may be sufficient to find its approximate value from the linearized equation (3.129):

$$\tilde{\lambda} = \frac{\int_{\mathcal{F}}\left(\rho_Z(T,Q) - \rho_U\right) dT dQ}{\int_{\mathcal{F}} \rho_Z(T,Q)\left(\rho_Z(T,Q) - \rho_U\right) dT dQ}. \quad (3.130)$$

Example Let a data matrix and the indicators of compactness based on the Euclidean and Manhattan distances be

$$U_{(3\times2)} = \begin{pmatrix} 0.1 & 0.8 \\ 0.5 & 0.4 \\ 0.9 & 0.2 \end{pmatrix}; \quad \rho_U^E = 0.506, \quad \rho_U^M = 0.933.$$

Consider projection onto the space $R_{(1)}^2$:

$$Z_{(2\times1)} = \mathbf{z}^{(2)} = T_{(2\times3)} U_{(3\times2)} \mathbf{q}(2),$$

where

$$\mathcal{T} = [T^-, T^+], \quad T^- = \begin{pmatrix} 0 & 0 & 0 \\ 0 & 0 & 0 \end{pmatrix}; \quad T^+ = \begin{pmatrix} 1 & 1 & 1 \\ 1 & 1 & 1 \end{pmatrix}$$

$$\mathcal{Q} = [\mathbf{q}^-, \mathbf{q}^+], \quad \mathbf{q}^- = \begin{pmatrix} -1 \\ -1 \end{pmatrix}, \mathbf{q}^+ = \begin{pmatrix} 1 \\ 1 \end{pmatrix}.$$

The vector $\mathbf{z}^{(2)}$ consists of the components

$$z_1 = q_1 \sum_{j=1}^{3} t_{1j} u_{j1} + q_2 \sum_{j=1}^{3} t_{1j} u_{j2}, \quad z_2 = q_1 \sum_{j=1}^{3} t_{2j} u_{j1} + q_2 \sum_{j=1}^{3} t_{2j} u_{j2}.$$

The two types of distances between the elements of $\mathbf{z}$ are:
—the Euclidean distance

$$\rho_Z^E(T, \mathbf{q}) = (z_1 - z_2)^2 = \left[q_1 \sum_{j=1}^{3} u_{j1}(t_{1j} - t_{2j}) - q_2 \sum_{j=1}^{3} u_{j2}(t_{1j} - t_{2j}) \right]^2 ;$$

—the Manhattan distance

$$\rho_Z^M(T, \mathbf{q}) = |z_1 - z_2| = |q_1 \sum_{j=1}^{3} u_{j1}(t_{1j} - t_{2j}) - q_2 \sum_{j=1}^{3} u_{j2}(t_{1j} - t_{2j})|.$$

The entropy-optimal PDFs (3.128) of the random projector matrices have the following form:
—for the Euclidean distance,

$$F_E^*(T, \mathbf{q} \,|\, \lambda) = \frac{\exp\left(-\lambda \, \rho_Z^E(T, \mathbf{q})\right)}{\mathbb{F}(\lambda)}, \tag{3.131}$$

$$\mathbb{F}_E(\lambda) = \int_{\mathcal{F}} \exp\left(-\lambda \rho_Z^E(T, \mathbf{q})\right) dT \, d\mathbf{q};$$

—for the Manhattan distance,

$$F_M^*(T, \mathbf{q} \,|\, \lambda) = \frac{\exp\left(-\lambda \, \rho_Z^M(T, \mathbf{q})\right)}{\mathbb{F}(\lambda)}, \tag{3.132}$$

$$\mathbb{F}_M(\lambda) = \int_{\mathcal{F}} \exp\left(-\lambda \rho_Z^M(T, \mathbf{q})\right) dB \, d\mathbf{q}.$$

The parameter λ satisfies the following equations for the Euclidean and Manhattan distances, respectively:

$$\int_{\mathcal{F}} \exp\left(-\lambda \rho_Z^E(T, \mathbf{q})\right) \left(\rho_Z^E(T, \mathbf{q}) - \delta \, \rho_U^E\right) dT \, d\mathbf{q} = 0, \tag{3.133}$$

$$\int_{\mathcal{F}} \exp\left(-\lambda \rho_Z^M(T, \mathbf{q})\right) \left(\rho_Z^M(T, \mathbf{q}) - \delta \, \rho_U^M\right) dT \, d\mathbf{q} = 0.$$

In equalities (3.131–3.133),

$$\mathcal{F} = \mathcal{T} \bigotimes \mathcal{Q}.$$

A qualitative picture of the morphology of the multidimensional PDFs (3.131, 3.132) can be obtained using some of their sections.
   a) *The projector matrix T is given:*

$$T^* = \begin{pmatrix} 1 & 0 & 1 \\ 0 & 1 & 0 \end{pmatrix}.$$

The corresponding sections of the unnormalized functions (3.131, 3.132) have the form

$$F_E^*(B^*, \mathbf{q}) \sim \exp\left[-\lambda\,(0.41q_1 - 0.28q_2)^2\right], \quad (q_1, q_2) \in [-1, 1],$$

and

$$F_M^*(B^*, \mathbf{q}) \sim \exp\left[-\lambda\,|0.41q_1 - 0.28q_2|\right], \quad (q_1, q_2) \in [-1, 1].$$

The graphs of these functions for different values λ are shown in Fig. 1(a, b, c) and Fig. 2(a, b, c).

b) The projector matrix T and the projector vector $\mathbf{q}^{(2)}$ are given by

$$T = \begin{pmatrix} 0 & 1 & 0 \\ t_{21} & 0 & 1 \end{pmatrix}, \qquad \mathbf{q}^{(2)} = \begin{pmatrix} q_1 \\ -1 \end{pmatrix}.$$

The corresponding sections of the unnormalized functions (3.131, 3.132) have the form

$$F_E^*(T^*, \mathbf{q}) \sim \exp\left[-\lambda\,(-0.1q_1t_{21} - 0.4q_1 + 0.8t_{21} + 0.2)^2\right],$$
$$q_1 \in [-1, 1], \quad t_{21} \in [0, 1],$$

and

$$F_M^*(B^*, \mathbf{q}) \sim \exp\left[-\lambda\,|-0.1q_1t_{21} - 0.4q_1 + 0.8t_{21} + 0.2|\right],$$
$$q_1 \in [-1, 1], \quad t_{21} \in [0, 1].$$

The graphs of these functions for different values λ are presented in Figs. 3a–c and Figs. 4a–c.

Note that these PDFs of random projectors are entropy-optimal for the corresponding data matrix and differ significantly from the Gaussian distribution recommended for random projection. For example, see [181].

3.7.3 Implementation of random projectors and their numerical characteristics

A standard approach to implement random projectors is to sample the entropy-optimal PDFs $P(Q), W(T)$ and $F(T, Q)$, which preserve the inter-vector distance of the data matrix in its projector matrices. As the result of sampling, the PDFs are transformed into a sequence of random projector matrices:

$$F^*(T, Q) \Rightarrow (T^{(1)}, Q^{(1)}), (T^{(2)}, Q^{(2)}), \dots, (T^{(N)}, Q^{(N)}), \dots \qquad (3.134)$$

A general method for generating random sequences with a given PDF was proposed in [36]. The balance property (see above) may be not satisfied for each separate element of this sequence, but it will surely hold "on the average." Therefore, the empirical projector for implementation can be the matrix

$$\bar{T} \approx \frac{1}{N} \sum_{i=1}^{N} T^{(i)}, \qquad \bar{Q} \approx \frac{1}{N} \sum_{i=1}^{N} Q^{(i)}. \qquad (3.135)$$

The exact values of the "average" projector matrices are given by

$$\bar{T} = \int_{\mathcal{F}} T\, F(T,Q)\, dT\, dQ, \qquad \bar{Q} = \int_{\mathcal{F}} Q\, F(T,Q)\, dT\, dQ. \qquad (3.136)$$

Although the integrals in these equalities are multidimensional, integrands are exponentials of polynomials. This often provides analytical formulas for calculating multidimensional integrals.

In addition, the problem of calculating multidimensional integrals can be solved by approximate methods, such as the Monte Carlo simulations and the polynomial approximation of integrands.

The PDF $F^*(T,Q)$ is defined on the bounded set $\mathcal{F}$ (3.122). Therefore, the projector for implementation can be the pair of matrices

$$(T^*, Q^*) = \arg \max_{\mathcal{F}} F(T,Q). \qquad (3.137)$$

3.7.4   Random projector matrices with given values of elements

1. The set of $(0,1)$-projector matrices and its ordering

Consider a projection procedure onto the space $R^r_{(m)}$ performed by a projector matrix $Q_{(s \times r)}$ with random elements taking values 0 or 1.

Write this matrix as a row vector of length sr. The number of different sequences composed of zeros and ones, i.e., the number of different row vectors, is $N = 2^{rs}$. For example, for $s = 3$ and $r = 1$, there are 8 vectors of this type:

$$000, 100, 010, 001, 110, 011, 101, 111;$$

for $s = 4$ and $r = 1$, 16 vectors of this type:

$$0000, 1000, 0100, 0010, 0001, 1100, 0110, 0011,$$

$$1001, 0101, 1010, 1111, 1110, 0111, 1011, 1101.$$

For a projector matrix $Q_{(s \times r)}$, there exists a finite set $\mathbb{Q}$ of its $(0,1)$-realizations:

$$\mathbb{Q} = \{Q^{(1)}, \ldots, Q^{(N)}\}, \qquad N = 2^{sr}. \qquad (3.138)$$

Suppose that the realizations are random, then their probabilistic properties can be characterized by a discrete distribution function (DDF) $W(\alpha)$ with a discrete support $\alpha \in \mathcal{A} = [1, N]$, where

$$W(\alpha) = \mathbf{w} = \{w_\alpha\, \alpha = \overline{1, N}\}. \qquad (3.139)$$

The projector matrix is given by

$$
Y_{(m \times r)}^{(\alpha)} = \begin{pmatrix} \mathbf{y}_{(1)}^{(\alpha)} \\ \cdots \\ \mathbf{y}_{(m)}^{(\alpha)} \end{pmatrix} = U_{(m \times s)} Q_{(s \times r)}^{(\alpha)}; \quad y_j^{(\alpha, i)} = \sum_{k=1}^{s} u_{ik} q_{kj}^{(\alpha)}, \quad (3.140)
$$

$$
i = \overline{1, m}; \; j = \overline{1, r}.
$$

The indicator of compactness of the set $\mathfrak{Y}^{(\alpha)}$ has the form

$$
\rho_Y^{(\alpha)}(Q) = \frac{2}{m(m-1)} \sum_{(\kappa, \beta)=1, \, \kappa \neq \beta}^{m} \varrho^{(\alpha)}(\mathbf{y}^{(\kappa)}, \mathbf{y}^{(\beta)} \mid Q_{(s \times r)}^{(\alpha)}). \quad (3.141)
$$

Order the sequence (3.138) in accordance with the sequence of inequalities

$$
0 < \rho_Y^{(1)}(Q) < \rho_Y^{(2)}(Q) < \cdots < \rho_Y^{(N)}(Q). \quad (3.142)
$$

2. MEE-estimate of DDFs

The MEE-estimate of the DDF $W(\alpha)$ is calculated as

$$
H(\mathbf{w}) = -\sum_{\alpha=1}^{N} w_\alpha \ln w_\alpha \Rightarrow \max, \quad (3.143)
$$

subject to the following constraints:
— the normalization conditions

$$
\sum_{\alpha=1}^{N} w_\alpha = 1; \quad (3.144)
$$

— the balance conditions for the indicators of compactness

$$
\sum_{\alpha=1}^{N} w_\alpha \, \rho_Y^{(\alpha)}(Q) = \rho_U. \quad (3.145)
$$

The problem (3.143)–(3.145) is a finite-dimensional constrained optimization problem with a concave objective function (information entropy) and linear equality constraints.

Consider the Lagrange function

$$
L(\mathbf{w}, \lambda) = H(\mathbf{w}) + \mu \left(1 - \sum_{\alpha=1}^{N} w_\alpha \right) + \lambda \left(\rho_U - \sum_{\alpha=1}^{N} w_\alpha \rho_Y^{(\alpha)}(Q) \right). \quad (3.146)
$$

From the first-order optimality conditions, for this function, it follows that the entropy-optimal discrete probability distribution is given by

$$w_\alpha^* = \frac{\exp\left(-\lambda \rho_Y^{(\alpha)}(Q)\right)}{\sum_{\alpha=1}^N \exp\left(-\lambda \rho_Y^{(\alpha)}(Q)\right)}, \qquad \alpha = \overline{1, N}, \tag{3.147}$$

where the parameter λ satisfies the equation

$$\sum_{\alpha=1}^N \exp(-\lambda \rho_Y^{(\alpha)}(Q))(\rho_Y^{(\alpha)}(Q) - \rho_U) = 0, \qquad -\infty < \lambda < \infty. \tag{3.148}$$

Introduce the change of variable $z = \exp(-\lambda)$. Then equation (3.148) can be written as

$$\varphi(z) = \sum_{\alpha=1}^N z^{h_\alpha} c_\alpha = 0, \ z \geq 0; \quad c_\alpha = (h_\alpha - \rho_U), \ h_\alpha = \rho_Y^{(\alpha)}. \tag{3.149}$$

Theorem 3.1 *Assume that there exists $\alpha = l$ such that*

$$c_\alpha < 0 \quad for \ \alpha = 1, \dots, l,$$
$$c_\alpha \geq 0 \quad for \ \alpha = l+1, \dots, N, \tag{3.150}$$

and, according to (3.142),

$$0 < h_1 < h_2 < \cdots < h_l < h_{l+1} < \cdots < h_N. \tag{3.151}$$

Then equation (3.149) has a unique positive solution $z^* > 0$.

Proof Using conditions (3.150, 3.151), write equation (3.149) as

$$\phi(z) = \frac{\sum_{\alpha=l+1}^N z^{h_\alpha} c_\alpha}{\sum_{\alpha=1}^l z^{h_\alpha} c_\alpha} = 1. \tag{3.152}$$

Due to (3.151), $h_{l+1} - h_1 > 0$, and

$$\phi(z) = z^{(h_{l+1}-h_1)} \frac{\sum_{\alpha=l+2}^N z^{h_\alpha} c_\alpha + c_{l+1}}{\sum_{\alpha=2}^l z^{h_\alpha} c_\alpha + c_1}. \tag{3.153}$$

Hence, $\phi(0) = 0$. Now represent the function $\phi(z)$ in the form

$$\phi(z) = z^{(h_N-h_l)} \frac{\sum_{\alpha=l+1}^{N-1} z^{h_\alpha-h_N} c_\alpha + c_N}{\sum_{\alpha=1}^{l-1} z^{h_\alpha-h_l} c_\alpha + c_l}. \tag{3.154}$$

As is easily checked, $\phi(\infty) = +\infty$. Consider its derivative:

$$\phi'(z) = \frac{\sum_{\beta=1}^l \sum_{\alpha=l+1}^N c_\beta c_\alpha (h_\alpha - h_\beta) z^{(h_\beta - h_\alpha)}}{\left(\sum_{\alpha=1}^l c_\alpha z^{h_\alpha}\right)^2} > 0. \tag{3.155}$$

Therefore, the function $\phi(z)$ is monotonically increasing, and equations (3.149), (3.151) have a unique positive solution $z^* > 0$. ■

3.7.5 Choice of appropriate projector matrix from $\mathbb{Q}$ (3.138)

The entropy-optimal DDF $W^*(\alpha) = w_\alpha^*$, $\alpha = \overline{1, N}$, (3.149) characterizes the probabilities of implementing the elements (projector matrices) from the set $\mathbb{Q}$ (3.138). However, for obtaining a particular projector matrix $Y_{(m \times r)}$, a particular projector matrix from the set $\mathbb{Q}$ is needed. Particular matrices can be chosen using different strategies as follows.

1. Projector matrix with maximum probability Let the difference $\Delta_\rho = \rho_Y^{(N)} - \rho_Y^{(1)}$ in (3.142) be sufficiently large. This means that the maximum value of probability in the DDF (3.149) is appreciably greater than the minimum one. Then the implemented projector matrix has the number

$$\alpha^* = \arg \max_{1 \leq \alpha \leq N} w_\alpha^*. \tag{3.156}$$

2. Average projector matrix If $\Delta_\rho = \rho_Y^{(N)} - \rho_Y^{(1)}$ is small, the sequence $\{\rho^{(\alpha)}\}$ increases with a low rate. In this case, the projector matrix for implementation should be obtained by averaging:

$$\tilde{\alpha} = \sum_{\alpha=1}^{N} \alpha w_\alpha^*, \qquad \alpha^* = \lfloor \tilde{\alpha} \rfloor, \tag{3.157}$$

where $\lfloor \bullet \rfloor$ denotes the integer part of a number $\bullet$.

3. Median projector matrix is given by

$$\sum_{\alpha=1}^{\alpha^*} w_\alpha^* = \sum_{\alpha=\alpha^*+1}^{N} w_\alpha^*. \tag{3.158}$$

4. Averaging of projector matrices This strategy consists of the following stages. In the first stage, subsets of projector matrices with definite properties are selected. In the second stage, a subset of the corresponding projector matrices is formed. Then the average projector matrix is calculated.

Whereas the second and third stages are related to standard calculations for mathematical statistics, the first one requires additional comments. The subsets of projector matrices are selected in accordance with rules reflecting their desired properties. Depending on the rule for selecting such subsets, the so-called variance and interquartile subsets of projector matrices will be studied below.

First, consider *the variance subsets*. For this, find the root mean square (RMS) value of the variable α:

$$\sigma_\alpha^* = \lfloor \tilde{\sigma} \rfloor, \qquad \tilde{\sigma} = \sqrt{D\alpha}, \qquad D\alpha = \sum_{\alpha=1}^{N} (\alpha^* - \alpha)^2 w_\alpha. \tag{3.159}$$

Define the variance interval

$$\mathcal{I}_\alpha = [\alpha^* - \sigma_\alpha^*, \alpha^* + \sigma_\alpha^*]. \tag{3.160}$$

The variance subset has the form

$$\tilde{\mathbb{Q}} = \{Q^{(\alpha^* - \sigma_\alpha^*)}, \ldots, Q^{(\alpha^* + \sigma_\alpha^*)}\}. \tag{3.161}$$

For each element of this subset, calculate the corresponding projector matrices:

$$\tilde{\mathcal{Y}} = \{Y_{(m \times r)}^{(\alpha^* - \sigma_\alpha^*)}, \ldots, Y_{(m \times r)}^{(\alpha^* + \sigma_\alpha^*)}\}. \tag{3.162}$$

In the last stage, find the average projector matrix on the variance set:

$$\bar{Y}_{(m \times r)}^D = \frac{1}{2\sigma_\alpha^*} \sum_{\alpha = \alpha^* - \sigma_\alpha^*}^{\alpha^* + \sigma_\alpha^*} Y_{(m \times r)}^{(\alpha)}. \tag{3.163}$$

Now analyze the formation of *the interquantile subsets*. The entropy-optimal probability distribution function (see (3.147)) has the form

$$F^*(\kappa) = \sum_{\alpha=1}^{\kappa} w_\alpha^*, \qquad \kappa \in [0, 1]. \tag{3.164}$$

Recall that the κ^*-quartile of a random variable α is the set $\mathcal{A}^{(\kappa^*)}$ of its values $[1, \kappa^*]$.

Consider two quartile sets, $\mathcal{A}^{(\kappa_1^*)}$ and $\mathcal{A}^{(\kappa_2^*)}$, where $\kappa_1^* > \kappa_2^*$. The subsets of projector matrices are associated with each of these sets:

$$\mathbb{Q}(\kappa*_1) = \{Q^{(1)}, \ldots, Q^{(\alpha(\kappa_1^*))}\},$$
$$\mathbb{Q}(\kappa*_2) = \{Q^{(1)}, \ldots, Q^{(\alpha(\kappa_2^*))}\}. \tag{3.165}$$

Construct the subset of projector matrices

$$\mathbb{Q}(\kappa_1^* - \kappa_2^*) = \{Q^{(\alpha(\kappa_2^*))}, \ldots, Q^{(\alpha(\kappa_1^*))}\}, \tag{3.166}$$

which is the $(\kappa_1^* - \kappa_2^*)$-interquantile subset $\mathcal{I}^{(\kappa_1^* - \kappa_2^*)}$ of the entropy-optimal projector matrices. The number of elements in this subsets is $M_{(\kappa_1^* - \kappa_2^*)} = \alpha(\kappa_1^*) - \alpha(\kappa_2^*)$.

For each element of the subset $\mathcal{Q}(\kappa_1^* - \kappa_2^*)$ (3.166), calculate the corresponding projector matrices

$$\mathcal{Y}(\kappa_1^* - \kappa_2^*) = \{Y_{(m \times r)}^{\alpha(\kappa_2^*)}, \ldots, Y_{(m \times r)}^{\alpha(\kappa_1^*)}\}. \tag{3.167}$$

The average projector matrix on the subset $\mathcal{I}^{(\kappa_1^* - \kappa_2^*)}$ is given by

$$\bar{Y}_{(m \times r)}^K = \frac{1}{M_{(\kappa_1^* - \kappa_2^*)}} \sum_{\alpha = \alpha(\kappa_2^*)}^{\alpha(\kappa_1^*)} Y_{(m \times r)}^\alpha. \qquad (3.168)$$

Averaging over interquantile sets yields projector matrices with given (κ_1, κ_2)-quantiles.

CHAPTER 4

RANDOMIZED PARAMETRIC MODELS

An important component of the RML procedures is a model with random parameters, which will be called *the randomized parametric model (RPM)*. In this way, we emphasize that the random parameters in this model are artificial and have optimal probabilistic characteristics. The optimality of the probabilistic characteristics of the RPM parameters is considered in terms of an quality functional and a set of admissible parameter values in a specific problem. As the RML procedures are used for a wide range of machine learning problems, we will study static and dynamics RPMs.

4.1 DEFINITION, CHARACTERISTICS AND CLASSIFICATION

A randomized parametric model (RPM) transforms input data into an *ensemble* of output data, which is generated by the model in accordance with given probability density functions (PDFs) of its parameters.

Consider a case in which the input data represent a one-dimensional array, i.e., a vector $\mathbf{x} \in R^n$, as illustrated in Fig. 4.1a. Its elements are some characteristics $x_1, \ldots, x_n$ of a studied object. An RPM has randomized parameters $\mathbf{a} \in R^r$ whose probabilistic properties are described by a PDF $P(\mathbf{a})$. An RPM generates an ensemble $\mathcal{Y}$ of vectors $\mathbf{y} \in R^m$ with the given PDF $P(\mathbf{a})$.

Next, let the input data be defined by a two-dimensional array, i.e., a matrix X composed of column vectors $\mathbf{x}^{(j)}$, $j = \overline{1, s}$ (see Fig. 4.1b). For example, such an array results from measuring the object's state vector at s different times. For each measured vector $\mathbf{x}^{(j)}$ from the input array,

DOI: 10.1201/9781003306566-4

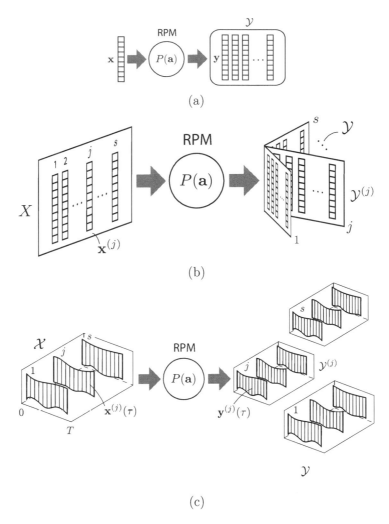

Figure 4.1

an RPM with the above-mentioned properties generates an ensemble $\mathcal{Y}^{(j)}$ of random vectors $\mathbf{y}^{(j)} \in R^m$ with the given PDF $P(\mathbf{a})$.

As a last example, consider a set $\mathcal{X}_\tau$ composed of three-dimensional input data arrays containing measurements of a vector function $\mathbf{x}^{(j)}(\tau)$, $j = \overline{1, s}$, where $\tau \in \mathcal{T} = [0, T]$. Here, for each measurement, an RPM generates three-dimensional ensembles $\mathcal{Y}^{(j)}$ of random vector functions $\mathbf{y}^{(j)}(t)$, $t \geq \tau$, with the given PDF $P(\mathbf{a})$.

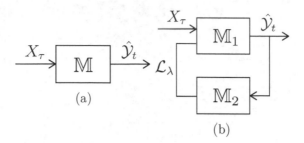

Figure 4.2

In all examples above, an RPM transforms deterministic input data into an ensemble of randomized output data. Such a transformation will be characterized by a mapping

$$\mathbb{M}(P(\mathbf{a}), \tau, t, T) =: \mathcal{X}_\tau \xrightarrow{P(\mathbf{a})} \hat{\mathcal{Y}}, \tag{4.1}$$

and a corresponding RPM will be assigned the "*input–output ensemble*" *[I–OE]* class; see Fig. 4.2a.

Sometimes, it is useful to structure the mapping $\mathbb{M}(P(\mathbf{a}), \tau, T)$. In particular, we can represent it as two mappings $\mathbb{M}_1(P_1(\mathbf{a}^{(1)}), \tau, \lambda, t, T)$ and $\mathbb{M}_2(P_2(\mathbf{a}^{(2)}), t, \lambda)$ within a feedback loop (see Fig. 4.2b), i.e.,

$$\mathbb{M}_1(P_1(\mathbf{a}^{(1)}), \tau, \lambda, t, T) =: (X_\tau, \hat{\mathcal{Z}}_\lambda) \xrightarrow{P_1(\mathbf{a}^{(1)})} \mathcal{Y};$$
$$\mathbb{M}_1(P_2(\mathbf{a}^{(2)}), t, \lambda) =: \hat{\mathcal{Y}} \xrightarrow{P_2(\mathbf{a}^{(2)})} \hat{\mathcal{Z}}_\lambda. \tag{4.2}$$

Here $(\lambda, \tau, t) \in \mathcal{T} = [0, T]$. The models described by the mappings (4.1, 4.2) will be classified as the "*(input, feedback)–output ensemble*" *[(I,FB)– OE]* RPMs.

These RPMs have randomized interval-time parameters, that is,

$$\mathbf{a} \in \mathcal{A} = [\mathbf{a}^-, \mathbf{a}^+], \qquad \mathbf{a}^\pm = \{a_1^\pm, \dots, a_r^\pm\}. \tag{4.3}$$

In addition, the RPMs of both classes will be divided into subclasses, in terms of their temporal properties and mathematical modeling methods. Based on the first classification attribute, there exist static and dynamic RPMs. For the static models, $\tau = t$ in the mapping (4.1), i.e., an ensemble generated by such a model at a time t depends on the data at this time only. In the dynamic RPMs, an ensemble $\hat{\mathcal{Y}}$ depends on the data X_τ, where $t_0 \le \tau \le t$.

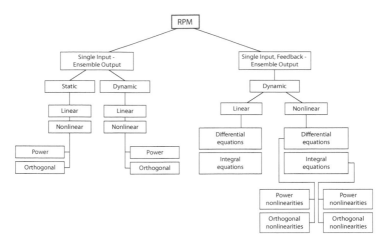

Figure 4.3

For further classification in terms of mathematical modeling methods, we will employ a standard approach by considering linear and nonlinear models, functions and functionals, differential and integral equations, and different approximations of nonlinearities. The classification tree is demonstrated in Fig. 4.3.

4.2 "SINGLE INPUT–ENSEMBLE OUTPUT" RANDOMIZED PARAMETRIC MODEL

4.2.1 Static models

For this class of models, an ensemble of output data at a time t depends on the input data at this time and also on a realization of the randomized parameters. Therefore, the index t can be omitted, and such mappings can be described using vector functions of the random argument $\mathbf{a}$.

Consider an input data array, which is characterized by a matrix $X = [x_{ij} \,|\, i = \overline{1,n},\, j = \overline{1,s}]$. This array consists of s measurements of an input vector $\mathbf{x}^{(j)} = \{x_1^{(j)}, \ldots, x_n^{(j)}\}$, where $j = \overline{1,s}$. The elements of this vector are the elements of column j in the matrix X.

A static RPM transforms the input matrix X into an output matrix $\hat{Y} = [\hat{y}_{kj} \,|\, k = \overline{1,m},\, j = \overline{1,s}]$. The columns of this matrix are the elements of the jth vector $\hat{\mathbf{y}}^{(j)} = \{\hat{y}_1^{(j)}, \ldots, \hat{y}_m^{(j)}\}$, where $j = \overline{1,s}$.

Introduce a matrix $A = [a_{hp} \,|\, h = \overline{1,m}, p = \overline{1,n}]$, composed of randomized parameters. The elements of this matrix are random variables taking values within the intervals

$$\mathcal{A}_{hp} = [a_{hp}^-, a_{hp}^+],\tag{4.4}$$

which form the $(m \times s)$-dimensional parallelepiped

$$\mathcal{A} = \bigotimes_{h=1}^{m} \bigotimes_{p=1}^{n} \mathcal{A}_{hp}.\tag{4.5}$$

By assumption, there exists a continuously differentiable PDF $P(A)$ on this set.

A relation between the input X and the output ensemble $\hat{\mathcal{Y}}$ of a static RPM is described by a vector function $\mathcal{F}$ that has the randomized parameters A with the PDF $P(A)$, i.e.,

$$\hat{\mathcal{Y}} = \mathcal{F}(X, A, P(A)).\tag{4.6}$$

In other words, for a fixed input data array X, a static RPM generates an ensemble $\hat{\mathcal{Y}}$ of random matrices $\hat{Y}$ with the above-mentioned properties.

An important special case of the RPMs (4.6) is the model with n inputs and one output. In this case, the model generates a vector $\hat{\mathbf{y}} = \{y_1^{(1)}, \ldots, y_m^{(s)}\}$, and its randomized parameters are characterized by a vector $\mathbf{a} = \{a_1, \ldots, a_r\}$ with interval-type elements

$$a_h \in \mathcal{A}_h = [a_h^-, a_h^+], \qquad \mathcal{A} = \bigotimes_{h=1}^{r} \mathcal{A}_h.\tag{4.7}$$

Then equality (4.6) takes the form

$$\hat{\mathcal{Y}} = \mathcal{F}(X, \mathbf{a}, P(\mathbf{a})),\tag{4.8}$$

where $\hat{\mathcal{Y}}$ denotes an ensemble of random vectors $\hat{\mathbf{y}}$.

1. Linear static RPMs. This class of the models involves a linear vector function $\mathbf{F}$ (4.6) of the form

$$\hat{\mathcal{Y}} = A X,\tag{4.9}$$

where A is a matrix of dimensions $(m \times n)$ with the randomized elements (4.4, 4.5) and the probabilistic properties characterized by a PDF $P(A)$.

If the RPM (4.6) has n inputs and one output, then the output ensemble consists of the values of a random variable

$$\hat{y} = \langle \mathbf{x}, \mathbf{a} \rangle.\tag{4.10}$$

2. Nonlinear static RPMs. The nonlinear RPMs form a significantly wider class than the linear ones. The nonlinear processes that occur in nature are modeled using appropriate laws. If a certain relation between an input and output of an object surely exists but its form is unknown, then a more of less universal approach lies in approximation. Here it is necessary to define a structure and parameters of an approximating function. As a result, to design a nonlinear static RPM means to search for appropriate parameter values. Consider some types of approximations and also the underlying hypotheses on the properties of the "Single Input–Ensemble Output" relations.

2.1. Analytical nonlinearities. By a prevailing hypothesis, there exists an analytical relation between the input and output matrices, X and Y. This hypothesis allows us to perform approximation using the *Taylor monomial*

$$\hat{y} = \sum_{h=1}^{R} B^{(h)} Z^{(h)}. \tag{4.11}$$

Here $B^{(h)}$ denotes a matrix of dimensions $(s \times \varphi(h))$ with interval-type randomized elements, and $Z^{(h)}$ is a matrix of dimensions $(\varphi(h)) \times r$ whose entries represent the products of degree h from the elements of the input matrix X. The matrices $B^{(h)}$ are defined on the intervals

$$\mathcal{B}_h = [\mathcal{B}_h^-, \mathcal{B}_h^+], \qquad h = \overline{1, R}. \tag{4.12}$$

For a one-dimensional static object (with one input and one output), equality (4.11) takes the simpler form

$$\hat{y} = \sum_{h=1}^{R} b_h x^h, \tag{4.13}$$

where the parameters $b_1, \ldots, b_R$ are randomized and independent. The domain of definition of the parameters $b_1, \ldots, b_R$ is given by

$$b_h \in \mathcal{B}_h = [b_h^-, b_h^+], \qquad h = \overline{1, R}. \tag{4.14}$$

2.2. Continuous nonlinearities. In fact, the existence of an analytical relation between an input and output of a static nonlinearity is a bold hypothesis. Following a more realistic approach, assume that the input is bounded and the nonlinearity $\Phi(x)$ has no discontinuities of the second kind (infinite jumps). Numerous approximation methods with orthogonal polynomials were developed on the basis of these hypotheses.

Table 4.1: Orthonormal polynomials.

Class	Interval	$w(x)$	Polynomial
Lagerra	$[0, \infty)$	$e^{(-x)}$	$1, 1-x, 1-2x+x^2/2, \ldots$
Hermit	$(-\infty, \infty)$	e^{-x^2}	$1, 2x, 4x^2-2, \ldots$
Legendre	$[-1, 1]$	1	$1, x, (3x^2-1)/2, \ldots$
Chebyshev	$[-1, 1]$	$1/\sqrt{1-x^2}$	$1, x, 2x^2-1, \ldots$

Recall some fundamental concepts of the corresponding methods. Consider a one-dimensional static object with a bounded input and a nonlinear characteristic that has no discontinuities of the second kind:

$$x^- \leq x \leq x^+.$$

By assumption, there exists a certain function $w(x)$ on this interval, which is often interpreted as a weight characteristic for the values of the variable x. Since the nonlinearity $\Phi(x)$ of the static object is considered on the same interval without discontinuities of the second kind,

$$\int_{x^-}^{x^+} w(x)\Phi^2(x)dx < \infty.$$

To obtain orthogonal polynomials, choose a system of power functions $1, x, x^2, \ldots$ and apply the following orthogonalization procedure:

$$\psi_0(x) = c_0,$$

$$\psi_1(x) = c_0 + c_1 x,$$

$$\psi_2(x) = c_0 + c_1 x + c_2 x^2,$$

$$\cdots = \cdots + \cdots + \cdots$$

The coefficients $c_0, c_1, c_2, \ldots$ in these formulas are adjusted so that

$$\int_{x^-}^{x^+} w(x)\psi_i^2(x)dx = 1, \qquad i = 0, 1, 2, \ldots$$

and

$$\int_{x^-}^{x^+} w(x)\psi_i(x)\psi_{i-1}(x)dx = 0, \qquad i = 1, 2, \ldots$$

This procedure yields a system of orthogonal polynomials. The weight function $w(x)$ defines the class of a corresponding orthonormal polynomial. Some classes are presented in Table 4.1.

Using the system of orthonormal polynomials, construct an approximating polynomial with the parameters $a_0, \ldots, a_R$ by

$$\hat{\Phi}(x) = \sum_{i=0}^{R} a_i \psi_i(x). \tag{4.15}$$

In contrast to the classical approach to approximation, the parameters of an approximating function will be treated as randomized ones taking values within an interval $[\mathbf{a}^-, \mathbf{a}^+]$.

4.2.2 Functional description of dynamic models

An output ensemble of dynamic models at a time t is characterized by a mapping $\mathbb{M}[P(\mathbf{a}), \tau, t]$ of the form (4.1) and depends on the input data on a time interval $[t - \tau, t]$ and a realization of the randomized parameters.

Denote by $\mathbf{x}(\tau)$ an input vector composed of independent observations $x_1(\tau), \ldots, x_n(\tau)$ that are defined on a time interval $\mathcal{T}_t = [0, t]$, where $\tau \in \mathcal{T}_t$. At each time $t > 0$, the model outputs a vector $\hat{\mathbf{y}}(t)$ composed of observations $\hat{y}_1(t), \ldots, \hat{y}_m(t)$, where $t \in \mathcal{T} = [0, T]$. A relation between the input $\mathbf{x}(\tau \,|\, \tau \in \mathcal{T}_t)$ and the output $\hat{\mathbf{y}}(t)$ is described by a functional $\mathcal{B}$, i.e.,

$$\hat{\mathbf{y}}(t) = \mathcal{B}(\mathbf{x}(\tau), \mathbf{a} \,|\, \tau \in \mathcal{T}_t), \qquad t \in \mathcal{T} = [0, T]. \tag{4.16}$$

Assume that this functional is parameterized by a vector $\mathbf{a} = \{a_1, \ldots, a_r\}$ of random elements that belong to intervals

$$\begin{aligned} \mathcal{A}_i &= [a_i^-, a_i^+], \qquad i \in [1, r]; \\ \mathcal{A} &= \prod_{i=1}^{r} \mathcal{A}_i. \end{aligned} \tag{4.17}$$

Three classes of the dynamic "Single Input–Ensemble Output" RPMs will be considered, namely, the linear ones, as well as the nonlinear ones with the power and polynomial nonlinearities.

4.2.3 Linear dynamic models

Consider a dynamic RPM with an input $\mathbf{x}(\tau)$, $\tau \in \mathcal{T}_t$ and an output $\hat{\mathbf{y}}(t)$, $t \in \mathcal{T}$. The input vector elements $x_i(\tau)$, $i = \overline{1, n}$, are *continuous time-varying functions*. A relation between the input and output of this RPM is described by a *linear* functional $\mathcal{B}$ with the parameters $\mathbf{a}$ (4.16), i.e.,

$$\hat{\mathbf{y}}(t) = \int_0^l B(t, \tau, \mathbf{a})\mathbf{x}(\tau)d\tau. \tag{4.18}$$

In this formula, the matrix $B(t, \tau, \mathbf{a})$ of dimensions $(m \times n)$ consists of pulse characteristics $[b_{ij}(t, \tau \mid \mathbf{a}^{(ij)}), \mid i = \overline{1, m}, j = \overline{1, n}]$ with random parameters $\mathbf{a}^{(ij)} = \{a_1^{(ij)}, \ldots, a_{k_{ij}}^{(ij)}\}$, where k_{ij} is the number of parameters for a pulse characteristic $b_{ij}(t, \tau)$.

According to a general concept, the parameters $\mathbf{a}^{(ij)}$ are interval-type, i.e.,

$$\mathbf{a}^{(ij)} \in \mathcal{A}^{(ij)} = [\mathbf{a}_-^{(ij)}, \mathbf{a}_+^{(ij)}], \quad \mathbf{a} \in \mathcal{A} = \prod_{i=1}^{m} \prod_{j=1}^{n} \mathcal{A}^{(ij)}. \tag{4.19}$$

The model (4.19) generates an ensemble $\hat{\mathcal{Y}}_\mathcal{T}$ of random trajectories $\hat{\mathbf{y}}(t)$, $t \in \mathcal{T}$. These pulse characteristics can be parameterized in different ways. Some of them will be discussed below.

1.1. Given functions with unknown parameters. As a rule, pulse characteristics are exponential functions or harmonic functions with exponential decay, i.e.,

$$b_{ij}(t, \tau) = \alpha_{ij}(t) \exp\left(-\beta_{ij}(t)\tau\right) \text{ or} b_{ij}(t, \tau) \exp\left(-\beta_{ij}\tau\right) \sin(\omega_{ij}(t)\tau + \phi_{ij}(t)), \tag{4.20}$$

where, e.g., time-varying parameters have the form

$$\begin{aligned} \alpha_{ij}(t) &= \alpha_{ij}^0 + \alpha_{ij}^1 t; & \beta_{ij}(t) &= \beta_{ij}^0 + \beta_{ij}^1 t, \\ \omega_{ij}(t) &= \omega_{ij}^0 \ln t; & \phi_{ij}(t) &= \phi_{ij}^0 + \phi_{ij}^1 t. \end{aligned} \tag{4.21}$$

In these expressions, $\alpha_{ij}^0, \alpha_{ij}^1, \beta_{ij}^0, \beta_{ij}^1, \omega_{ij}^0, \phi_{ij}^0$ and ϕ_{ij}^1 are random variables that take values within corresponding intervals.

1.2. Approximation by given family of functions with unknown weights. Consider a family $\mathcal{R}_v$ of v deterministic functions, further referred to as the *basis* functions:

$$\mathcal{R}_v = \{\phi_1(t, \tau), \ldots, \phi_v(t, \tau)\}. \tag{4.22}$$

Then the pulse characteristics are written as

$$b_{ij}(t, \tau) = \sum_{k=1}^{K_{ij}} b_k^{(ij)} \phi_k(t, \tau), \tag{4.23}$$

where the coefficients $b_k^{(ij)}$ are randomized independent variables that take values within some intervals.

To a large extent, a proper choice of basis functions depends on the properties of an object under study. Here a widespread approach is to employ the families of basis functions formed by the exponential functions $\exp(\alpha(t, \tau))$, the polynomial-exponential functions $\beta t^p \exp(\alpha(t, \tau))$ (with integers p), or the polynomial-exponential-harmonic functions $\beta t^p \exp(\alpha(t, \tau)) \sin(\omega(t, \tau))$.

Example 3.1 Consider a linear dynamic system with random time-varying parameters. The input vector consists of two elements $\{x_1(\tau), x_2(\tau)\}$, which are continuous functions on the time interval $\mathcal{T}$. The output vector has one element of the form

$$y_1(t) = \int_0^t (b_{11}(t, \tau)x_1(\tau) + b_{12}(t, \tau)x_2(\tau)) \, d\tau.$$

The pulse characteristics in this equality are given by

$$b_{11}(t, \tau) = A \exp(-\alpha(t)\tau),$$

$$b_{12}(t, \tau) = C(t) \exp(-\beta\tau).$$

This model incorporates the constant parameters $A = 2.0$ and $\beta = 0.5$, as well as the random time-varying parameters $\alpha(t) = \alpha_0 + \alpha_1 t$ and $C(t) = C_0 + C_1 t$, where α_0, α_1, C_0 and C_1 are independent random variables taking values within the following intervals:

$$-0.1 \leq \alpha_0 \leq 0.2; \quad -0.05 \leq \alpha_1 \leq 0.10;$$

$$1.0 \leq C_0 \leq 2.0; \quad -0.5 \leq C_1 \leq 0.5.$$

They obey the uniform distributions on the corresponding intervals, i.e., the PDFs are given by

$$p(\alpha_0) = \begin{cases} 3.3 & \text{for } -0.1 \leq \alpha_0 \leq 0.2, \\ 0 & \text{for } -0.1 > \alpha_0 > 0.2; \end{cases} \quad p(\alpha_1) = \begin{cases} 6.6 & \text{for } -0.05 \leq \alpha_1 \leq 0.10, \\ 0 & \text{for } -0.05 > \alpha_1 > 0.10; \end{cases}$$

$$p(C_0) = \begin{cases} 1.0 & \text{for } 1.0 \leq C_0 \leq 2.0, \\ 0 & \text{for } 1.0 > C_0 > 2.0; \end{cases} \quad p(C_1) = \begin{cases} 1,0 & \text{for } -0.5 \leq C_1 \leq 0.5, \\ 0 & \text{for } -0.5 > C_1 > 0.5. \end{cases}$$

The pulse characteristics $b_{11}(t, \tau)$ and $b_{12}(t, \tau)$ are functions of two random variables, as illustrated in Figs. 4.4a and 4.4b. Their graphical representations are the corresponding surface ensembles $\mathcal{B}_{11}(t, \tau)$ and $\mathcal{B}_{12}(t, \tau)$. At the time $t = 0$, the pulse characteristics depend on the variable τ only. The cutsets of these surfaces, i.e., the ensembles $\mathcal{B}_{11}(0, \tau)$ and $\mathcal{B}_{12}(0, \tau)$ are generated by the random parameters α_0 and C_0, respectively.

If the elements of the matrix B have random time-invariable parameters, the expression (4.18) takes the form

$$\mathbf{y}(t) = \int_0^t B(t - \tau) \mathbf{x}(\tau) d\tau. \tag{4.24}$$

The elements of the matrix B, which has dimensions $(m \times n)$, are the pulse characteristics $b_{ij}(\tau)$ with the independent randomized time-invariable parameters.

Example 3.2 For the linear dynamic system of Example 3.1, let the pulse characteristics be given by

$$b_{11}(\tau) = A \exp(-\alpha\tau),$$

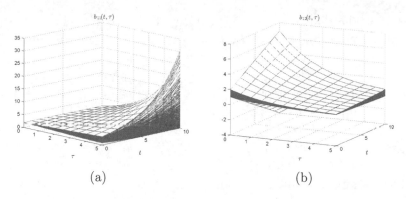

(a)   (b)

Figure 4.4

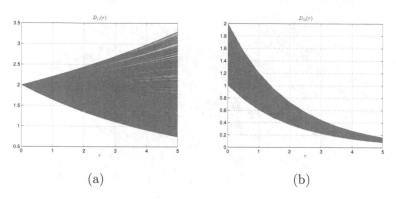

(a)   (b)

Figure 4.5

$$b_{12}(\tau) = C \exp(-\beta\tau).$$

In this model, $A = 2.0$ and $\beta = 0.5$ act as the constant parameters, whereas α and C are the random parameters taking values within the intervals

$$-0.1 \leq \alpha \leq 0.2 \qquad \text{and} \qquad 1.0 \leq C \leq 2.0.$$

The parameters are independent and have the uniform distributions, i.e.,

$$p(\alpha) = \begin{cases} 3.3 & \text{for } -0.1 \leq \alpha \leq 0.2, \\ 0 & \text{for } -0.1 > \alpha > 0.2; \end{cases} \qquad p(C) = \begin{cases} 1.0 & \text{for } 1.0 \leq C \leq 2.0, \\ 0 & \text{for } 1.0 > C > 2.0. \end{cases}$$

Figures 4.5a and 4.5b demonstrate ensembles $\mathcal{B}_{11}(\tau)$ and $\mathcal{B}_{12}(\tau)$ of the pulse characteristics $b_{11}(\tau)$ and $b_{12}(\tau)$, respectively.

1.3. Discrete input functions. In most cases, the input and output of an object are measured at discrete times with a fixed step Δ. An object operates on finite time intervals, and hence the available information about its input and output is represented as one-dimensional numerical data arrays. These arrays are used to form the vectors of lattice functions, $\mathbf{x}[k\Delta] = \mathbf{x}[k]$ and $\hat{\mathbf{y}}[k\Delta] = \hat{\mathbf{y}}[k]$ for the object's input and output, respectively. For details, the reader is referred to [176].

In a linear approximation, a dynamic relation between the input and output lattice function vectors, $\mathbf{x}[k]$ and $\hat{\mathbf{y}}[k]$, can be described by a matrix $B[k, q]$ composed of the lattice pulse characteristics $b_{ij}[k, q]$ for the RPM with time-varying parameters or by a matrix composed of the lattice pulse characteristics $b_{ij}[q]$ for the RPM with constant parameters. In both statements, the RPM contain independent random parameters taking values within given intervals. The pulse characteristics are assumed to have the so-called finite memory. For the RPMs with time-varying parameters, this means that

$$b_{ij}[k, q] = \begin{cases} b_{ij}[k, q] & \text{if } 0 \leq l \leq L_{ij} \text{ and } k \geq 0, \\ 0 & \text{if } l > L_{ij} \text{ and } k \geq 0; \end{cases} \quad (4.25)$$

for the RPM with constant parameters,

$$b_{ij}[q] = \begin{cases} b_{ij}[q] & \text{if } 0 \leq l \leq L_{ij}, \\ 0 & \text{if } l > L_{ij}. \end{cases} \quad (4.26)$$

Then the RPM with discrete inputs and outputs is described by the following formulas:

- for the time-varying parameters,

$$\hat{\mathbf{y}}[k] = \sum_{q=0}^{L^*} B[k, q]\,\mathbf{x}[q], \qquad k \geq L^*; \quad (4.27)$$

- for the constant parameters,

$$\hat{\mathbf{y}}[k] = \sum_{q=0}^{L^*} B[q]\,\mathbf{x}[k - q], \qquad k \geq L^*. \quad (4.28)$$

Here $L^* = \max_{ij} L_{ij}$.

In contrast to the linear RPMs with continuous input functions, in this case the parameters can be the ordinates of the corresponding pulse characteristics.

Example 3.3 Consider a one-dimensional RPM with discrete input and memory of capacity 2, i.e.,

$$\hat{y}[k] = \sum_{q=0}^{1} b_{1,1}[q]\,x[k - q], \qquad k \in [1, 10].$$

Table 4.2

k	1	2	3	4	5	6	7	8	9	10
$x[k]$	2.5	3.5	5.7	2.8	2.1	1.9	1.7	2.1	3.1	4.4
$x[k-1]$	1.8	2.5	3.5	5.7	2.8	2.1	1.9	1.7	2.1	3.1

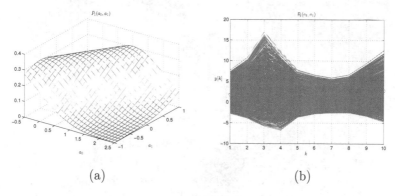

(a)   (b)

Figure 4.6

Write this representation as

$$\hat{y}[k] = a_0\, x[k] + a_1\, x[k-1], \qquad a_0 = b_{1,1}[0];\ a_1 = b_{1,1}[1].$$

The parameters a_0 and a_1 are random and take values within the intervals

$$\mathcal{A}_0 = [-0.5,\, 2.5] \qquad \text{and} \qquad \mathcal{A}_1 = [-1.0,\, 6.0].$$

The PDF of the parameters has the following form:
  –for the dependent parameters,

$$P(a_0, a_1) = \begin{cases} \frac{1}{2\sqrt{\pi}} \exp\left(-\frac{a_0^2 + a_1^2 - 2a_0 a_1}{8}\right) & \text{if } (a_0, a_1) \in \mathcal{A}_0 \times \mathcal{A}_1, \\ 0 & \text{if } (a_0, a_1) \notin \mathcal{A}_0 \times \mathcal{A}_1; \end{cases}$$

  –for the independent parameters,

$$P(a_0, a_1) = \begin{cases} \frac{1}{2\sqrt{\pi}} \exp\left(-\frac{a_0^2 + a_1^2}{8}\right) & \text{if } (a_0, a_1) \in \mathcal{A}_0 \times \mathcal{A}_1, \\ 0 & \text{if } (a_0, a_1) \notin \mathcal{A}_0 \times \mathcal{A}_1. \end{cases}$$

The RPM input on the interval $[1, 10]$ is given in Table 4.2.

Figures 4.6-4.7 show the PDFs $P_1(a_0, a_1)$, $P_2(a_0, a_1)$ and the corresponding ensembles $\mathcal{Y}_1$, $\mathcal{Y}_2$, respectively.

4.2.4 Nonlinear dynamic models with power nonlinearities

Most of real objects are intrinsically nonlinear. A nonlinear object may demonstrate "almost" linear properties under a small variation of its

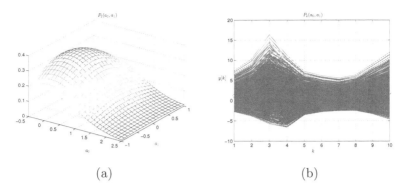

$$(a) \qquad\qquad\qquad\qquad (b)$$

Figure 4.7

input. An approach to describe the nonlinear properties of real objects in terms of the "Single Input–Ensemble Output" models is based on their approximation using functional power series. The elements of such series represent power functionals of different structure that depends on the nonlinear properties of an object and also on the properties of input functions. Particularly, if the object's nonlinearities are *continuous functions* and the object's input signals are *continuous time-varying functions*, then the RPM of such an object can be defined by a functional Volterra series [177, 184].

Consider a dynamic system that operates on a time interval $\mathcal{T}_t = [0, t]$ and is characterized by a continuous functional $\mathcal{B}(\mathbf{x}(\tau)) \,|\, \tau \in \mathcal{T}_t)$. Let the input vector $\mathbf{x}(\tau) = \{x_1(\tau), \ldots, x_n(\tau)\}$ be belonging to the space $\mathcal{X}_{\mathcal{T}_T}$ of continuous time-varying functions. According to [100, 146, 184, 185], the functional $\mathcal{B}(\mathbf{x}(\tau))$ (equivalently, the RPM output $\hat{\mathbf{y}}(t)$) can be represented as a convergent functional series, i.e.,

$$\hat{\mathbf{y}}(t) = \mathcal{B}(\mathbf{x}(\tau)) \,|\, \tau \in \mathcal{T}_t) =$$

$$= \sum_{h=1}^{\infty} \int_0^t \cdots \int_0^t W^{(h)}(t - \tau_1, \ldots, t - \tau_h) \times$$

$$\times \mathbf{x}^{(h)}(\tau_1, \ldots, \tau_h) d\tau_1 \cdots d\tau_h. \qquad (4.29)$$

Here $\mathbf{x}^{(h)}(\tau_1, \ldots, \tau_h)$ denotes the vector composed of the lexicographically ordered products $\{x_{i_1}(\tau_{j_1}) \cdots x_{i_n}(\tau_{j_h})\}$, where $(i_1, \ldots, i_n) \in \overline{1, n}$ and $(j_1, \ldots, j_h) \in \overline{1, h}$. The dimension of the vector $\mathbf{x}^{(h)}$ is

$$l(n, h) = n^h. \qquad (4.30)$$

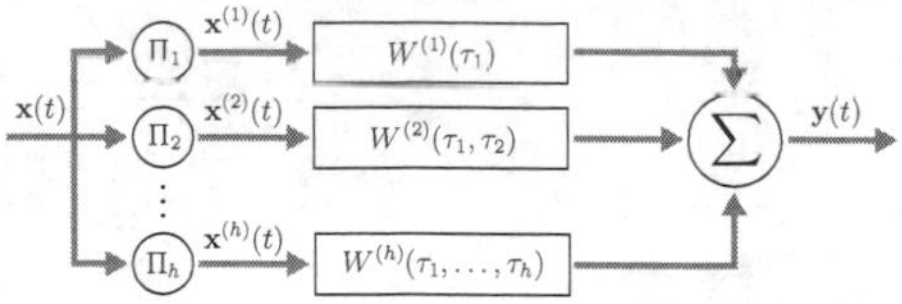

Figure 4.8

The matrix $W^{(h)}(t_1, \ldots, t_h)$ has dimensions $(m \times l(m, h))$, and its elements are multidimensional weight functions (pulse characteristics) $w_{k_h}^{(h)}(t_1, \ldots, t_h)$, where $k_h = \overline{1, l(n, h)}$.

Figure 4.8 illustrates the structure of the functional series (4.28). The blocks $\prod_h$ in Fig. 4.8 indicate the lexicographical ordered products from the elements of the vector $\mathbf{x}$ of power h.

The weight functions contain random parameters $\mathbf{a}^{(h)}$ of the interval type, i.e.,

$$\mathbf{a}^{(h)} \in \mathcal{A}^{(h)} = \{\mathbf{a}_-^{(h)} \leq \mathbf{a}^{(h)} \leq \mathbf{a}_+^{(h)}\}. \tag{4.31}$$

Since the parameters are random, they have a PDF $P^h(\mathbf{a}^{(h)})$ on corresponding intervals.

Example 3.4 Consider the quadratic dynamic system (4.2.4) with two inputs $\mathbf{x}(\tau) = \{x_1(\tau), x_2(\tau)\}$ and two outputs $\hat{\mathbf{y}}(t) = \{y_1(t), y_2(t)\}$, which are continuous functions. A relation between them is described by a continuous functional $\mathcal{B}$ with random parameters, i.e.,

$$\hat{\mathbf{y}}(t) = \int_0^t W^{(1)}(t - \tau)\mathbf{x}(\tau)d\tau + \int_0^t \int_0^t W^{(2)}(t - \tau_1, t - \tau_2)\mathbf{x}^{(2)}(\tau_1, \tau_2)d\tau_1 d\tau_2.$$

Here

$$\mathbf{x}^{(2)}(\tau_1, \tau_2) = \{x_1(\tau_1)x_1(\tau_2), x_1(\tau_1)x_2(\tau_2), x_1(\tau_2)x_2(\tau_1), x_2(\tau_1)x_2(\tau_2)\}.$$

The matrices of the weight functions (pulse characteristics) have the following elements:

- the linear block is characterized by the four weight functions

$$w_{ij}^{(1)}(\tau) = A_{ij} \exp(-\alpha_{ij}\tau), \quad (i, j) = 1, 2;$$

- the quadratic block is characterized by the eight two-dimensional weight functions

$$w_{ij}^{(2)}(\tau_1, \tau_2) = B_{ij} \exp(-\gamma_{ij}\tau_1 - \rho_{ij}\tau_2), \quad i = 1, 2; \; j = 1, 2, 3, 4.$$

For obtaining the graphical representations of the random weight functions, consider some of them in the linear $(w_{11}^{(1)}(\tau), w_{22}^{(1)}(\tau))$ and nonlinear $(w_{21}^{(2)}(\tau_1, \tau_2), w_{13}^{(2)}(\tau_1, \tau_2))$ blocks.

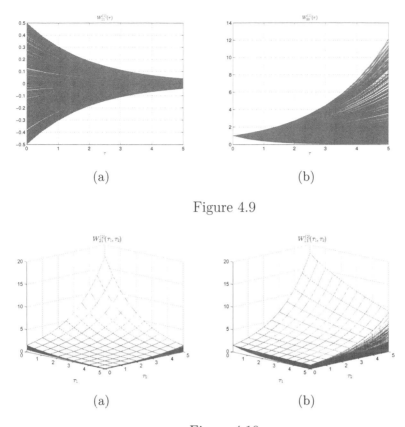

Figure 4.9

Figure 4.10

The linear block has the constant parameters $A_{22} = 1.0$, $\alpha_{11} = 0.5$ and also the random parameters $A_{11} \in [-0.5, 0.5]$, $\alpha_{22} \in [-0.5, 1.0]$ with the uniform distribution on the corresponding intervals. The ensembles $\mathcal{W}_{11}^{(1)}$ and $\mathcal{W}_{22}^{(1)}$ are shown in Figs. 4.9a and 4.9b.

The nonlinear block has the constant parameters $\gamma_{21} = 0.5$, $B_{13} = 1.5$ and also the random parameters $B_{21} \in [0.5, 1.5]$, $\rho_{21} \in [-0.5, 0.5]$, $B_{13} = 1.5$, $\gamma_{13} \in [0.1, 0.8]$ and $\rho_{13} \in [-0.5, 0.5]$. The PDFs of all random parameters are uniform on the corresponding intervals. The random surface ensembles $\mathcal{W}_{21}^{(2)}$ and $\mathcal{W}_{13}^{(2)}$ are demonstrated in Figs. 4.10a and 4.10b.

The functional $\mathcal{B}$ (4.2.4) can be often written as a functional series with structured weight matrices (pulse characteristics). Under structuring we mean a sequence of linear dynamic and nonlinear static operations. The block diagram of an RPM in which each channel represents a

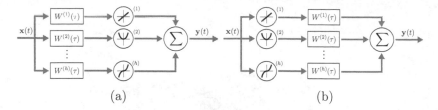

$$(a) \qquad\qquad\qquad (b)$$

Figure 4.11

serial connection of a linear dynamic block and a power nonlinearity is shown in Fig. 4.11a. The output of this model is defined by the following modification of the functional Volterra series:

$$\hat{\mathbf{y}}(t) = \mathcal{B}(\mathbf{x}(\tau)) \,|\, \tau \in \mathcal{T}_t) =$$

$$= \sum_{h=1}^{\infty} \left(\int_0^t W^{(h)}(t-\tau) \,\mathbf{x}(\tau) \, d\tau \right)^h. \tag{4.32}$$

For the one-dimensional RPM, the last-mentioned equality takes the form

$$\hat{y}(t) = \sum_{h=1}^{\infty} \int_0^t \cdots \int_0^t \prod_{i=1}^{h} \left(w(t-\tau_i) \, x(\tau_i) d\tau_i \right). \tag{4.33}$$

The block diagram in Fig. 4.11b corresponds to an RPM with the inverse sequence of the linear and nonlinear operations in each channel: first the input undergoes a power transformation and then a dynamic linear transformation. The RPM output is described by another modification of the functional Volterra series:

$$\hat{\mathbf{y}}(t) = \sum_{h=1}^{\infty} \int_0^t W(t-\tau) \mathbf{x}^{(h)}(\tau) \, d\tau. \tag{4.34}$$

Here, the vector $\mathbf{x}^{(h)}(\tau)$ consists of the lexicographically ordered elements formed by all possible products of the elements of the vector $\mathbf{x}(\tau)$.

Example 3.5 Consider a one-dimensional RPM with a structured quadratic nonlinearity of the following form:

$$\hat{y}_1(t) = \int_0^t \int_0^t w_1^{(2)}(t-\tau_1) \, w_2^{(2)}(t-\tau_2) \, x_1(\tau_1) \, x_1(\tau_2) \, d\tau_1 \, d\tau_2.$$

In this case, the quadratic pulse characteristic is also structured—it represents the product of the pulse characteristics:

$$w^{(2)}(\tau_1, \tau_2) = w_1^{(2)}(\tau_1) \, w_2^{(2)}(\tau_2).$$

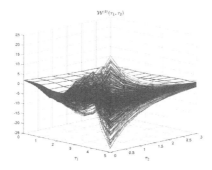

$$\mathcal{W}^{(2)}(\tau_1, \tau_2)$$

Figure 4.12

The pulse characteristics have the form

$$w_1^{(2)}(\tau) = A \exp(-\alpha\tau)\cos(\omega\tau); \quad w_2^{(2)}(\tau) = C \exp(-\beta\tau).$$

Hence,

$$w^{(2)}(\tau_1, \tau_2) = A\,C\,\exp(-\alpha\tau_1 - \beta\tau_2)\cos(\omega\tau_2).$$

The constant parameters are $AC = 2.0$, whereas the random parameters α, β and ω obey the uniform distributions on the intervals

$$\alpha \in [-0.5, 0.5]; \quad \beta \in [0.8, 1.6]; \quad \omega \in [1, 2].$$

The ensemble $\mathcal{W}^{(2)}$ is shown in Fig. 4.12.

Consider an RPM with a finite memory K; the input and output of this model are measured at discrete times t_k with step Δ. Letting $k = [t_k/\Delta]$,[1] write equality (4.2.4) as

$$\hat{\mathbf{y}}[k] = \sum_{h=1}^{\infty} \sum_{(n_1,\ldots,n_h)=0}^{K} W^{(h)}[n_1,\ldots,n_h]\,\mathbf{x}^{(h)}[k-n_1,\ldots,k-n_h], \quad (4.35)$$

$$k \geq K.$$

The vector $\mathbf{x}^{(h)}[k-n_1,\ldots,k-n_h]$ has the elements $x_{i_1}[k-n_1]\cdots x_{i_n}[k-n_h]$, which are the lexicographically ordered all possible products of $(i_1,\ldots,i_n) = \overline{1,n}$ elements of the vector $\mathbf{x}$ at the times $(n_1,\ldots,n_h) = \overline{0,K}$. The dimension of the vector $\mathbf{x}^{(h)}[k-n_1,\ldots,k-n_h]$ is

$$l(n,h) = \frac{(n\,K + h - 1)!}{h!(n\,k - 1)!}. \quad (4.36)$$

---

[1] Here $[t_k/\Delta]$ denotes the integer part of t_k/Δ.

The parameter K defines the memory capacity of the RPM, i.e.,

$$
W^{(h)}[n_1, \ldots, n_h] = \begin{cases} W^{(h)}[n_1, \ldots, n_h] & \text{if all } (n_1, \ldots, n_h) \leq K, \\ 0 & \text{if at least one } n_j > K. \end{cases} \quad (4.37)
$$

The elements of the matrix $W^{(h)}[n_1, \ldots, n_h]$ are discrete random weight functions (pulse characteristics) $w_{i,k_h}^{(h)}[n_1, \ldots, n_h]$, $i = 1, m$; $k_h = \overline{1, l(n, h)}$. They reflect the dynamic and nonlinear properties of the RPM. Note that the nonlinear properties are formulated in terms of multidimensional functions (pulse characteristics), not in terms of structurally representable nonlinear elements.

The pulse characteristics $w_{i,k_h}^{(h)}[n_1, \ldots, n_h]$ have the interval type, i.e.,

$$
w_{i,k_h}^{(h)}[n_1, \ldots, n_h] \in \mathcal{W}_{i,k_h}^{(h)} = \left[\tilde{w}_{i,k_h}^{(h,-)}[n_1, \ldots, n_h], \tilde{w}_{i,k_h}^{(h,+)}[n_1, \ldots, n_h] \right].
$$
$$(4.38)$$

In this expression, $\tilde{w}_{i,k_h}^{(h,\pm)}[n_1, \ldots, n_h]$ are given functions that describe the left and right limits of the above intervals. For example, they can be defined as the exponential functions

$$
\tilde{w}_{i,k_h}^{(h,-)}[n_1, \ldots, n_h] = \prod_{j=1}^{h} \beta_{(i,k_h)}^{(h,-)}(j) \exp\left(-\alpha_{(i,k_h)}^{(h,-)}(j)\, n_j \right),
$$

$$
\tilde{w}_{i,k_h}^{(h,+)}[n_1, \ldots, n_h] = \prod_{j=1}^{h} \beta_{(i,k_h)}^{(h,+)}(j) \exp\left(-\alpha_{(i,k_h)}^{(h,+)}(j)\, n_j \right). \quad (4.39)
$$

A convenient approach is to treat their ordinates as the RPM parameters. This allows us to transform the nonlinear RPM to its linear analog of considerably higher dimension. The transformation procedure for the one-dimensional RPM (with one input and one output) is given by

$$
\hat{y}[k] = \sum_{h=1}^{H} \sum_{K}^{} (n_1, \ldots, n_h) = 0 w^{(h)}[n_1, \ldots, n_h] \prod_{i=1}^{h} x[k - n_i], \quad k \geq K.
$$
$$(4.40)$$

Introduce a lexicographical ordering of the variables $\{n_1, \ldots, n_h\}$ by reassigning numbers from 0 to $t_h = (K + 1)^h - 1$ to the resulting sets. Then

$$
\{n_1, \ldots, n_h\} \rightarrow i^{(h)}, \qquad i^{(h)} \in [0, t_h]. \quad (4.41)
$$

In accordance with this numbering scheme, define the following indexes for the random parameters that correspond to the values of the pulse characteristics at the discrete times $0, 1, \ldots, K$:

$$a_{i^{(h)}}^{(h)} \rightarrow w^{(h)}[n_1, \ldots, n_h], \qquad i^{(h)} \in [0, t_h]. \tag{4.42}$$

Construct a random vector

$$\mathbf{a}(h) = \{a_0^{(h)}, \ldots, a_{t_h}^{(h)}\}. \tag{4.43}$$

The elements of the vector $\mathbf{a}(h)$—see (4.38)—take values within the intervals

$$\mathcal{A}^{(h)}(i^{(h)}) = \left[a_-^{(h)}(i^{(h)}), \, a_+^{(h)}(i^{(h)})\right]. \tag{4.44}$$

By analogy with (4.41), introduce a lexicographical ordering of the variables $k - n_1, \ldots, k - n_h$, where k is a fixed parameter and the indexes $n_1, \ldots, n_h$ take values within the interval $[0, K]$.

For each fixed k, renumber the resulting sets as follows:

$$\{k - n_1, \ldots, k - n_h\} \rightarrow (k, i^{(h)}). \tag{4.45}$$

Then the lexicographically ordered products of the variables $x[k - n_1], \ldots, x[k - n_h]$ with fixed k form the vector

$$\mathbf{x}^{(h)}[k] = \left\{x_{k,0}^{(h)}, \ldots, x_{k,t_h}^{(h)}\right\}. \tag{4.46}$$

Consequently, the expression (4.40) can be written as

$$\hat{y}[k] = \sum_{h=1}^{H} \langle \mathbf{a}^{(h)}, \mathbf{x}^{(h)}[k] \rangle, \tag{4.47}$$

where $\langle \bullet, \bullet \rangle$ denotes the scalar product of vectors $\bullet$ and $\bullet$.

Consider a one-dimensional RPM with structured nonlinearities (see Fig. 4.11a) and the following discrete time-varying functions as its input:

$$\hat{y}[k] = \sum_{h=1}^{H} \left(\sum_{n=0}^{K} w^{(h)}[n]x[k-n]\right)^h, \qquad k \geq K. \tag{4.48}$$

The pulse characteristics $w^{(h)}[n]$ have a finite memory K and random parameters. Again, as the latter choose their ordinates, i.e.,

$$w^{(h)}[n] = a_n^{(h)}, \qquad n = \overline{0, K}. \tag{4.49}$$

Construct the vector

$$\mathbf{a}^{(h)} = \{a_0^{(h)}, \ldots, a_K^{(h)}\}. \tag{4.50}$$

The parameters $a_n^{(h)}$ are random, independent, and take their values within the intervals

$$\mathcal{A}_n^{(h)} = [-A_n^{(h)}, A_n^{(h)}], \quad A_n^{(h)} = \beta_n^{(h)} \exp(-\alpha_n^{(h)}). \tag{4.51}$$

The PDFs $p_n^{(h)}(a_n^{(h)})$, $n = \overline{0, K}$, are defined on these intervals.

Get back to equality (4.48) and introduce the following notations:

- the index set

$$\mathcal{I}_h(K) = \{i_n : \sum_{n=0}^{K} i_n = h, \, i_n = \overline{0, K}\}; \tag{4.52}$$

- the vector

$$\mathbf{i} = \{i_0, \ldots, i_K\}; \tag{4.53}$$

- the function

$$C_{(k,i_n)}(x) = \frac{1}{i_n!} x^{i_n}[k - n]. \tag{4.54}$$

Then formula (4.48) can be written as

$$\hat{y}[k] = \sum_{h=1}^{H} V_k^{(h)}(\mathbf{a}^{(h)}, x),$$

$$V_k^{(h)}(\mathbf{a}^{(h)}, x) = \sum_{\mathbf{i} \in \mathcal{I}_h(K)} \prod_{n=0}^{K} C_{(k,i_n)}(x) \, [a_n^{(h)}]^{i_n}. \tag{4.55}$$

Hence, unlike the RPM (4.40), the model under consideration is nonlinear with power nonlinearities.

4.2.5 Nonlinear dynamic models with polynomial nonlinearities

In machine learning problems, it is often necessary to deal with dynamic objects whose observable output has jumps at certain times or under certain values of its input. In other cases, there exists prior information about the object's properties that defines its structure as illustrated in Fig. 4.13a or Fig. 4.13b. The nonlinear blocks in these diagrams are described by orthonormal polynomials of different classes (see Table 4.1).

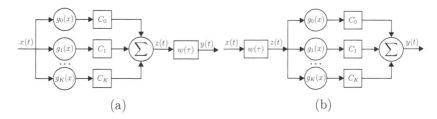

(a)   (b)

Figure 4.13

Consider an RPM with the structure presented in Fig. 4.13a. Using orthogonal polynomials $g_k(x)$, let us write this RPM in the following form:

$$\hat{y}(t) = \sum_{k=0}^{K} c_k \int_0^t w(t - \tau \,|\mathbf{b})\, g_k(x(\tau))d\tau, \qquad (4.56)$$

where the coefficients $c_0, \ldots, c_K$ are independent random variables of the interval type, i.e.,

$$c_k \in \mathcal{C}_k = [c_k^-, c_k^+], \qquad k = \overline{0, K}. \qquad (4.57)$$

The parameters $\mathbf{b}$ of the pulse characteristic $w(t \,|\mathbf{b})$ are randomized and belong to the intervals

$$\mathbf{b} \in \mathcal{B} = [\mathbf{b}^-, \mathbf{b}^+]. \qquad (4.58)$$

Example 3.6 Consider a nonlinear RPM (Fig. 4.13a) in which the randomized model of the nonlinear element is based on the four Hermite polynomials from Table 4.1, i.e.,

$$z(t) = c_0 + c_1 x(t) + c_2(x^2 - 1) + c_3(x^3(t) - 3x(t)).$$

The coefficients c_0, c_1, c_2 and c_3 are random, independent and take values within the interval $\mathcal{C} = [c^-, c^+]$, where $c^- = -1$ and $c^+ = 1$. All the coefficients have the same PDF

$$f(c) = \begin{cases} A \exp(-c^2) & \text{if } -1 \le c < 1, \\ 0 & \text{if } -1 > c \ge 2. \end{cases}$$

The normalization constant is $A = 1.71$. Then the joint PDF takes the form

$$F(c_0, \ldots, c_3) = A^4 \exp(-c_0^2 - c_1^2 - c_2^2 - c_3^2).$$

The pulse characteristic of the linear dynamic element is given by

$$w(\tau) = \exp(-\alpha\tau),$$

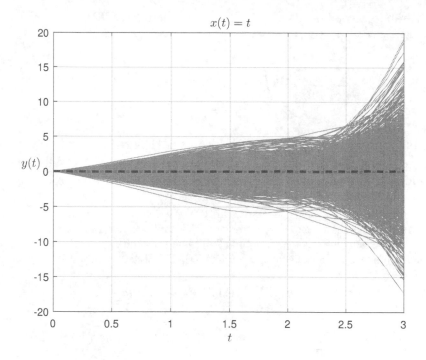

Figure 4.14

where α denotes a random parameter with the uniform distribution on the interval $[-0.5, 0.5]$, i.e., its PDF has the form

$$b(\alpha) = \begin{cases} 1 & \text{if } -1 \le \alpha < 1, \\ 0 & \text{if } -1 > \alpha \ge 1. \end{cases}$$

Therefore, the RPM is described by the equation

$$\hat{y}(t) = \sum_{i=0}^{3} c_i \int_0^t \exp(-\alpha(t - \tau))\, H_i(x(\tau)) d\tau.$$

Here $H_i(x)$ stand for the Hermite polynomials, c_i and α are the random parameters, and $x(t) = 1$. Figure 4.14 demonstrates the ensemble of all trajectories, as well as the mean and median trajectories, which match one another.

Now get back to the block diagram in Fig. 4.13b. Obviously, the linear and nonlinear dynamic elements of the object have interchanged their places. The object's input is a function $x(t)$. Let us construct the RPM of this object using the pulse characteristic with random parameters $w(\tau \mid \mathbf{b})$ for the linear element and a collection of orthogonal polynomials with random coefficients for the nonlinear element. Then the RPM can be written as

$$\hat{y}(t) = \sum_{k=1}^{K} \sum_{i=0}^{k} c_k a_{ki} \int_0^t \cdots \int_0^t w(t - \tau_0|\mathbf{b}) \cdots w(t - \tau_i|\mathbf{b}) \times$$

$$\times x(\tau_0) \cdots x(\tau_i) d\tau_1 \cdots d\tau_i. \tag{4.59}$$

In this equation, a_{ki} denote the coefficients of the orthonormal polynomials (see Table 4.1); $c_1, \ldots, c_K$ are the independent random coefficients of the randomized model of the nonlinear element. They are of the interval type (4.57).

Example 3.7 Consider a nonlinear RPM with the structure presented in Fig. 4.13b, where the randomized model of the nonlinear element is given by the four Hermite polynomials (see Example 3.6). The pulse characteristic of the linear dynamic element is the same as in Example 3.6. The RPM output has the form

$$y(t) = c_0 + c_1 \int_0^t \exp(-\alpha(t - \tau))x(\tau)d\tau +$$

$$+ c_2 \left(\int_0^t \int_0^t \exp(-\alpha(t - \tau_1)) \exp(-\alpha(t - \tau_2))x(\tau_1)x(\tau_2)d\tau_1 d\tau_2 - 1 \right) +$$

$$+ c_3 \int_0^t \int_0^t \int_0^t \exp(-\alpha(t - \tau_1)) \exp(-\alpha(t - \tau_2)) \exp(-\alpha(t - \tau_3)) \times$$

$$\times x(\tau_1)x(\tau_2)x(\tau_3)d\tau_1 d\tau_2 d\tau_3 -$$

$$- 3c_3 \int_0^t \exp(-\alpha(t - \tau))x(\tau)d\tau.$$

The ensemble of all trajectories, as well as the mean and median trajectories, are shown in Fig. 4.15.

If the input and output of this RPM is characterized by the lattice functions, then the following equations hold:

- for the RPM in Fig. 4.13a,

$$\hat{y}[n] = \sum_{k=0}^{K} \sum_{i=0}^{k} c_k a_{ki} \sum_{m=0}^{n} w[n - m \mid \mathbf{b}] x^i[m]; \tag{4.60}$$

- for the RPM in Fig. 4.13b,

$$\hat{y}[n] = \sum_{k=1}^{K} \sum_{i=0}^{k} c_k a_{ki} \sum_{(m_0,\ldots,m_i)=0}^{n} \prod_{j=0}^{i} w[n - m_j \mid \mathbf{b}] x[m_j]. \tag{4.61}$$

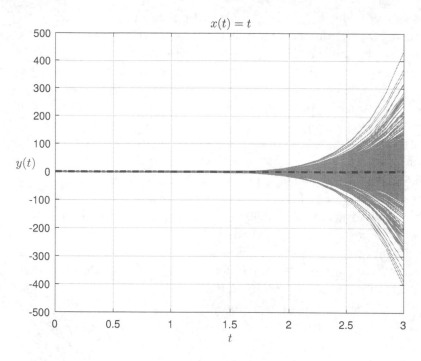

Figure 4.15

In these formulas, the coefficients c_k are the independent random variables distributed on the corresponding intervals (4.57) in accordance with the given PDFs. The parameters $b_1, \ldots, b_s$ of the pulse characteristic are also random, independent and take values within the corresponding intervals with the given PDF.

4.2.6 Randomized neural networks

Neural networks are a kind of parameterized models used in machine learning procedures. A structural foundation of a neural network is the Kolmogorov theorem [94], which states that a continuous function of n variables on the nonnegative unit cube can be written as the superposition of one-variable functions with summation:

$$f(x_1, \ldots, x_n) = \sum_{k=1}^{(2n+1)} h_k \left(\sum_{i=1}^{n} \phi_{ik}(x_i) \right); \qquad (4.62)$$

here the variables $h_1, \ldots, h_{2n+1}$ depend on the function f, whereas the functions $\phi_{ik}(x_i)$ do not. This theorem claims the existence of such a representation but gives no technical details to find it.

On the one hand, the universalism of neural networks can be explained by a structural analogy with the Kolmogorov theorem; on the other, by a simplified (statistical) analogy with operation of a biological neuron [122]. The general structure of the neuron's statistical model is illustrated in Fig. 4.16a. It consists of three serially connected blocks, namely, synaptic coefficients $\mathbf{w} = \{w_1, \ldots, w_n\}$, an adder and a nonlinear transformation block with an activation function $\phi(u, \mathbf{b})$, where $\mathbf{b}$ are the activation function parameters. Figure 4.16b shows an example of this model, in which the nonlinear transformation block has an activation function $\phi(u, \Delta)$ with a threshold Δ. In the general case, the activation function $\phi(u)$ may contain several parameters, further denoted by $b_1, \ldots, b_q$. Besides the function $\phi(u)$ in Fig. 4.16b, the following activation functions are widespread:

—the linear activation function (see Fig. 4.17a)

$$\phi(u) = \begin{cases} 0 & \text{if } u \leq \Delta, \\ u & \text{if } u \geq \Delta; \end{cases} \tag{4.63}$$

—the sigmoidal activation function (see Fig. 4.17b)

$$\phi(u) = \frac{1}{1 + \exp(u - \Delta)}; \tag{4.64}$$

—the tangent activation function (see Fig. 4.17c)

$$\phi(u) = \tan(u). \tag{4.65}$$

A network with the structure presented in Fig. 4.16b is called a single-layer neural network. A serial connection of single-layer neural networks yields a multilayer neural network. Its internal layers are called hidden whereas the last one is called the output layer. Most applications involve one hidden layer, as illustrated in Fig. 4.18. The input layer has a set of synaptic coefficients $w_{11}, \ldots w_{nh}$; the hidden layer, $g_{11}, \ldots, g_{hs}$.

Thus, single- and multilayer neural networks (Figs. 4.16 and 4.18) are statistical systems, i.e., the network's input at a current time theoretically depends on its input at this time. For example, the output of a single-layer neural network has the form

$$y(t) = S\left(\mathbf{x}(t) \mid \mathbf{w}, \mathbf{b}\right), \tag{4.66}$$

where $\mathbf{w}$ and $\mathbf{b}$ are network parameters.

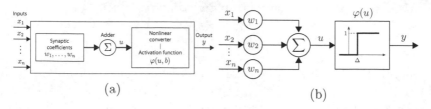

Figure 4.16

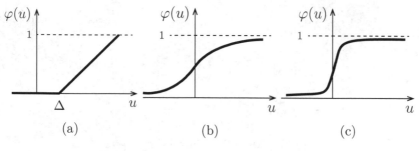

Figure 4.17

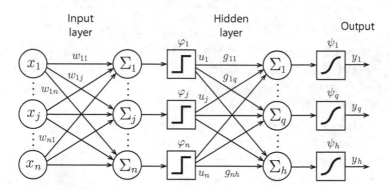

Figure 4.18

By assumption, a randomized neural network incorporates random parameters that take values within given intervals, i.e.,

$$\mathbf{w} \in \mathcal{W} = [\mathbf{w}^-, \mathbf{w}^+]; \qquad \mathbf{b} \in \mathcal{B} = [\mathbf{b}^-, \mathbf{b}^+]. \qquad (4.67)$$

The probabilistic properties of the parameters on these intervals are characterized by probability density functions $P_w(\mathbf{w})$ and $P_b(\mathbf{b})$. The output of a randomized neural network represents an ensemble of the values of the random variable y.

Example 3.7 Consider a randomized neural network with the structure shown in Fig. 4.16 and the sigmoidal activation function. The equations of this network have the form

$$u = \sum_{i=1}^{3} w_i x_i, \qquad y = \frac{1}{1 + \exp(u - \Delta)},$$

where $\Delta = 0.5$ and the coefficients $w_i \in [-0.5, 1.5]$ are random and independent with the PDF

$$P_w(\mathbf{w}) = \prod_{i=1}^{3} p_w(w_i),$$

where

$$p_w(w_i) = \begin{cases} \frac{1}{2} & \text{for } -0.5 \geq w_i \leq 1.5, \\ 0 & \text{for } -0.5 \leq w_i \geq 1.5. \end{cases}$$

A probability density of the output of the network generated in the course of 1000 Monte Carlo simulations with the PDF $P_w(\mathbf{w})$ can be observed in Fig. 4.19. For this ensemble, the mean value and variance are $m = 0.55094$ and $\sigma = 0.053719$.

4.3 "(SINGLE INPUT, FEEDBACK)–ENSEMBLE OUTPUT" DYNAMIC RANDOMIZED PARAMETRIC MODEL

4.3.1 Definition and structure

For a structural design of many dynamic RPMs, it is insufficient to define an input-output relation. In such cases, we have to reflect some internal mechanisms of the processes or objects under modeling. A mathematical framework that can be used to describe these mechanisms involves different classes of equations. Their general structure is represented by a certain block with a parameterized mapping $\mathbb{G}$ and a feedback loop characterized by a parameterized mapping $\mathbb{C}$, see Fig. 4.20.

In this figure, $\mathbf{x} \in R^n$ denotes the model input; $\mathbf{y} \in R^m$ is the model output; finally, $\mathbf{z} \in R^d$ gives the feedback block output.

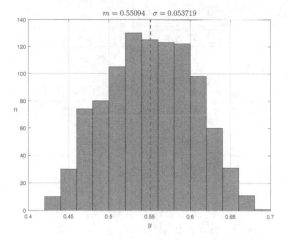

Figure 4.19

The mappings $\mathbb{G}$ and $\mathbb{C}$ describe the internal mechanisms of the processes running in an object. Using a functional description for these mappings, introduce the following notations:

$$\begin{aligned} \mathbb{G} &= \mathcal{G}\left(\mathbf{x}(\tau), \mathbf{z}(\tau), \,|\, \mathbf{a}; \, \tau \in \mathcal{T}\right), \\ \mathbb{C} &= \mathcal{C}\left(\mathbf{y}(\tau), \,|\, \mathbf{c}; \, \tau \in \mathcal{T}\right). \end{aligned} \tag{4.68}$$

The functionals $\mathcal{G}$ and $\mathcal{C}$ contain random parameters $\mathbf{a} \in R^r$ and $\mathbf{c} \in R^h$, respectively, with independent elements that belong to the intervals

$$\mathbf{a} \in \mathcal{A} = [\mathbf{a}^-, \mathbf{a}^+], \qquad \mathbf{c} \in \mathcal{S} = [\mathbf{c}_-, \mathbf{c}^+]. \tag{4.69}$$

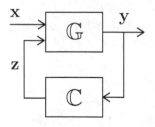

Figure 4.20

Consequently, the [(SI,FB)–EO] RPM can be written as

$$\begin{aligned}
\mathbf{y}(t) &= \mathcal{G}\left(\mathbf{x}(\tau), \mathbf{z}(\tau)\right), |\mathbf{a}; \tau \in \mathcal{T}), \\
\mathbf{z}(t) &= \mathcal{C}\left(\mathbf{y}(\tau), |\mathbf{c}; |\tau \in \mathcal{T}\right).
\end{aligned} \tag{4.70}$$

The functionals in these equations may have different mathematical models, depending on the declared properties of objects under analysis.

4.3.2 Linear dynamic models

The linear [(SI,FB)–EO] RPM have the structure illustrated in Fig. 4.20, except for its blocks are described by *linear functionals $\mathcal{G}$ and $\mathcal{C}$ with random parameters*. Assume that the input vector consists of *continuous time-varying functions* and the parameters are time-invariable. Then *the integral form of the linear [(SI,FB)–EO] RPM* can be written as

$$\begin{aligned}
\mathbf{y}(t) &= \int_{\mathcal{T}} G(t - \tau \,|\, \mathbf{a}) \, \mathbf{x}(\tau) \, d\tau + \int_{\mathcal{T}} F(t - \tau \,|\, \mathbf{b}) \, \mathbf{z}(\tau) \, d\tau, \\
\mathbf{z}(t) &= \int_{\mathcal{T}} C(t - \tau) \, \mathbf{y}(\tau) \, d\tau.
\end{aligned} \tag{4.71}$$

Recall that these equalities involve vectors $\mathbf{x} \in R^n$, $\mathbf{y} \in R^m$ and $\mathbf{z} \in R^d$, as well as the matrices $G(t \,|\, \mathbf{a}) = [g_{ij}(t \,|\, \mathbf{a}), i = \overline{1, m}, j = \overline{1, n}]$; $F(t \,|\, \mathbf{b}) = [f_{il}(t \,|\, \mathbf{b}), i = \overline{1, m}], l = \overline{1, d}$; $C(t \,|\, \mathbf{b}) = [c_{h,k}(t \,|\, \mathbf{c}), h = \overline{1, d}, k = \overline{1, m}]$. As their elements the matrices contain the pulse characteristics $g_{ij}(t \,|\, \mathbf{a})$, $f_{il}(t \,|\, \mathbf{b})$ and $c_{h,k}(t \,|\, \mathbf{c})$ with random parameters described by the vectors $\mathbf{a}, \mathbf{b}$ and $\mathbf{c}$; the elements of these vectors are independent and take values within the intervals

$$\mathbf{a} \in \mathcal{A} = [\mathbf{a}^-, \mathbf{a}^+], \quad \mathbf{b} \in \mathcal{B} = [\mathbf{b}^-, \mathbf{b}^+], \quad \mathbf{c} \in \mathcal{C} = [\mathbf{c}^-, \mathbf{c}^+]. \tag{4.72}$$

The [(SI,FB)–EO] RPM can be transformed into the [SI–EO] RPM. In our case, a most convenient approach is to employ the Fourier transform. First, eliminate the vector $\mathbf{z}$ in (4.71) to get the equation

$$\mathbf{y} = \int_{\mathcal{T}} G(t - \tau \,|\, \mathbf{a}) \, \mathbf{x}(\tau) \, d\tau + \int_{\mathcal{T}} F(t - \tau \,|\, \mathbf{b}) \int_{\mathcal{T}_\tau} C(t - \lambda \,|\, \mathbf{c}) \, \mathbf{y}(\lambda) \, d\tau d\lambda. \tag{4.73}$$

Using the Fourier transform [43], write the solution of this equation in the frequency domain (i.e., in terms of image functions):

$$\begin{aligned}
\mathbf{Y}(s) &= K(s \,|\, \mathbf{a}, \mathbf{b}, \mathbf{c}) \, \mathbf{X}(s), \\
K(s \,|\, \mathbf{a}, \mathbf{b}, \mathbf{c}) &= \left(I - F(s \,|\, \mathbf{b}) C(s \,|\, \mathbf{c})\right)^{(-1)} G(s \,|\, \mathbf{a}).
\end{aligned} \tag{4.74}$$

Here $s = \alpha + \iota\omega$ and $\iota = \sqrt{-1}$, whereas α and ω are real variables. Transition from the frequency domain to the time domain yields the following integral equation for an equivalent [SI–EO] RPM:

$$\mathbf{y}(t) = \int_{\mathcal{T}} K(t - \tau \,|\, \mathbf{a}, \mathbf{b}, \mathbf{c}) \,\mathbf{x}(\tau) \, d\tau, \qquad (4.75)$$

where the matrix

$$K(t \,|\, \mathbf{a}, \mathbf{b}, \mathbf{c}) = \mathcal{L}^{(-1)}\{K(s \,|\, \mathbf{a}, \mathbf{b}, \mathbf{c})\}. \qquad (4.76)$$

Example 3.8 Consider a one-dimensional [(SI,FB)–EO] RPM with the structure shown in Fig. 4.20:

$$y(t) = \int_0^t \exp(-\lambda(t - \tau))(x(\tau) - z(\tau)) \, d\tau -$$

$$- \int_0^t \exp(-\lambda(t - \tau)) \left(\int_0^\tau \exp(-\alpha(\tau - \mu)) \, y(\mu) \, d\mu \right) d\tau.$$

The parameters λ and μ are random, independent, and belong to the intervals

$$\lambda \in [-1, 2], \quad \mu \in [0.1, 0.5].$$

The PDFs of these parameters have the form

$$p(\lambda) = \begin{cases} 1/3 & \text{if } -1 \leq \lambda < 2, \\ 0 & \text{if } -1 \geq \lambda \geq 2, \end{cases} \quad u(\mu) = \begin{cases} 0.25 & \text{if } 0.1 \geq \mu < 0.5, \\ 0 & \text{if } 0.1 \leq \mu \geq 0.5. \end{cases}$$

Consider *the differential form of the linear [(SI,FB)–EO] RPMs*. Introduce the following notations: $\mathbf{x}(t) = \{x_1(t), \ldots, x_n\}$ as the input vector; $\mathbf{z}(t) = \{z_1(t), \ldots, z_d\}$ as the state vector; $\mathbf{y}(t) = \{y_1(t), \ldots, y_m(t)\}$ as the output vector; $A = [a_{ij} \,|\, (i,j) = \overline{1,d}]$ as a transition matrix; $B = [b_{kl} \,|\, k = \overline{1,d}, l = \overline{1,n}]$ as the input matrix; finally, $C = [c_{ht} \,|\, h = \overline{1,m}, t = \overline{1,d}]$. The elements of these matrices are independent random variables that take values within the intervals

$$\begin{aligned} a_{ij} &\in \mathcal{A}_{ij} = [a_{ij}^-, a_{ij}^+], & (i,j) = \overline{1,d}; \\ b_{kl} &\in \mathcal{B}_{kl} = [b_{kl}^-, b_{kl}^+], & k = \overline{1,d}, \, l = \overline{1,n}; \\ c_{ht} &\in \mathcal{C}_{ht} = [c_{ht}^-, c_{ht}^+], & h = \overline{1,m}, \, t = \overline{1,d}. \end{aligned} \qquad (4.77)$$

Due to independence, perform the direct product of these intervals for defining the matrix intervals

$$\mathcal{A} = [\mathcal{A}^-, \mathcal{A}^+], \quad \mathcal{B} = [\mathcal{B}^-, \mathcal{B}^+], \quad \mathcal{C} = [\mathcal{C}^-, \mathcal{C}^+]. \qquad (4.78)$$

As a result, the differential form of the linear [(SI,FB)–EO] RPM is given by

$$\begin{aligned} \frac{d\mathbf{z}(t)}{dt} &= A\mathbf{z}(t) + B\mathbf{x}(t), \quad \mathbf{z}(0) = \mathbf{z}^0, \\ \mathbf{y}(t) &= C\,\mathbf{z}(t), \end{aligned} \qquad (4.79)$$

where the matrices A, B, and C have random elements of the interval type (4.78).

The linear [(SI,FB)–EO] RPM can be written in the input–output representation using the fundamental solution matrix [85]

$$W(A|t - \tau) = \exp[A(t - \tau)]. \tag{4.80}$$

In this case,

$$\mathbf{y}(t) = CW(A|t)\mathbf{z}^0 + C \int_0^t W(A|t - \tau)B\mathbf{x}(\tau)d\tau. \tag{4.81}$$

Let the input, output and feedback vectors consist of the lattice functions $\mathbf{x}[s], \mathbf{z}[s]$ and $\mathbf{y}[s]$, respectively, and also let the matrices G, F and C consist of the corresponding lattice pulse characteristics $g_{ij}[s], \, | \, i = \overline{1, n}, \, j = \overline{1, m}; \, f_{il}[s], \, | \, i = \overline{1, n}, \, l = \overline{1, d}; \, c_{hk}[m], \, | \, h = \overline{1, d}, \, k = \overline{1, m}$. In addition, assume that the input functions have discrete time, i.e., are *lattice functions*. Under all these hypotheses, the integral form (4.71) can be transformed into the additive one, i.e.,

$$\mathbf{y}[s] = \sum_{t=1}^{s} G[t, \, | \, \mathbf{a}] \, \mathbf{x}[s - t] + \sum_{t=1}^{s} F[t, \, | \, \mathbf{b}] \, \mathbf{z}[s - t],$$

$$\mathbf{z}[s] = \sum_{t=1}^{s} C[t, \, | \, \mathbf{c}] \, \mathbf{y}[s - t]. \tag{4.82}$$

As a rule, the lattice pulse characteristics have finite memory: they are nonzero if $s \in [0, M]$. In this case, equations (4.82) take the following form:

$$\mathbf{y}[s] = \sum_{t=1}^{M} G[t, \, | \, \mathbf{a}] \, \mathbf{x}[s - t] + \sum_{t=1}^{M} F[t, \, | \, \mathbf{b}] \, \mathbf{z}[s - t],$$

$$\mathbf{z}[s] = \sum_{t=1}^{M} C[t, \, | \, \mathbf{c}] \, \mathbf{y}[s - t], \qquad s \geq M. \tag{4.83}$$

4.3.3 Nonlinear dynamic models with power nonlinearities

Consider an object whose [(SI,FB)–EO] RPM has the structure shown in Fig. 4.20. Let the functionals $\mathcal{G}$ and $\mathcal{C}$ be continuous. Then they can be represented using the framework of functional Volterra series [184]; see Section 3.1.

This framework will be adopted for the one-dimensional [(SI,FB)–EO] RPM (Fig. 4.20), which is described by the system of equations

$$\begin{aligned} y(t) &= \mathcal{G}\left(\varepsilon(\tau)\right), \, | \, \mathbf{a}; \, \tau \in \mathcal{T}\right), \\ z(t) &= \mathcal{C}\left(y(\tau), \, | \, \mathbf{c}; \, | \, \tau \in \mathcal{T}\right), \\ \varepsilon(t) &= x(t) - z(t). \end{aligned} \tag{4.84}$$

Assume that the time-varying functions $x(t), z(t)$, and $y(t)$, as well as the functionals $\mathcal{G}(\varepsilon, \, | \, \mathbf{a})$ and $\mathcal{C}(y, \, | \, \mathbf{c})$, are continuous. In this case, they can be expressed as the functional series with the corresponding weight functions (pulse characteristics), i.e.,

$$y(t) = \sum_{i=1}^{\infty} \int_{\tau_1 \in \mathcal{T}} \cdots \int_{\tau_i \in \mathcal{T}} g_i(t - \tau_1, \ldots, t - \tau_i \mid \mathbf{a}^{(i)}) \times$$
$$\varepsilon(\tau_1) \cdots \varepsilon(\tau_i) \, d\tau_1 \cdots d\tau_i,$$

$$z(t) = \sum_{j=1}^{\infty} \int_{\lambda_1 \in \mathcal{T}} \cdots \int_{\lambda_j \in \mathcal{T}} c_j(t - \lambda_1, \ldots, t - \lambda_j \mid \mathbf{c}^{(j)}) \times$$
$$\times y(\lambda_1) \cdots y(\lambda_j) \, d\lambda_1 \cdots d\lambda_j, \qquad (4.85)$$

$$\varepsilon(t) = x(t) - z(t).$$

Here the pulse characteristics have random parameters, namely, the random independent vectors $\mathbf{a}^{(i)}$ and $\mathbf{c}^{(j)}$ with independent elements taking values within the intervals

$$\mathbf{a}^{(i)} \in \mathcal{A}^{(i)} = [\mathbf{a}_-^{(i)}, \mathbf{a}_+^{(i)}], \qquad i \in [1, \infty),$$
$$\mathbf{c}^{(j)} \in \mathcal{S}^{(j)} = [\mathbf{c}_-^{(j)}, \mathbf{c}_+^{(j)}], \qquad j \in [1, \infty). \qquad (4.86)$$

As a matter of fact, there exists a certain relation between the [(SI,FB)–EO] RPM (4.3.3) and the [SI–EO] RPM (4.2.4) with continuous nonlinearities. Recall that the latter model is described by a continuous functional $\mathcal{B}$, which can be also written as a functional Volterra series, i.e.,

$$\mathbf{y}(t) = \mathcal{B}(\mathbf{x}(\tau), \mid \tau \in \mathcal{T}) =$$
$$+ \sum_{k}^{\infty} \int_{\tau_1 \in \mathcal{T}} \cdots \int_{\tau_k \in \mathcal{T}} w_k(t - \tau_1, \ldots, t - \tau_k) \times$$
$$\times x(\tau_1) \cdots x(\tau_k) \, d\bar{\tau}. \qquad (4.87)$$

Define the pulse characteristics in this formula so that equations (4.3.3) turn into identities [146]. First, substitute the last-mentioned equation into the first one to obtain an integral equation in the variable $y(t)$:

$$y(t) = \int_{\tau_1 \in \mathcal{T}} g_1(t - \tau_1 \mid \mathbf{a})[x(\tau_1) - \int_{\lambda_1 \in \mathcal{T}} c_1(\tau_1 - \lambda_1 \mid \mathbf{c}) \, y(\lambda_1) \, d\lambda_1 +$$
$$+ \int_{\lambda_1 \in \mathcal{T}} \int_{\lambda_2 \in \mathcal{T}} c_2(t - \lambda_1, t - \lambda_2 \mid \mathbf{c}) \, y(\lambda_1) y(\lambda_2) \, d\lambda_1 d\lambda_2 + \ldots \qquad (4.88)$$

Then construct the solution in the form of the functional series (4.3.3): substitute it into the left- and right-hand sides of equation (4.3.3), collecting and comparing the terms with the same powers of the variable x. These manipulations lead to the following system of integral equations for the cores $w_k(\tau_1, \ldots, \tau_k) = w_k(\tau_1, \ldots, \tau_k \,|\, \mathbf{a}, \mathbf{c})$ of (4.3.3):

$$
\int_{\tau_1 \in \mathcal{T}} w_1(t - \tau_1 \,|\, \mathbf{a}, \mathbf{c})\, x(\tau_1)\, d\tau_1 = \int_{\tau_1 \in \mathcal{T}} g_1(t - \tau_1 \,|\, \mathbf{a})\, [x(\tau_1)\, d\tau_1 -
$$

$$
- \int_{\lambda_1 \in \mathcal{T}_{\tau_1}} c_1(\tau_1 - \lambda_1 \,|\, \mathbf{c}) \int_{\mu_1 \in \mathcal{T}_{\lambda_1}} w_1(\lambda_1 - \mu_1 \,|\, \mathbf{a}, \mathbf{c}) \times
$$

$$
\times x(\mu_1)\, d\tau_1 d\lambda_1 d\mu_1, \tag{4.89}
$$

$$
\int_{\tau_1 \in \mathcal{T}} \int_{\tau_2 \in \mathcal{T}} w_2(t - \tau_1, t - \tau_2 \,|\, \mathbf{a}, \mathbf{c})\, x(\tau_1) x(\tau_2)\, d\tau_1 d\tau_2 =
$$

$$
- \int_{\tau_1 \in \mathcal{T}} g_1(t - \tau_1 - \tau_2 \,|\, \mathbf{a}) \int_{\lambda_1 \in \mathcal{T}_{\tau_1}} \int_{\lambda_2 \in \mathcal{T}_{\tau_2}} c_2(\tau_1 - \lambda_1, \tau_2 - \lambda_2 \,|\, \mathbf{c}) \times
$$

$$
\times \int_{\mu_1 \in \mathcal{T}_{\lambda_1}} \int_{\mu_2 \in \mathcal{T}_{\lambda_2}} w_1(\lambda_1 - \mu_1, \,|\, \mathbf{a}, \mathbf{c}) w_1(\lambda_2 - \mu_2, \,|\, \mathbf{a}, \mathbf{c}) \times
$$

$$
\times x(\mu_1) x(\mu_2)\, d\tau_1 d\tau_2 d\lambda_1 d\lambda_2 d\mu_1 d\mu_2 -
$$

$$
- \int_{\tau_1 \in \mathcal{T}} \int_{\tau_2 \in \mathcal{T}} g_2(t - \tau_1, t - \tau_2, \,|\, \mathbf{a}) \times
$$

$$
\times \int_{\lambda_1 \subset \mathcal{T}_{\tau_1}} \int_{\lambda_2 \subset \mathcal{T}_{\tau_2}} c_1(\tau_1 - \lambda_1) c_1(\tau_2 - \lambda_2, \,|\, \mathbf{c}) \times
$$

$$
\times \int_{\mu_1 \in \mathcal{T}_{\lambda_1}} \int_{\mu_2 \in \mathcal{T}_{\lambda_2}} w_1(\lambda_1 - \mu_1, \,|\, \mathbf{a}, \mathbf{c}) \times
$$

$$
w_1(\lambda_2 - \mu_2, \,|\, \mathbf{a}, \mathbf{c})\, d\tau_1 d\tau_2 d\lambda_1 d\lambda_2 d\mu_1 d\mu_2, \cdots.
$$

This system consists of homogeneous (with respect to the powers of x) integral equations with a recursive structure. Using the multidimensional Fourier transform (see the details in [42,178]), write the final expressions

$$
W(s_1, \,|\, \mathbf{a}, \mathbf{c}) = \frac{G_1(s_1, \,|\, \mathbf{a})}{1 + G_1(s_1, \,|\, \mathbf{a}) C(s_1, \,|\, \mathbf{c})},
$$

$$
W(s_1, s_2, \,|\, \mathbf{a}, \mathbf{c}) =
$$

$$
= \frac{-G_1(s_1 + s_2, \,|\, \mathbf{a})\, C_2(s_1, s_2, \,|\, \mathbf{c})\, W_1(s_1, \,|\, \mathbf{a}, \mathbf{c}) W_1(s_2, \,|\, \mathbf{a}, \mathbf{c})}{1 + .G_1(s_1 + s_2, \,|\, \mathbf{a}) C_1(s_1 + s_2, \,|\, \mathbf{c})} +
$$

$$
+ \frac{G_2(s_1, s_2, \,|\, \mathbf{a}) W_1(s_2, \,|\, \mathbf{a}, \mathbf{c}) [C_1(s_2, \,|\, \mathbf{c}) W_1(s_1, \,|\, \mathbf{a}, \mathbf{c}) - 2C_1(s_2, \,|\, \mathbf{c})]}{1 + G_1(s_1 + s_2, \,|\, \mathbf{a}) C_1(s_1 + s_2), \,|\, \mathbf{c}},
$$

$$
\tag{4.90}
$$

$$\cdots$$

Here $s_1, \ldots, s_i$ are the complex parameters of the multidimensional Fourier transform.

Therefore, the [(SI,FB)–EO] RPM (4.3.3) can be written as the [SI–EO] RPM (4.2.4). This transformation may be fruitful in machine learning problems for randomized dynamic regression models.

4.3.4 Nonlinear dynamic models with polynomial nonlinearities

In the previous subsection, we have studied the procedure of using the orthogonal polynomials in the [SI–EO] RPMs. Now consider this procedure subject to the [(SI,FB)–EO] RPMs. For the [(SI,FB)–EO] RPM with the structure shown in Fig. 4.21a,

$$
\begin{aligned}
y(t) &= \int_{\mathcal{T}} w(t - \tau) \sum_{k=0}^{K} w_k g_k(\varepsilon(\tau)) \, d\tau, \\
z(t) &= \int_{\mathcal{T}} c(t - \tau) \sum_{m=0}^{M} v_m c_m(y(\tau)) \, d\tau, \\
\varepsilon(t) &= x(t) - z(t).
\end{aligned}
\tag{4.91}
$$

Another [(SI,FB)–EO] RPM (see Fig. 4.21b) is described by the following system of equations:

$$
\begin{aligned}
y(t) &= \sum_{k=0}^{K} w_k g_k(u(\tau)) \, d\tau, \\
u(t) &= \int_{\mathcal{T}} w(t - \tau) \, \varepsilon(\tau) \, d\tau, \\
e(t) &= \int_{\mathcal{T}} c(t - \tau) \, y(\tau) \, d\tau, \tag{4.92} \\
z(t) &= \sum_{m=0}^{M} v_m c_m(e(\tau)) \, d\tau, \tag{4.93} \\
\varepsilon(t) &= x(t) - z(t).
\end{aligned}
$$

The parameters of these RPMs, $\mathbf{w}$ and $\mathbf{v}$, are random, independent and take values within the intervals

$$
\mathbf{w} \in \mathcal{W} = [\mathbf{w}^-, \mathbf{w}^+], \qquad \mathbf{v} \in \mathcal{V} = [\mathbf{v}^-, \mathbf{v}^+].
\tag{4.94}
$$

Recall that the orthogonal polynomials can be applied for approximating the continuous nonlinear functions and, moreover, the discontinuous nonlinearities, which forms their distinctive feature. However, the orthogonal property of the polynomials holds only on a certain set (the domain of definition) and in a certain functional space.

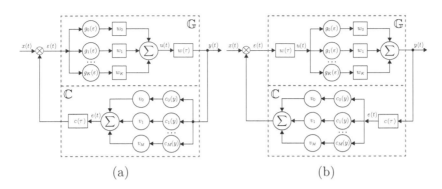

(a)   (b)

Figure 4.21

The usage procedure of the orthogonal polynomials from Section 3.2 is naturally extended to the [(SI,FB)–EO] RPMs.

Example 3.9 Consider the [(SI,FB)–EO] RPM with the block diagram shown in Fig. 4.22a. The corresponding equations have the form

$$y(t) = \int_0^t w(t-\tau)\phi(\varepsilon(\tau))d\tau,$$

$$z(t) = \psi\left(\int_0^t c(t-\tau)y(\tau)d\tau\right),$$

$$\varepsilon(t) = x(t) - z(t).$$

The nonlinear elements of this model are described by the Hermite polynomials

$$H_1(\varepsilon) = \varepsilon, \qquad H_3(\varepsilon) = \varepsilon^3 - 3\varepsilon,$$

$$\phi(\varepsilon) = w_1 H_1(\varepsilon) + w_3 H_3(\varepsilon), \quad \psi(e) = v_1 H_1(e) + v_3 H_3(e).$$

The coefficients w_1, w_3, v_1 and v_3, acting as parameters, represent independent random variables distributed on the intervals

$$w_1 \in [-1,1], \; w_3 \in [-0.1, 0.1], \quad v_1 \in [-2,2], \; v_3 \in [-1,1].$$

Their PDFs are uniform on the corresponding intervals. The properties of the dynamic elements are reflected by the pulse characteristics

$$w(\tau) = \exp(-0.5\tau)\cos(2\tau), \qquad c(\tau) = \tau\exp(-0.5\tau).$$

The input function has the form $x(t) = 2\sin(2t)$.

The ensemble of all trajectories, as well as the mean and median trajectories and the variance pipe, are shown in Fig. 4.22b.

4.4 PROBABILISTIC CHARACTERISTICS OF RANDOMIZED PARAMETERS AND ENSEMBLES

The models used in the RML procedures have realizations of *randomized parameters*, which generate the *ensembles* of random objects (vectors,

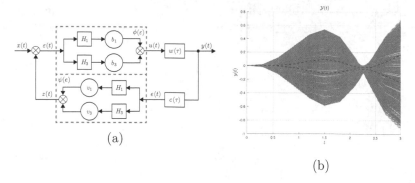

(a)

(b)

Figure 4.22

trajectories, events) at the output of these models. The properties of both are described in terms of appropriate probabilistic characteristics.

1. Probabilistic characteristics of randomized parameters

The RPMs under study involve randomized interval-type parameters with desired probabilistic properties. We will discriminate between

- the *original* parameters $\mathbf{a} = \{a_1, \ldots, a_n\}$, which take values within the n-dimensional parallelepiped $\mathcal{A}$ with arbitrary-size faces, i.e.,

$$\mathcal{A} = [\mathbf{a}^-, \mathbf{a}^+] = \bigotimes_{i=1}^{n} \mathcal{A}_i = [a_i^-, a_i^+], \qquad \text{and} \qquad (4.95)$$

- the *relative nonnegative* parameters $\mathbf{u} = \{u_1, \ldots, u_n\}$, which take values within the n-dimensional nonnegative parallelepiped $\mathcal{U}$ with the unit faces, i.e.,

$$\mathcal{U} = [\mathbf{0}, \mathbf{1}]. \qquad (4.96)$$

The relative nonnegative parameters are related to the original ones by

$$\begin{aligned} a_i &= a_i^- + u_i \delta_i, \quad \delta_i = a_i^+ - a_i^-, \qquad i = \overline{1, n}, \\ \mathbf{a} &= \mathbf{a}^- + \Delta \mathbf{u}, \qquad \Delta = \mathrm{diag}[(a_i^+ - a_i^-), i = \overline{1, n}]. \quad (4.97) \end{aligned}$$

The model parameters of these types undergo randomization in accordance with the desired probabilistic characteristics, more specifically, in accordance with the PDFs, $P(\mathbf{a})$ for the original parameters $\mathbf{a}$ and $W(\mathbf{u})$ for the relative nonnegative parameters $\mathbf{u}$. In the general case, the elements of the vectors $\mathbf{a}$ and $\mathbf{u}$ can be related to each other. Their

PDFs represent multivariable functions of general form. If the parameters are independent, then the corresponding PDFs have the multiplicative structure:

$$P(\mathbf{a}) = \prod_{i=1}^{n} P_i(a_i), \qquad W(\mathbf{u}) = \prod_{i=1}^{n} w_i(u_i). \qquad (4.98)$$

For making presentation compact, let us introduce the following terminological abbreviations:

- RPM-Orig for the randomized parametric models with the original parameters $\mathbf{a}$ and the *PDF* $P(\mathbf{a})$;

- RPM-Rel for the randomized parametric models with the relative nonnegative parameters $\mathbf{u}$ and the *PDF* $W(\mathbf{u})$.

2. Empirical probabilistic characteristics of ensembles

The randomized models output the ensembles of random vectors (static models), the ensembles of time-varying functions, i.e., trajectories (dynamic models), or the ensembles of random events. For each model, these ensembles appear as the result of generating its the randomized parameters. The ensembles can be described in quantitative terms using the well-known probabilistic characteristics of mathematical statistics.

Consider an arbitrary ensemble $\hat{\mathcal{Y}}_T$ of trajectories $\hat{y}(t)$ on a finite time interval $\mathcal{T} = [0, T]$, that is, generated by the one-dimensional RPM-Orig model with a given PDF. An example of such an ensemble is presented in Fig. 4.23. Assume that it contains N trajectories. Denote by $\hat{y}(t^*)$ its cutset under a fixed value $t = t^*$. This cutset forms a collection of .values (realizations) of the random variable $\hat{y}(t^*)$, which will be used for empirical estimation.

A) The evolution of mean trajectory. The mean value of the random variable $\hat{y}(t^*)$ is given by

$$\bar{y}(t^*) = \frac{1}{N} \sum_{k=1}^{N} \hat{y}^{(k)}(t^*), \qquad (4.99)$$

where $\hat{y}^{(k)}(t^*)$ denote the realizations of the random variable $\hat{y}(t^*)$.

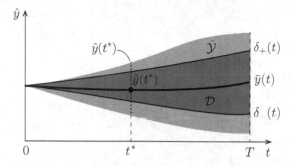

Figure 4.23

Calculating these means for different times $t^* \in \mathcal{T}$, we obtain a collection of empirical estimates $\bar{y}(t_u^*)$, $u \in \overline{1, U}$. They can be approximated using a continuous curve $\bar{y}(t)$, which forms the mean trajectory [104].

B) *The evolution of the empirical variance pipe.* Take the ensemble $\hat{\mathcal{Y}}_T$ illustrated in Fig. 4.23, and consider its cutset $\hat{y}(t^*)$. Define the random variable

$$D(t^*) = [\hat{y}(t^*) - \bar{y}(t^*)]^2. \tag{4.100}$$

The mean of this random variable (i.e., the variance of the random variable $\hat{y}(t^*)$) is given by

$$\bar{D}(t^*) = \frac{1}{N-1} \sum_{k=1}^{N} D^{(k)}(t^*), \tag{4.101}$$

where $D^{(k)}(t^*)$ denote the realizations of the random variable $D(t^*)$ (4.100).

Introduce the standard deviation of the random variable $\hat{y}(t^*)$ from its mean $\bar{y}(t^*)$:

$$\sigma(t^*) = \sqrt{\bar{D}(t^*)}. \tag{4.102}$$

Under an assumption that t^* takes values within the interval $\mathcal{T}$, let us approximate the resulting points using a continuous curve—the standard deviation trajectory $\sigma(t)$.

Define two trajectories, $\delta_\pm = \bar{y}(t) \pm \sigma(t)$. The domain

$$\mathcal{D} = \{\hat{y} : \delta_-(t) \leq \hat{y}(t) \leq \delta_+(t)\} \tag{4.103}$$

will be called *the standard deviation pipe.* Figure 4.23 demonstrates the standard deviation pipe $\mathcal{D}$ about the mean trajectory $\bar{y}(t)$.

C) The evolution of the empirical probability density function. The empirical estimates of the PDFs can be constructed using the random variable realizations $\hat{y}(t^*)$ at a fixed time t^*.

The empirical PDF, i.e., the histogram $p[\hat{y}(t^*)]$, is constructed in accordance with the well-known rules [104]. Denote by $\hat{y}^-(t^*)$ and $\hat{y}^+(t^*)$ the left and right limits, respectively, for the random variable $\hat{y}(t^*)$. Next, divide the interval $[\hat{y}^-(t^*), \hat{y}^+(t^*)]$ into J subintervals η_j, $(j = \overline{1, J})$ of the length $(\hat{y}^+(t^*) - \hat{y}^-(t^*))/J$. The empirical probability density function is the frequency with which the values of the random variable $\hat{y}(t^*)$ get into the subinterval η_j, $(j = \overline{1, J})$, i.e.,

$$p_j = \frac{M_j}{M}, \qquad (4.104)$$

where M_j gives the number of values $\hat{y}(t^*)$ within the subinterval η_j and M is the total number of trajectories in the ensemble. Figure 4.24 provides an example of the histogram $p[\hat{y}(t^*)]$. The evolution of the histogram can be observed by varying the time t^*. As a rule, an ensemble contains very many trajectories. Therefore, the histogram can be approximated with good accuracy using a continuous curve $\hat{P}[\hat{y}(t^*)]$.

D). The evolution of the empirical median trajectory. Often the empirical PDF has no symmetry. In such cases, the median trajectory seems to be more informative. Consider the PDF $\hat{P}[\hat{y}(t^*)]$, and define a value $z(t^*)$ such that

$$\int_{\hat{y}^-(t^*)}^{z(t^*)} \hat{P}(x)dx = \int_{z(t^*)}^{\hat{y}^-(t^*)} \hat{P}(x)dx. \qquad (4.105)$$

The value $z(t^*)$ of the random variable $\hat{y}(t^*)$ is called the median for the time t^*. Performing these calculations for all $t^* \in \mathcal{T}$, obtain a continuous curve, which will be called the median trajectory of the ensemble.

E). The evolution of the α-quantiles of the empirical PDF. For a fixed time t^*, introduce the empirical probability density function

$$\hat{F}_{t^*}(y) = \int_{\hat{y}^-(t^*)}^{y} \hat{P}_{t^*}(v)dv. \qquad (4.106)$$

According to the definition given in [104], the α_{t^*}-quantile ($\alpha_{t^*} \in [0, 1]$)

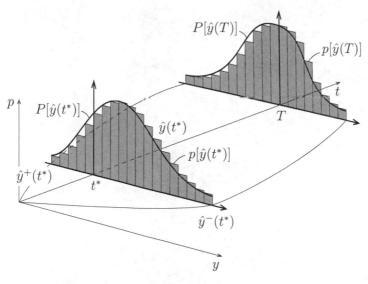

Figure 4.24

of the distribution $\hat{F}_{t^*}(y)$ of a random variable y is the number $y_{\alpha_{t^*}}$ satisfying the inequalities

$$Pb\{y \leq y_{\alpha_{t^*}}\} \leq \alpha_{t^*} \qquad (4.107)$$

and

$$Pb\{y > y_{\alpha_{t^*}}\} \leq 1 - \alpha_{t^*}, \qquad (4.108)$$

where Pb denotes probability. The α_{t^*}-quantile is illustrated in Fig. 4.25b. By analogy with items A)–C), consider different cutsets $t^* \in \mathcal{T}$ of the ensemble to get the α_t-quantile trajectory, see Fig. 4.25b.

F). The evolution of the empirical confidence intervals and confidence levels. The α_{t^*}-quantile allows us to define an interval (known as *the confidence interval*) that covers the values of the random variable $\hat{y}(t^*)$ with the probability α_{t^*} (*the confidence level*).[2]

By definition, the confidence interval

$$C_{t^*} = [y_{1-\alpha_{t^*}/2}, y_{1+\alpha_{t^*}/2}] \qquad (4.109)$$

---

[2]In mathematical statistics, these concepts were suggested and developed by J. Neyman [123].

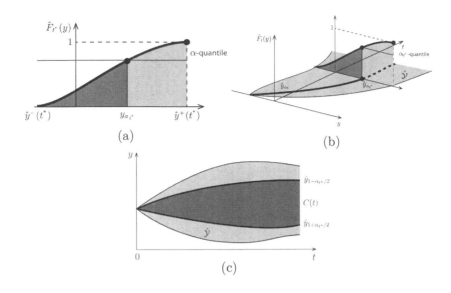

Figure 4.25

covers the values of the random variable $\hat{y}(t^*)$ with the confidence level

$$P[\hat{y}(t^*) \in C_{t^*}] = \alpha_{t^*}. \qquad (4.110)$$

Letting $t^* = t \in \mathcal{T}$ in both formulas, construct the confidence domain $C(t)$, as illustrated in Fig. 4.25c.

3. The moment characteristics of the ensemble

Consider a one-dimensional ensemble $\hat{\mathcal{Y}}_T$ consisting of random trajectories $\hat{y}(t\,|\,\mathbf{a})$, where $t \in \mathcal{T}$. The parameters $\mathbf{a}$ are randomized with a given PDF $P(\mathbf{a})$. Let $t = t^*$ be fixed. Then $\hat{y}(t^*\,|\,\mathbf{a})$ represents a function of the random argument $\mathbf{a}$.

Define its k*th moments* by

$$M_k(t^*) = \mathcal{M}\{y(t^*\,|\,\mathbf{a})\} = \int_{\mathcal{A}} y^k(t^*\,|\,\mathbf{a})P(\mathbf{a})d\mathbf{a}. \qquad (4.111)$$

Introduce *the k-means* of these moments in the form

$$m_k(t^*) = M_k^{1/k}(t^*). \qquad (4.112)$$

For $k = 1$, the moment coincides with its mean, i.e., $M_1 = m_1$.

Assuming t^* to be a variable that takes values within the interval $\mathcal{T}$, find the kth moment trajectory $M_k(t)$ and the k-mean trajectory $m_k(t)$.

In some applications, the object's output vector consists of the elements acting as the average functionals with random parameters. For example, consider option trading [15,182]. Each of r financial derivatives contains p random parameters $a_1, \ldots, a_p$, which are realized in the course of trading with some PDF $P(\mathbf{a})$. The relations between the financial derivatives and parameters are described by functionals $\mathcal{G}_1(\mathbf{a}), \ldots, \mathcal{G}_r(\mathbf{a})$. As a rule, they belong to the class of power functionals; in the elementary statement, these functionals have the form

$$\mathcal{G}_1(\mathbf{a}) = \sum_{k=1}^{p} c_k a_k,$$

$$\mathcal{G}_2(\mathbf{a}) = \sum_{k=1}^{p} c_k a_k + \sum_{k_1=1}^{r} \sum_{k_2=1}^{r} c_{k_1 k_2} a_{k_1} a_{k_2},$$

$$\cdots$$

$$\mathcal{G}_r(\mathbf{a}) = \sum_{k=1} c_k a_k + \sum_{k_1,k_2} c_{k_1,k_2} a_{k_1} a_{k_2} + \cdots + \sum_{k_1,\ldots,k_r} c_{k_1,\ldots,k_r} a_{k_1} \cdots a_{k_r}.$$

The more efficient models that are used to predict future mean prices involve the integral power functionals; their cores depend on the time variables, $c(\tau), c(\tau_1, \tau_2), \ldots$.

In both statements, the relations under consideration give the mean prices of the financial derivatives, i.e.,

$$\int_{\mathcal{A}} \mathcal{G}_1(\mathbf{a}) P(\mathbf{a}) d\mathbf{a}, \int_{\mathcal{A}} \mathcal{G}_2(\mathbf{a}) P(\mathbf{a}) d\mathbf{a}, \ldots$$

These mean price formulas represent the combinations of the weighted moments of the random vector $\mathbf{a}$. In the option trading models, the object's output $\mathbf{y}$ is the predicted mean prices for the corresponding financial derivatives.

CHAPTER 5

ENTROPY-ROBUST ESTIMATION PROCEDURES FOR RANDOMIZED MODELS AND MEASUREMENT NOISES

The main stage of each randomized machine learning procedure is to estimate the probability density functions. Estimation involves input and output data arrays. According to a common hypothesis, output arrays contain errors due to imperfect measurements. The core of the estimation procedures consists in a model with *randomized* interval-type parameters, which can be *original or relative*. A static model with randomized parameters generates an ensemble of random vectors, whereas a dynamic model an ensemble of random trajectories. They are characterized in terms of different-order moments, which serve for balancing the model's output with its input data.

DOI: 10.1201/9781003306566-5

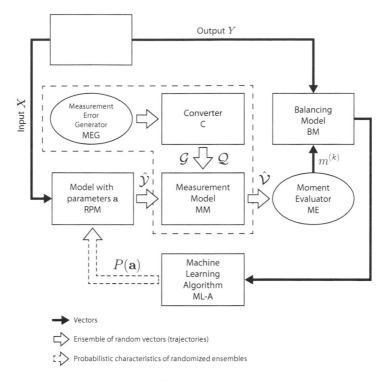

Figure 5.1

The PDF estimation problem is formulated as the problem to find the "best" functions under maximum uncertainty. The latter is measured using the information entropy of the model and measurement noises. Maximizing the entropy, we construct the entropy-optimal PDFs of the model parameters and the "worst" PDFs of the noises in the sense of this criterion.

5.1 STRUCTURE OF ENTROPY-ROBUST ESTIMATION PROCEDURE

Consider the general phenomenological structure of the estimation procedure. It represents a sequence of stages shown in Fig. 5.1. This procedure yields the estimated PDFs of the model parameters and noises—$P^*(\mathbf{a})$ and $Q(\bar{\xi})$, respectively.

1. Data source (DS) generates an array of input data vectors $\mathbf{x}(t_1), \ldots, \mathbf{x}(t_s)$ and a corresponding array of output data vectors $\mathbf{y}(t_1), \ldots, \mathbf{y}(t_s)$, where s denotes the number of measurements, $\mathbf{x} \in R^n$ and $\mathbf{y} \in R^m$. Sometimes, it seems convenient to write these data arrays in the form of matrices $X = [\, x_{ij} \,|\, i = \overline{1,n}; \; j = \overline{1,s}\,]$ and $Y = [\, y_{ij} \,|\, i = \overline{1,m}; \; j = \overline{1,s}\,]$. By a standard assumption, input data are measured perfectly and output data with errors, and this aspect should be taken in account for estimation procedure design.

2. Randomized parametric model (RPM) generally represents a dynamic system with randomized *original* parameters $\mathbf{a} \in R^r$ that take values within the r-dimensional parallelepiped

$$\mathbf{a} \in \mathcal{A} = [\mathbf{a}^-, \mathbf{a}^+] \in R^r, \quad \mathcal{A} = \bigcup_{i=1}^{r} \mathcal{A}_i, \quad \mathcal{A}_i = [a_i^-, a_i^+]. \qquad (5.1)$$

The probabilistic properties of the randomized original parameters are described by the PDF $P(\mathbf{a})$. The models of this class will be referred to as the *RPM-Orig* models.

In many cases, it is reasonable to use nonnegative *relative* randomized parameters $\mathbf{u} \in \mathcal{I} = [\mathbf{0}, \mathbf{1}]; \; \mathcal{I} \in R_+^r$. Recall that they have the following relation to the original parameters:

$$\mathbf{a} = \mathbf{a}^- + A\,\mathbf{u}, \quad \mathbf{u} \in \mathcal{I}, \qquad (5.2)$$

where

$$A = \operatorname{diag}[a_i^+ - a_i^- \; i = \overline{1,r}]. \qquad (5.3)$$

The probabilistic properties of the randomized relative parameters are characterized in terms of the PDF $W(\mathbf{u})$. The models of this class will be mentioned as the *RPM-Rel* models.

For *each realization* of the randomized parameters, the models generate trajectories with the given PDFs, which have the following functional descriptions:

- for the RPM-Orig model,

$$\mathbf{y}[t_j \,|\, P(\mathbf{a})] = \mathcal{B}[\mathbf{x}(t_j, \tau), \mathbf{a}, P(\mathbf{a}); (t_j, \tau) \in \mathcal{T}], \qquad (5.4)$$
$$j = \overline{1,s};$$

- for the RPM-Rel model,

$$\mathbf{z}[t_j \,|\, W(\mathbf{u})] = \mathcal{D}[\mathbf{x}(t_j, \tau), \mathbf{a}^- + A\,\mathbf{u}, W(\mathbf{u}); (t_j, \tau) \in \mathcal{T}], j = \overline{1,s}.$$

In formulas (5.4, 5.5), $\mathcal{B}$ and $\mathcal{D}$ are integral functionals.

At the discrete measurement times, the randomized models output the ensembles of corresponding trajectories, i.e.,

- for the RPM-Orig model,

$$\mathcal{Y}[t_j \,|\, P(\mathbf{a})] = \{\mathbf{y}[t_j \,|\, P(\mathbf{a})\}, \qquad t_j \in \mathcal{T}, \, j = \overline{1, s}, \qquad (5.5)$$

- for the RPM-Rel model,

$$\mathcal{Z}[t_j \,|\, W(\mathbf{u})] = \{\mathbf{z}[t_j \,|\, W(\mathbf{u})], \qquad t_j \in \mathcal{T}, \, j = \overline{1, s}. \qquad (5.6)$$

3. Measurement error generator (MEG) is intended to simulate the measurement errors in output data. This device produces the ensembles of random vectors with certain probabilistic properties. It consists of a vector randomizer (VR) with a standard PDF and a converter (C) of these vectors into the ones with the given PDFs. The standard PDF is the density of the random vectors that have the uniform distribution on the unit nonnegative m-dimensional cube. The converter transforms these vectors into the random vectors with a PDF that is defined by the machine learning of the model.

Thus, the MEG outputs the random noise vectors $\bar{\xi}(t_j) = \bar{\xi}^{(j)}$, ($j = \overline{1, s}$), where $\bar{\xi}^{(j)} \in R^m$. In the same way, describe the noise arrays that appear as the result of data accumulation using a matrix $\Xi = [\,\xi_{ij} \,|\, i = \overline{1, m}; \, j = \overline{1, s}\,]$.

The elements of the noise vectors are assumed to be independent and belonging to the intervals

$$\bar{\xi}^{(j)} \in \mathcal{K}_j = [\bar{\xi}_-^{(j)}, \bar{\xi}_+^{(j)}], \quad j = \overline{1, s},$$

$$\Xi \in \mathcal{K} = \bigotimes_{j=1}^{s} \mathcal{K}_j. \qquad (5.7)$$

It is often useful to introduce the relative nonnegative vectors $\bar{\eta}^{(j)}$ that take values within the unit nonnegative m-dimensional cube $\mathcal{E}_j = [\mathbf{0}, \mathbf{1}]$. In addition, define the auxiliary variable matrix $E = [\,\eta_{ij} \,|\, i = \overline{1, m}; \, j = \overline{1, s}\,]$. The transformation of the random vectors $\bar{\xi}^{(j)}$ into the vectors $\eta^{(j)}$ runs as follows:

$$\bar{\xi}^{(j)} = \bar{\xi}_-^{(j)} + K_j \,\bar{\eta}^{(j)}, \quad K_j = \text{diag}\,[\xi_{i,+}^{(j)} - \xi_{i,-}^{(j)}], \qquad (5.8)$$

$$i = \overline{1, m}; j = \overline{1, s}.$$

The probabilistic properties of the noises are characterized by the PDF $Q_j(\xi^{(j)})$ defined on $\mathcal{K}_j$ (5.7) or by the PDF $G_j(\bar{\eta}^{(j)})$ defined on the unit cube $\mathcal{E}$. If these density functions are specified, then the MEG produces the ensembles of random vectors that simulate

- the original noises

$$\mathcal{Q}_j[Q_j(\bar{\xi}^{(j)})] = \{\bar{\xi}^{(j)}[Q_j(\bar{\xi}^{(j)})]\}, \qquad j = \overline{1,s}; \qquad (5.9)$$

- the nonnegative relative noises

$$\mathcal{G}_j[G_j(\bar{\eta}^{(j)})] - \{\bar{\eta}^{(j)}[G_j(\bar{\eta}^{(j)})]\}, \qquad j = \overline{1,s}. \qquad (5.10)$$

4. Measurement model (MM) simulates the influence of the measurement errors on the estimation results. By assumption, this influence is described well by the additive measurement model. The ensembles of the measurement model outputs have the following form:

- for the RPM-Orig model,

$$\mathcal{V}_j[P(\mathbf{a}), Q_j(\bar{\xi}^{(j)})] = \mathcal{Y}_j[P(\mathbf{a})] + \mathcal{Q}_j[Q_j(\bar{\xi}^{(j)})], \qquad (5.11)$$
$$j = \overline{1,s};$$

- for the RPM-Rel model,

$$\mathcal{F}_j[W(\mathbf{u}), G_j(\bar{\eta}^{(j)})] = \mathcal{Z}_j[W(\mathbf{u})] + \bar{\xi}_{-}^{(j)} + K_j\,\mathcal{G}_|[G_j(\bar{\eta}^{(j)})], j = \overline{1,s}.$$

5. Moment evaluator (ME) performs online calculations for the moment characteristics of the ensembles generated by the measurement model.

For the randomized models, the ensembles consist of the following random trajectories:

- for the RPM-Orig model,

$$\mathbf{v}(t_j) = \mathbf{y}[t_j \mid P(\mathbf{a})] + \bar{\xi}^{(j)}[Q_j(\bar{\xi}^{(j)})]; \qquad (5.12)$$

- for the RPM-Rel model,

$$\mathbf{f}(t_j) = \mathbf{z}[t_j \mid W(\mathbf{u})] + \bar{\xi}_{-}^{(j)} + K_j\bar{\eta}^{(j)}[G_j(\bar{\eta}^{(j)})]. \qquad (5.13)$$

As the numerical characteristics of the ensembles $\mathcal{V}_j$ and $\mathcal{F}_j$, adopt the vectors composed of the k-means of the kth moments (see (4.111, 4.112)), i.e.,

- for the RPM-Orig model,

$$\mathbf{m}^{(k)}[P(\mathbf{a}), Q_j(\bar{\xi}^{(j)}), j] = \left\{ [M_i^{(k)}(j)]^{1/k}, \, | \, i = \overline{1,m} \right\},$$ (5.14)

$$j = \overline{1,s};$$

- for the RPM-Rel model,

$$\mathbf{n}^{(k)}[W(\mathbf{u}), G_j(\bar{\eta}^{(j)}, j)] = \left\{ [N_i^{(k)}(j)]^{1/k}, \, | \, i = \overline{1,m} \right\},$$ (5.15)

$$j = \overline{1,s}.$$

In these formulas, the kth moments of the corresponding vector elements are given by

$$M_i^{(k)}(j) = \int_{\mathcal{A}} y_i^k(t_j, \, | \, \mathbf{a}) P(\mathbf{a}) d\mathbf{a} + \int_{\Xi_j} [\xi_i^{(j)}]^k Q_j(\bar{\xi}^{(j)}) d\xi_i^{(j)},$$

$$N_i^{(k)}(j) = \int_{\mathcal{U}} z_i^k(t_j, \, | \, \mathbf{u}) W(\mathbf{u}) d\mathbf{u} + [\xi^{(j)})_-]_i^k +$$

$$+ K_{ji} \int_{\mathcal{E}} [\eta_i^{(j)}]^k G_j(\bar{\eta}^{(j)}) d\eta_i^{(j)}.$$ (5.16)

6. Balancing model (BM) is used to compare available output data $\mathbf{y}$ with the accepted characteristics of ensembles $\mathcal{V}$ or $\mathcal{F}$.

For the randomized models, the balance constraints have the form

$$\mathbf{m}^{(k)}[P(\mathbf{a}), Q_j(\bar{\xi}^{(j)}), j] = \mathbf{y}(t_j), \qquad j = \overline{1,s};$$ (5.17)

$$\mathbf{n}^{(k)}[W(\mathbf{u}), G_j(\bar{\eta}^{(j)}, j)] = \mathbf{y}(t_j), \qquad j = \overline{1,s}.$$ (5.18)

7. Machine learning algorithm (ML-A) is intended to form optimal estimates for the PDFs of the parameters and measurement noises. As repeatedly mentioned earlier, the randomized machine learning procedure has to estimate the PDFs under the maximum uncertainty characterized by information entropy. Thus, the ML-A is searching for its constrained maximum on an admissible set in the form of the balance constraints (5.17, 5.18).

The information entropy functional $\mathcal{H}$ is defined over the class of continuously differentiable PDFs $P(\mathbf{a}), Q_1(\bar{\xi}^{(1)}), \ldots, Q_s(\bar{\xi}^{(s)})$, which describe the probabilistic properties of the randomized parameters and noises simulating the measurement errors, respectively. The entropy functional generalizes the finite-dimensional concept of information entropy.

In 1865, German physicist R. Clausius introduced a physical quantity to characterize a thermodynamic system, called *entropy*.[1] According to its meaning, the entropy S gives a quantitative assessment for any changes in a thermodynamic system, whose macrostate is characterized by a variation of the quantity of heat ΔQ and a corresponding variation of the temperature ΔT. The ratio of these physical quantities defines the thermodynamic entropy

$$S = \frac{\Delta Q}{\Delta T}.$$

In 1877, L. Boltzmann established a relation between the thermodynamic entropy S (or the physical entropy E) and the statistical weight Z for the macrostate $(\Delta Q, T)$ of a thermodynamic system. Later on, using the hypothesis of equiprobable microstates, M. Planck wrote this relation in the form

$$S = k \ln Z = E.$$

Here $k = 1.38 \cdot 10^{-23}$ denotes the Boltzmann constant. The statistical weight of the macrostate $(\Delta Q, T)$ is measured by the number of all possible microstates (combinations) that make up this macrostate. Recall that these terms apply to the so-called (mono- or polyatomic) gas model, which considers a system (macrosystem) of very many indistinguishable particles (molecules). The space of the system's macrostates consists of m nonintersecting subsets with given capacities $G_1, \ldots, G_m$ (i.e., the number of states that can be occupied by the particles). Behaving stochastically, the particles are distributed in a random way over the states in these subsets, with the same probability and independently of each other. Let each particle be uniquely labeled, and list the resulting subsets for all labeled particles; in fact, this list is the system's microstate. Of course, the list under consideration is of a huge volume. The sublists with the same number of particles N_k in the kth subset can be extracted. The number of particles N_k is an element of the system's macrostate $\mathbf{N} = \{N_1, \ldots, N_m\}$. The number of elements in a sublist (the number of microstates that induce a given macrostate) is called its statistical weight $Z = Z(N_k)$.

To find a relation between the statistical weight Z and the macrostate characteristic N_k (the number of particles that fill the kth subset), we will need a certain generalization of the macrosystem model. This generalization involves information about the prior probabilities $a_1, \ldots, a_m$ with which the particles are distributed over the subsets of the macrostate space, as well as information about the properties of states in these subsets. Note that in the gas model the prior probabilities are assumed to be equal and the state properties are not described.

---

[1] Greek tropē 'transformation.'

Consider a macrosystem in which each state can be occupied by one particle only (otherwise, the state is unfilled). The numbers of such states in the subsets are $G_1, \ldots, G_m$. The distribution mechanism has the following organization. Each particle may *occupy N_k states in the kth subset with the prior probability a_k* randomly and independently of the other particles; hence, $(G_k - N_k)$ *states remain free with the prior probability* $(1 - a_k)$. Denote this event by $\{N_k, (G_k - N_k)\}$. Due to the independence of all particles in this distribution mechanism, the probability of the event makes up

$$a_k^{N_k} (1 - a_k)^{(G_k - N_k)}.$$

The number of filled states N_k can be achieved in different distributions of the particles over the kth subset, i.e., different microstates. The number of such microstates $Z(N_k)$ is the number of N_k-combinations of G_k particles:

$$C_{N_k}^{G_k} = \frac{G_k!}{N_k! (G_k - N_k)!}.$$

The event $\{N_k, (G_k - N_k)\}$ can be realized *either* by the 1st microstate, *or* by the 2nd microstate, ... , *or* by the $C_{N_k}^{G_k}$th microstate. Therefore, the event $\{N_k, (G_k - N_k)\}$ occurs with the probability

$$P_k(N_k) = \frac{G_k!}{N_k! (G_k - N_k)!} a_k^{N_k} (1 - a_k)^{(G_k - N_k)}.$$

Because all participants of the distribution mechanism are independent, the probability of the corresponding macrostate of this macrosystem is given by

$$P(\mathbf{N}) = \prod_{k=1}^{m} P_k(N_k) = \prod_{k=1}^{m} \frac{G_k!}{N_k! (G_k - N_k)!} a_k^{N_k} (1 - a_k)^{(G_k - N_k)}.$$

According to Planck's definition, physical entropy is proportional to the logarithm of the probability distribution function, since the statistical weight is an unnormalized distribution. Therefore,

$$S \sim ln P(\mathbf{N}).$$

Then it follows that

$$S \sim \sum_{k=1}^{m} \ln G_k! - \ln(G_k - N_k)! - \ln N_k! + N_k \ln a_k + (G_k - N_k) \ln(1 - a_k).$$

As the values N_k are large, employ Stirling's approximation

$$\ln N! = N(\ln N - 1).$$

In this case, the right-hand side of the expression for S can be written as the sum of a constant and a function of the macrostate:

$$S \sim C + H_F(\mathbf{N}).$$

The function

$$H_F(\mathbf{N}) = -\sum_{k=1}^{m} \left[N_k \frac{N_k}{\tilde{a}_k} + (G_k - N_k) \ln(G_k - N_k) \right], \quad \tilde{a}_k = \frac{a_k}{1 - a_k},$$

is called *the generalized Fermi information entropy*. For this function, the domain of definition is the m-dimensional parallelepiped $\mathcal{N} = [\mathbf{0}, \mathbf{G}]$, i.e., the set of admissible macrostates. If the parameters $\mathbf{a}$ and $\mathbf{G}$ satisfy the conditions

$$G_k a_k = \text{const}, \qquad N_k \ll G_k,$$

then $H_F(\mathbf{N})$ can be transformed into the entropy function

$$H_B(\mathbf{N}) = -\sum_{k=1}^{m} N_k \ln \frac{N_k}{a_k G_k e},$$

which is called *the generalized Boltzmann information entropy*. This function is defined on the nonnegative orthant $\mathcal{N} \in R_+^m$, which represents the set of admissible macrostates.

The random distribution of the indistinguishable particles over the subsets in the state space consumes r types of resources (energy, time, finances, etc.). The available quantities of these resources are $q_1, \ldots, q_r$. Resource consumption is characterized by consumption functions $\Phi_1(\mathbf{t}, \mathbf{N}), \ldots, \Phi_r(\mathbf{t}, \mathbf{N})$, which have parameters $\mathbf{t}$ (specific consumption rates) and depend on the macrostate vector $\mathbf{N}$.

As the distribution process requires resources, it is possible to realize only those macrostates that satisfy the conditions

$$\Phi_i(\mathbf{t}, \mathbf{N}) = q_i, \quad i = \overline{1, s}; \qquad \Phi_{i+s}(\mathbf{t}, \mathbf{N}) = q_{i+s}, \quad i = \overline{1, r - s}.$$

These constraints define the so-called resource set $\mathcal{D}$. The intersection of this set with the corresponding sets of admissible macrostates gives *the admissible set*

$$\mathcal{F} = \mathcal{N} \bigcap \mathcal{D}.$$

Using the above concepts and terminology, we can state *the generalized variational principle* as follows. *The macrostate that is realized in the macrosystem corresponds to the constrained maximum of the generalized information entropy*, i.e.,

$$H(\mathbf{N}) \Rightarrow \max \text{ subject to } \mathbf{N} \in \mathcal{F}.$$

5.2 ENTROPY-ROBUST ESTIMATION ALGORITHMS FOR PROBABILITY DENSITY FUNCTIONS

The estimation algorithms of the PDFs form the core of the randomized machine learning procedure. They are oriented towards the corresponding classes of the randomized parametric models, the RPM-Orig and RPM-Rel models. The two classes of randomized models have technical differences. First, some estimation algorithms for the RPM-Orig model will be considered, including the general problem statement and the structure of the resulting solutions. Then these estimation algorithms will be adapted to the RPM-Rel model.

5.2.1 Estimation algorithms for RPM-Orig model

According to (5.4, 5.12), the observable output of the RPM-Orig model at a time instant t_j is described by

$$\mathbf{v}(t_j) = \mathbf{y}[t_j \mid P(\mathbf{a})] + \bar{\xi}^{(j)}[Q_j(\bar{\xi}^{(j)})],$$
$$\mathbf{y}[t_j \mid P(\mathbf{a})] = \mathcal{B}[\mathbf{x}((t_j, \tau_l), \mid l = \overline{1, (j-1)}), \mathbf{a}], \tag{5.19}$$
$$j = \overline{1, s}.$$

Here, $\mathcal{B}$ denotes a matrix functional of dimensions $(m \times n)$ that characterizes an existing relation between the model's output $\mathbf{y}(t_j)$ and the array composed of the input vectors $\mathbf{x}(t_j, \tau_1), \dots, \mathbf{x}(t_j, \tau_{(j-1)})$. The measurement errors are simulated by the noise vectors $\bar{\xi}^{(j)}$. The RPM-Orig model output (5.2.1) depends on the randomized parameters $\mathbf{a}$ with a PDF $P(\mathbf{a})$ and also on the randomized noise vectors $\bar{\xi}^{(j)}$ with a PDF $Q_j(\bar{\xi}^{(j)})$; see (5.2.1). The available prior information about these PDFs (if any) is represented in the form of prior PDFs $P_0(\mathbf{a})$ and $Q_j^0(\bar{\xi}^{(j)})$. By assumption, the PDFs belong to the class $\mathbb{C}^1$ of all continuously differentiable functions.

For the RPM-Orig model (5.4), the entropy-robust estimation problem of the PDFs of its parameters and measurement noises is formulated as follows:

- maximize the generalized Boltzmann information entropy functional [149]

$$\mathcal{H}[P(\mathbf{a}), \bar{Q}(\xi)] = -\int_{\mathcal{A}} P(\mathbf{a}) \ln \frac{P(\mathbf{a})}{P_0(\mathbf{a})} \, d\mathbf{a} -$$
$$- \sum_{j=1}^{s} \int_{\Xi_j} Q_j(\bar{\zeta}^{(i)}) \ln \frac{Q_j(\bar{\xi}^{(j)})}{Q_j^0(\bar{\xi}^{(j)})} d\bar{\xi}^{(j)} \to \min_{P, Q_1, \dots, Q_s} \tag{5.20}$$

- subject to:

–the normalization conditions for the PDFs, i.e.,

$$\int_{\mathcal{A}} P(\mathbf{a})d\mathbf{a} = 1, \quad \int_{\Xi_j} Q_j(\bar{\xi}^{(j)})\, d\bar{\xi}^{(j)} = 1, \quad j = \overline{1, s}, \qquad (5.21)$$

and

–the balance constraints for the k-means of the output of the RPM-Orig model (5.14) and the measurements of the object's output, i.e.,

$$\mathbf{m}^{(k)}[P(\mathbf{a}), Q_j(\bar{\xi}^{(j)}), j] = \mathbf{y}(t_j), \quad j = \overline{1, s}. \qquad (5.22)$$

Due to (5.13), the vector $\mathbf{m}^{(k)}[P(\mathbf{a}), Q_j(\bar{\xi}^{(j)}), j]$ in these equality constraints has the components

$$m_i^{(k)}[j] = \left[\int_{\mathcal{A}} P(\mathbf{a}) y_i^k[t_j \mid P(\mathbf{a})]\, d\mathbf{a} \right]^{1/k} +$$

$$+ \left[\int_{\Xi_j} Q_j(\bar{\xi}^{(j)})\, [\xi_i^{(j)}]^k\, d\xi_i^{(j)} \right]^{1/k}, \qquad (5.23)$$

$$i = \overline{1, m},\, j = \overline{1, s}.$$

Consider the case in which the balance constraints involve the 1-mean of the RPM-Orig model output:

$$\left[\int_{\mathcal{A}} P(\mathbf{a}) \mathbf{y}[t_j \mid P(\mathbf{a})]\, d\mathbf{a} \right] + \left[\int_{\Xi_j} Q_j(\bar{\xi}^{(j)})\, \bar{\xi}^{(j)}\, d\bar{\xi}^{(j)} \right] = \mathbf{y}(t_j), \qquad (5.24)$$

$$j = \overline{1, s}.$$

In this optimization problem, the functional constraints extract the subset $\Omega[P(\mathbf{a}), Q(\bar{\xi})]$ of all admissible functions, where $Q(\bar{\xi}) = \{Q_1(\bar{\xi}^{(1)}), \dots, Q_s(\bar{\xi}^{(s)})\}$, from the class $\mathbb{C}^1$. In fact, the problem (5.2.1)–(5.24) is a functional entropy-linear programming problem [149]. It belongs to the class of the Lyapunov-type optimization problems [77, 174], in which the admissible set of functions is described by a system of linear integral functionals.

For the RPM-Orig model, the entropy-robust RML algorithm has the form

$$\left[P(\mathbf{a}), Q(\bar{\xi}) \right]^* = \arg \max_{[P(\mathbf{a}), Q(\bar{\xi})]} \left\{ \mathcal{H}, \mid [\mathcal{P}(\mathbf{a}), \mathcal{Q}(\bar{\xi})] \in \Omega \right\}, \qquad (5.25)$$

where the admissible set Ω is given by the normalization conditions (5.21) and the empirical balance constraints (5.24).

5.2.2 Estimation algorithms for RPM-Rel model

In the RPM-Rel model, the randomized parameters include the nonnegative relative parameters $\mathbf{u}$ and the measurement noises $\{\bar{\eta}^{(1)}, \ldots, \bar{\eta}^{(s)}\}$. According to (5.13), the observable output of the RPM-Rel model at a time instant t_j is described by

$$\mathbf{f}(t_j) = \mathbf{z}[t_j \,|\, W(\mathbf{u})] + \bar{\xi}_-^{(j)} + K_j \bar{\eta}^{(j)} [G_j(\bar{\eta}^{(j)})],$$
$$\mathbf{z}[t_j \,|\, W(\mathbf{u})] = \mathcal{D}[\mathbf{x}(t_j, \tau), \mathbf{a}^- + A\,\mathbf{u}, W(\mathbf{u}); (t_j, \tau) \in \mathcal{T}], \qquad (5.26)$$
$$j = \overline{1, s}.$$

Here, $\mathcal{D}$ denotes a matrix functional of dimensions $(m \times n)$ that characterizes an existing relation between the model's output $\mathbf{z}(t_j)$ and the array composed of the input vectors $\mathbf{x}(t_j, \tau_1), \ldots, \mathbf{x}(t_j, \tau_{(j-1)})$. The RPM-Rel model output (5.2.2) depends on the randomized relative parameters $\mathbf{u}$ with a PDF $W(\mathbf{u})$ and also on the randomized relative noise vectors $\bar{\eta}^{(j)}$ with a PDF $G_j(\bar{\eta}^{(j)})$; see (5.2.2). The available prior information about these PDFs (if any) is represented in the form of prior PDFs $W_0(\mathbf{u})$ and $G_j^0(\bar{\eta}^{(j)})$. By assumption, the PDFs belong to the class $\mathbb{C}^1$ of all continuously differentiable functions.

For the RPM-Rel model (5.5), the entropy-robust estimation problem of the PDFs of its parameters and measurement noises is formulated as follows:

- maximize the generalized Boltzmann information entropy functional

$$\mathcal{H}[W(\mathbf{u}), \bar{G}(\eta)] = -\int_{\mathcal{U}} W(\mathbf{u}) \ln \frac{W(\mathbf{u})}{W_0(\mathbf{u})}\, d\mathbf{u} -$$
$$- \sum_{j=1}^{s} \int_{\mathcal{E}_j} G_j(\bar{\eta}^{(j)}) \ln \frac{G_j(\bar{\eta}^{(j)})}{G_j^0(\bar{\eta}^{(j)})} d\bar{\eta}^{(j)} \Rightarrow \min_{W, \bar{G}} \qquad (5.27)$$

- subject to:

  −the normalization conditions for the PDF, i.e.,

$$\int_{\mathcal{U}} W(\mathbf{u}) d\mathbf{u} = 1, \qquad \int_{\mathcal{E}_j} G_j(\bar{\eta}^{(j)})\, d\bar{\eta}^{(j)} = 1, \quad j = \overline{1, s}, \qquad (5.28)$$

  and

  −the balance constraints for the k-means of the output of the RPM-Rel model (5.15) and the measurements of the object's output, i.e.,

$$\mathbf{n}^{(k)}[W(\mathbf{u}), G_j(\bar{\eta}^{(j)}), j] = \mathbf{y}(t_j), \qquad j = \overline{1, s}. \qquad (5.29)$$

Due to (5.15), the vector $\mathbf{n}^{(k)}[W(\mathbf{u}), G_j(\bar{\eta}^{(j)}), j]$ in these equality constraints has the components

$$
n_i^{(k)}[j] = \left[\int_{\mathcal{U}} W(\mathbf{u}) z_i^k[t_j \mid W(\mathbf{u})] \, d\mathbf{u} \right]^{1/k} +
$$

$$
\left[\int_{\mathcal{E}_j} G_j(\bar{\eta}^{(j)}) \, [\eta_i^{(j)}]^k \, d\eta_i^{(j)} \right]^{1/k}, \qquad (5.30)
$$

$$
i = \overline{1, m}, \; j = \overline{1, s}.
$$

Consider the case in which the balance constraints involve the 1-mean of the RPM-Rel model output:

$$
\left[\int_{\mathcal{U}} W(\mathbf{u}) \mathbf{z}[t_j \mid W(\mathbf{u})] \, d\mathbf{u} \right] + \left[\int_{\mathcal{E}_j} G_j(\bar{\eta}^{(j)}) \, \bar{\eta}^{(j)} \, d\bar{\eta}^{(j)} \right] = \mathbf{y}(t_j), \qquad (5.31)
$$

$$
j = \overline{1, s}.
$$

Like in the optimization problem presented above, the functional constraints extract the subset $\Psi[W(\mathbf{u}), G(\bar{\eta})]$ of all admissible functions, where $G(\bar{\eta}) = \{G_1(\bar{\eta}^{(1)}), \ldots, G_s(\bar{\eta}^{(s)})\}$, from the class $\mathbb{C}^1$. The problem (5.27-5.31) is also of the Lyapunov type.

For the RPM-Rel model, the entropy-robust RML algorithm has the form

$$
\left[W(\mathbf{u}), \bar{G}(\bar{\eta}) \right]^* = \arg \max_{[W(\mathbf{u}), \bar{G}(\bar{\eta})]} \{ \mathcal{H}, \mid [W(\mathbf{u}), G(\bar{\eta})] \in \Psi \}, \qquad (5.32)
$$

where the admissible set Ψ is given by the normalization conditions (5.28) and the empirical balance constraints (5.31).

5.2.3 Estimation algorithms for RPM-F model with measurement errors of input and output

The previous sections have studied cases of measurement errors figuring in the output only. However, in many applications, there are input data collections containing errors as well. Hence, the general RML procedure should be modified accordingly.

If errors occur in the input data, the observed output of the RPM-F model takes the following form:

$$
\mathbf{v}(t_j) = \mathbf{y}[t_j \mid F(\mathbf{a}, \bar{\mu}^{(j)})] + \bar{\xi}[Q_j(\bar{\xi}^{(j)})], \qquad (5.33)
$$

$$
\mathbf{y}[t_j \mid F(\mathbf{a}, \bar{\mu}^{(t_j)})] = \mathcal{B}[\mathbf{x}(t_j, \tau) + \bar{\mu}^{(t_j)}].
$$

Here $\bar{\mu}^{(j)}$ denotes the vector of input noises at a time instant t_j, which are of the interval type:

$$\bar{\mu}^{(j)} \in \mathcal{M} = [\mu^-, \mu^+] \quad \text{for all } t_j. \tag{5.34}$$

From the RPM-F model equations (5.33) it follows that the random elements are the parameters $\mathbf{a}$ and the input noise $\bar{\mu}$. Their probabilistic properties are characterized by a joint probability density function $F(\mathbf{a}, \bar{\mu}^{(t_j)})$. If there exists some information about this function, it is represented in the form of a prior PDF $F_0(\mathbf{a}, \bar{\mu}^{(t_j)})$.

For the RPM-F model (5.33), the entropy-robust estimation problem of the PDFs of its parameters and input and output measurement noises is formulated as follows:

- maximize the generalized Boltzmann information entropy functional

$$\mathcal{H}[F(\mathbf{a}, \bar{\mu}), \bar{Q}(\bar{\xi})] = -\int_{\mathcal{R}} F(\mathbf{a}, \bar{\mu}) \ln \frac{F(\mathbf{a}, \bar{\mu})}{F_0(\mathbf{a}, \bar{\mu})} \, d\mathbf{a} \, d\bar{\mu} -$$
$$- \sum_{j=1}^{s} \int_{\Xi_j} Q_j(\bar{\xi}^{(j)}) \ln \frac{Q_j(\bar{\xi}^{(j)})}{Q_j^0(\bar{\xi}^{(j)})} d\bar{\xi}^{(j)} \Rightarrow \max_{F, \bar{Q}} \tag{5.35}$$

- subject to:

—the normalization conditions for the PDFs, i.e.,

$$\int_{\mathcal{R}} F(\mathbf{a}, \bar{\mu}) \, d\mathbf{a} \, d\bar{\mu} = 1, \qquad \int_{\Xi_j} Q_j(\bar{\xi}^{(j)}) \, d\bar{\xi}^{(j)} = 1, \quad j = \overline{1, s}; \tag{5.36}$$

—the balance constraints for the k-means of the observed output (5.33) and the real data, i.e.,

$$\mathbf{m}^{(k)}[F(\mathbf{a}, \bar{\mu}), Q_j(\bar{\xi}^{(j)}), j] = \mathbf{y}(t_j), \qquad j = \overline{1, s}. \tag{5.37}$$

Due to (5.15), the vector $\mathbf{m}^{(k)}[F(\mathbf{a}, \bar{\mu}), Q_j(\bar{\xi}^{(j)}), j]$ in these equality constraints has the components

$$m_i^{(k)}[j] = \left[\int_{\mathcal{R}} F(\mathbf{a}, \bar{\mu}) \mathbf{y}[t_j \mid F(\mathbf{a}, \bar{\mu}^{(t_j)})] \, d\mathbf{a} d\bar{\mu} \right]^{1/k} +$$
$$\left[\int_{\Xi_j} Q_j(\bar{\xi}^{(j)}) \, [\xi_i^{(j)}]^k \, d\xi_i^{(j)} \right]^{1/k}, \tag{5.38}$$
$$\imath = \overline{1, m}, \ j = \overline{1, s}.$$

Consider the case in which the balance constraints involve the 1-mean of the RPM-F model output:

$$m_i^{(1)}[j] = \int_{\mathcal{R}} F(\mathbf{a}, \bar{\mu}) \mathbf{y}[t_j \mid F(\mathbf{a}, \bar{\mu}^{(t_j)})] \, da d\bar{\mu} +$$

$$\int_{\Xi_j} Q_j(\bar{\xi}^{(j)}) [\xi_i^{(j)}] d\xi_i^{(j)}, \qquad (5.39)$$

$$i = \overline{1, m}, \ j = \overline{1, s}.$$

Like in the optimization problem presented above, the functional constraints extract the subset $\Gamma[F(\mathbf{a}, \bar{\mu}), Q(\bar{\xi})]$ of all admissible functions, where $Q(\bar{\xi}) = \{Q_1(\bar{\xi}^{(1)}), \dots, Q_s(\bar{\xi}^{(s)})\}$, from the class $\mathbb{C}^1$. The problem (5.2.3—5.2.3) is also of the Lyapunov type.

For the RPM-F model with input and output data errors, the entropy-robust RML algorithm has the form

$$\left[F(\mathbf{a}, \bar{\mu}), \bar{Q}(\bar{\xi})\right]^* = \arg \max_{[F(\mathbf{a}, \bar{\mu}), \bar{Q}(\bar{\xi})]} \left\{\mathcal{H}, \mid [F(\mathbf{a}, \bar{\mu}), \bar{Q}(\bar{\xi})] \in \Gamma\right\}, \quad (5.40)$$

where the admissible set Γ is given by the normalization conditions (5.36) and the empirical balance constraints (5.2.3).

5.3 OPTIMALITY CONDITIONS FOR LYAPUNOV-TYPE PROBLEMS

A comprehensive study of the Lyapunov-type functional programming problems, including a possible application of Lagrange's method of multipliers, can be found in the monographs [77,174]. This section will briefly describe a mathematical framework based on the cited results and the Gateaux derivative [114].

A minimization problem with integral nonlinear functionals (a Lyapunov-type problem) can be written as

$$\mathcal{J}[f(\mathbf{x})] = \int_{\mathcal{X}} \Phi[f(\mathbf{x})] \, d\mathbf{x} \Rightarrow \min_{f(\mathbf{x})} \qquad (5.41)$$

subject to the constraints

$$\mathcal{G}_i[f(\mathbf{x})] = \int_{\mathcal{X}} \Psi_i[f(\mathbf{x})] \, d\mathbf{x} = 0, \qquad i = \overline{1, m}. \qquad (5.42)$$

Here $\mathbf{x} \in R^n$, the functions $\Phi[\bullet], \Psi_1[\bullet], \dots, \Psi_m[\bullet]$ and also the desired function $f(\bullet)$ that minimizes the functional $\mathcal{J}[\bullet]$ belong to the class $\mathbb{C}^1$ of all continuously differentiable functions.

Define the Lagrange functional by

$$\mathcal{L}[f(\mathbf{x})] = \int_{\mathcal{X}} \mathcal{J}[f(\mathbf{x})] - \sum_{i=1}^{m} \lambda_i \int_{\mathcal{X}} \Psi_i[f(\mathbf{x})] \, d\mathbf{x}. \qquad (5.43)$$

The first-order optimality condition claims that the variation of this functional with respect to the function $f(\mathbf{x})$ is zero, i.e.,

$$\delta\mathcal{L} = 0. \tag{5.44}$$

Because the function under variation belongs to the class $\mathbb{C}^1$, condition (5.44) is equivalent to the zero value of the Gateaux derivative of the functional $\mathcal{L}$.

Recall the definition of the Gateaux derivative. Consider a functional $\mathcal{E}[w(\mathbf{u})]$, where $w(\mathbf{u}) \in \mathbb{C}^1$ is a scalar function, $\mathbf{u} \in R^n$. Take an arbitrary function $h(\mathbf{u}) \in \mathbb{C}^1$, and introduce a linear combination $w^*(\mathbf{u}) + \alpha h(\mathbf{u})$, where α denotes a real variable and $w^*(\mathbf{u})$ is a certain fixed function from $\mathbb{C}^1$ that possesses some necessary properties (e.g., minimizes the functional $\mathcal{E}$). For a chosen function $h(\mathbf{u})$, the functional $\mathcal{E}$ becomes a function of the real variable α. The Gateaux derivative of the functional $\mathcal{E}$ is given by

$$D\mathcal{E} = \frac{d}{d\alpha}\mathcal{E}[w^*(\mathbf{u}) + \alpha h(\mathbf{u})]|_{\alpha=0}. \tag{5.45}$$

Getting back to the Lagrange functional (5.43), consider an arbitrary function $h(\mathbf{x})$ from the class $\mathbb{C}^1$ and the real variable α. Denote by $f^*(\mathbf{x})$ the function that yields minimum in the problem (5.41, 5.42). Represent an arbitrary function $f(\mathbf{x})$ as the linear combination

$$f(\mathbf{x}) = f^*(\mathbf{x}) + \alpha h(\mathbf{x}). \tag{5.46}$$

Under fixed $f^*(\mathbf{x})$ and $h(\mathbf{x})$, the Lagrange functional (5.43) turns out to be a function of the real variable α, i.e.,

$$\mathcal{L} = \int_X \Phi\left[f^*(\mathbf{x}) + \alpha h(\mathbf{x})\right]d\mathbf{x} - \sum_{i=1}^{m}\lambda_i \int_X \Psi_i\left[f^*(\mathbf{x}) + \alpha h(\mathbf{x})\right]d\mathbf{x}. \tag{5.47}$$

Then the first-order optimality condition for the Lagrange functional takes the form

$$D\mathcal{L} = \int_X h(\mathbf{x})\left(\frac{d\Phi(y)}{dy} - \sum_{i=1}^{m}\lambda_i\frac{d\Psi_i(y)}{dy}\right)d\mathbf{x} = 0, \tag{5.48}$$

where y is a scalar argument of the functions Φ and $\Psi_1, \ldots, \Psi_m$. For this integral expression to vanish, a necessary condition is that its integrand takes zero value for any function $h(\mathbf{x}) \in \mathbb{C}^1$.

Consequently, *the first-order optimality condition in the problem (5.41, 5.42)* can be written as

$$\frac{d\Phi(y)}{dy} = \sum_{i=1}^{m}\lambda_i\frac{d\Psi_i(y)}{dy}. \tag{5.49}$$

Here the derivatives of the functions $\Phi, \Psi_1, \ldots, \Psi_m$ depend on $f^*(\mathbf{x})$, the function that minimizes the functional (5.41), i.e.,

$$d\Phi(y)dy = \phi[f^*(\mathbf{x})], \qquad d\Psi_i(y)dy = \psi_i[f^*(\mathbf{x})], \quad i = \overline{1, m}. \qquad (5.50)$$

Hence, equality (5.49) can be transformed into

$$\phi[f^*(\mathbf{x})] = \sum_{i=1}^{m} \lambda_i \psi_i[f^*(\mathbf{x})]. \qquad (5.51)$$

This equality defines a parameterization of the optimal function $f^*(\mathbf{x})$ by the Lagrange multipliers $\lambda_1, \ldots, \lambda_m$. In some cases, the parameterization can be expressed in an analytical form. Particularly, let the function Φ in (5.41, 5.42) be quadratic, i.e.,

$$\Phi[f(\mathbf{x})] = a_0 + a_1 f(\mathbf{x}) + a_2 f^2(\mathbf{x}), \qquad (5.52)$$

and also let the functions $\Psi_1, \ldots, \Psi_m$ be linear, i.e.,

$$\Psi_i[f(\mathbf{x})] = b_0^i + b_1^i f(\mathbf{x}), \qquad i = \overline{1, m}, \qquad (5.53)$$

which leads to the functional quadratic programming problem. Then

$$\phi[f^*(\mathbf{x})] = a_1 + a_2 f^*(\mathbf{x}), \qquad \psi_i[f^*(\mathbf{x})] = b_1^i, \qquad (5.54)$$

and the optimality conditions (5.51) yield the following parametric representation for the optimal function f^*:

$$f^*(\mathbf{x}) = \sum_{i=1}^{m} \lambda_i b_1^i. \qquad (5.55)$$

The role of parameters is played by the Lagrange multipliers $\lambda_1, \ldots, \lambda_m$.

5.4 OPTIMALITY CONDITIONS AND STRUCTURE OF ENTROPY-OPTIMAL PROBABILITY DENSITY FUNCTIONS

The entropy-robust estimation problems (5.2.1–5.22) and (5.27–5.29) are constrained maximization problems of an integral functional in functions from the class $\mathbb{C}^1$ on an admissible set defined by integral equalities. To establish optimality conditions and perform a structural analysis of the resulting solution (optimal PDFs), we will take advantage of Lagrange's method of multipliers and the Gateaux derivatives.

5.4.1 Randomized models of the RPM-Orig class with output errors

Consider the problem (5.2.1–5.22). Define the Lagrange functional in the form

$$\mathcal{L}[P(\mathbf{a}), Q(\bar{\xi})] = \mathcal{H}[P(\mathbf{a}), Q(\bar{\xi})] + \lambda p[P(\mathbf{a})] + \langle \bar{\mu}, \mathbf{q}[Q(\bar{\xi})] \rangle +$$
$$+ \langle \bar{\theta}^{(j)}, \mathbf{1}^{(j)}[P(\mathbf{a}), Q(\bar{\xi})] \rangle, \qquad (5.56)$$

where $\lambda, \mu_1, \ldots, \mu_s$, and $\theta_1, \ldots, \theta_s$ denote the Lagrange multipliers for the corresponding constraints, i.e.,

$$p[P(\mathbf{a})] = 1 - \int_A P(\mathbf{a}) d\mathbf{a} = 0, \qquad (5.57)$$

$$\mathbf{q}[Q(\bar{\xi}] = \{q_1[Q_1(\bar{\xi}^{(1)}], \ldots, q_s[Q_s(\bar{\xi}^{(s)}]\}, \qquad (5.58)$$

$$q_j[Q_j(\bar{\xi}^{(j)}] = 1 - \int_{\div_|} Q_j(\bar{\xi}^{(j)}) d\bar{\xi}^{(j)} = 0,$$

$$\mathbf{1}^{(j)}[P(\mathbf{a}), Q(\bar{\xi})] = \mathbf{y}^{(j)} - \left[\int_A P(\mathbf{a}) \hat{\mathbf{y}}(t_j, \mathbf{a}) \, d\mathbf{a} \right] +$$
$$+ \int_{\Xi_j} Q_j(\bar{\xi}^{(j)}) \, \bar{\xi}^{(j)} \, d\bar{\xi}^{(j)} = 0, \qquad (5.59)$$

$$j = \overline{1, s}.$$

To obtain optimality conditions for the Lagrange functional, use the Gateaux derivative-based method described in Section 4.3.

Let the PDFs $P^*(\mathbf{a})$ and $Q_1^*(\bar{\xi}^{(1)}), \ldots, Q_s^*(\bar{\xi}^{(s)})$ satisfy the first-order optimality conditions for the functional (5.56). Consider arbitrary functions $h(\mathbf{a})$ and $g_1(\bar{\xi}^{(1)}), \ldots, g_s(\bar{\xi}^{(s)})$, all from the class $\mathbb{C}^1$, as well as real variables α and $\beta_1, \ldots, \beta_s$. Write the PDFs in (5.56) in the form

$$P(\mathbf{a}) = P^*(\mathbf{a}) + \alpha h(\mathbf{a}); \quad Q_j(\bar{\xi}^{(j)}) = Q_j^*(\bar{\xi}^{(j)}) + \beta_j g_j(\bar{\xi}^{(j)}), \qquad (5.60)$$
$$j = \overline{1, s}.$$

Then the Lagrange functional becomes a function of the real variables $\mathbb{W} = \{\alpha; \beta_1, \ldots, \beta_s\}$. The Gateaux derivatives in these variables are given by

$$\frac{\partial \mathcal{L}}{\partial \alpha}|_{\mathbb{W}=0} = - \int_A h(\mathbf{a})(\ln \frac{P^*(\mathbf{a})}{P_0(\mathbf{a})} + 1 + \lambda +$$
$$+ \sum_{j=1}^{s} \langle \bar{\theta}^{(j)}, \mathbf{B}^{(j)}[\mathbf{x}(t_j, \tau_h, \mid h = \overline{1, (j-1)}), \mathbf{a}] \rangle) d\mathbf{a}, \qquad (5.61)$$

$$\frac{\partial \mathcal{L}}{\partial \beta_j}|_{\mathbb{W}=0} = - \int_{\Xi_j} g_j(\bar{\xi}^{(j)}) \left[\ln \frac{Q_j^*(\bar{\xi}^{(j)})}{Q_j^0(\bar{\xi}^{(j)})} + 1 + \mu_j + \langle \bar{\theta}^{(j)}, \bar{\xi}^{(j)} \rangle \right] d\bar{\xi}^{(j)},$$
$$j = \overline{1, s}.$$

For these integrals to vanish for any functions $h(\mathbf{a}), g_1(\bar{\xi}^{(j)}), \ldots, g_s(\bar{\xi}^{(j)}) \in \mathbb{C}^1$, a necessary condition is that the bracketed expressions all take zero value, i.e.,

$$\ln \frac{P^*(\mathbf{a})}{P_0(\mathbf{a})} + 1 + \lambda + \sum_{j=1}^{s} \langle \bar{\theta}^{(j)}, \mathbf{B}^{(j)}[\mathbf{x}(\tau_h, \mid h = \overline{1, (j-1)}), \mathbf{a}] \rangle = 0,$$

$$\ln \frac{Q_j^*(\bar{\xi}^{(j)})}{Q_j^0(\bar{\xi}^{(j)})} + 1 + \mu_j + \langle \bar{\theta}^{(j)}, \bar{\xi}^{(j)} \rangle = 0, \qquad j = \overline{1, s}. \tag{5.62}$$

The solution of these equations within the Lagrange multipliers λ, $\mu_1, \ldots, \mu_s$, and $\bar{\theta}^{(1)}, \ldots, \bar{\theta}^{(s)}$ has the form

$$P^*(\mathbf{a}) = P_0(\mathbf{a}) \times$$

$$\times \exp\left[-1 - \lambda - \sum_{j=1}^{s} \langle \bar{\theta}^{(j)}, \mathbf{B}^{(j)}[\mathbf{x}((t_j, \tau_h), \mid h = \overline{1, (j-1)}), \mathbf{a}] \rangle\right],$$

$$Q_j^*(\bar{\xi}^{(j)}) = Q_j^0(\bar{\xi}^{(j)}) \exp\left[-1 - \mu_j - \langle \bar{\theta}^{(j)}, \bar{\xi}^{(j)} \rangle\right], \tag{5.63}$$

$$j = \overline{1, s}.$$

The Lagrange multipliers λ and $\mu_1, \ldots, \mu_s$ can be eliminated using the normalization conditions (5.21). Then formulas (5.4.1) yield the following entropy-optimal estimates for the PDFs of the RPM-Orig model parameters and measurement noises:

$$P^*(\mathbf{a}) = P_0(\mathbf{a}) \frac{\exp\left[-\sum_{j=1}^{s} \langle \bar{\theta}^{(j)}, \mathbf{B}^{(j)}[\mathbf{x}((t_j, \tau_h), \mid h = \overline{1, (j-1)}), \mathbf{a}] \rangle\right]}{\mathcal{P}(\bar{\theta}^{(1)}, \ldots, \bar{\theta}^{(s)})},$$

$$Q_j^*(\bar{\xi}^{(j)}) = Q_j^0(\bar{\xi}^{(j)}) \frac{\exp\left[-\langle \bar{\theta}^{(j)}, \bar{\xi}^{(j)} \rangle\right]}{\mathcal{Q}_j(\bar{\theta}^{(j)})}, \qquad j = \overline{1, s}. \tag{5.64}$$

Here

$$\mathcal{P}(\bar{\theta}^{(1)}, \ldots, \bar{\theta}^{(s)}) = \int_{\mathcal{A}} P_0(\mathbf{a}) \times$$

$$\times \exp\left[-\sum_{j=1}^{s} \langle \bar{\theta}^{(j)}, \mathbf{B}^{(j)}[\mathbf{x}((t_j, \tau_h), \mid h = \overline{1, (j-1)}), \mathbf{a}] \rangle\right] d\mathbf{a},$$

$$\mathcal{Q}_j(\bar{\theta}^{(j)}) = \int_{\Xi} Q_j^0(\bar{\xi}^{(j)}) \exp\left[-\langle \bar{\theta}^{(j)}, \bar{\xi}^{(j)} \rangle\right] d\bar{\xi}^{(j)}, \tag{5.65}$$

$$j = \overline{1, s}.$$

Clearly, the entropy-optimal PDFs are parameterized by the Lagrange multipliers and represent exponential functions. However, they may have any form, depending on the parameters $\mathbf{a}$. Figures 5.2–5.5 demonstrate

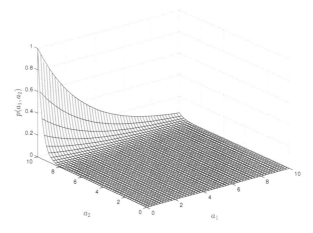

Figure 5.2: $p(a_1, a_2) = \exp(-0.3a_1 + 2.5a_2)$.

some examples of the estimated PDFs of the model parameters. For the linear RPM-Orig model, the estimated PDFs are always exponential functions; see Fig. 5.2. Note that their form is affected by the input and output measurements.

For the nonlinear RPMs, the estimates have a wider variety of structural forms. For example, the estimated PDFs for the RPM-Orig models with the quadratic, quadratic-linear and cubic nonlinearities are shown in Figs. 5.3–5.5.

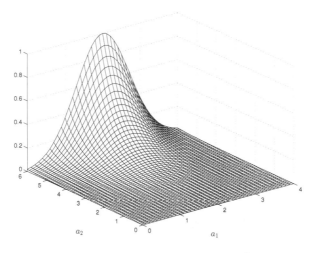

Figure 5.3: $p(a_1, a_2) = exp(4a_1 - a_1^2 + a_2)$.

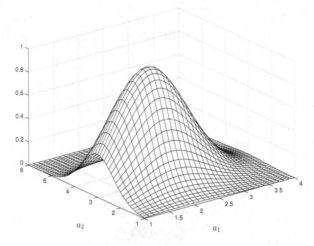

Figure 5.4: $p(a_1, a_2) = exp(4a_1 - a_1^2 + 6a_2 - a_2^2)$.

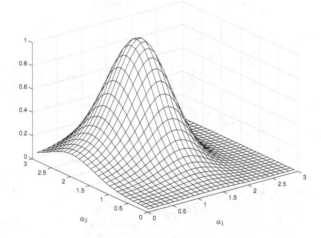

Figure 5.5: $p(a_1, a_2) = exp(0.5a_1 + 3.5a_2 + 2a_1^2 - a_1^3 - 0.2a_2^2 - 0.2a_2^3)$.

5.4.2 Randomized models of the RPM-Rel class with output errors

To construct optimality conditions for this class of the randomized models, introduce similar notations as (5.57–5.4.1), i.e.,

- the Lagrange multiplier λ for the first constraint in (5.21), writing it in the form

$$w[W(\mathbf{u})] = 1 - \int_{\mathcal{U}} W(\mathbf{u}) d\mathbf{u}; \qquad (5.66)$$

- the vector $\bar{\mu} = \{\mu_1, \ldots, \mu_s\}$ composed of the Lagrange multipliers for the second group of constraints in (5.21), writing them in the form

$$\mathbf{g}[G(\bar{\eta})] = \{g^{(1)}[G_1(\bar{\eta}^{(1)})], \ldots, g^{(s)}[G_s(\bar{\eta}^{(s)})]\},$$

$$g^{(j)}[G_j(\bar{\eta}^{(j)})] = 1 - \int_{\mathcal{E}} G_j(\bar{\eta}^{(j)}) \, d\bar{\eta}^{(j)}, \qquad (5.67)$$

$$j = \overline{1, s};$$

- the vectors $\bar{\theta}^{(j)} = \{\theta_1^{(j)}, \ldots, \theta_m^{(j)}\}$, $j = \overline{1, s}$, composed of the Lagrange multipliers for the constraints (5.22), writing them in the form

$$\mathbf{h}^{(j)}[W(\mathbf{u}), G_j(\bar{\eta}^{(j)})] = \mathbf{y}(t_j) -$$

$$- \int_{\mathcal{U}} W(\mathbf{u}) \mathbf{D}^{(j)}[\mathbf{x}((t_j, \tau_h), \mid h = \overline{1, (j-1)}), \mathbf{a}^- + A\,\mathbf{u}] \, du -$$

$$+ \int_{\mathcal{E}_j} G_j(\bar{\eta}^{(j)}) \, \bar{\eta}^{(j)} \, d\bar{\eta}^{(j)}, \quad j = \overline{1, s}. \qquad (5.68)$$

Using these notations, define the Lagrange functional

$$\mathcal{L}[W(\mathbf{u}), G(\bar{\eta})] = \mathcal{H}[W(\mathbf{u}), G(\bar{\eta})] + \lambda w[W(\mathbf{u})] + \langle \bar{\mu}, \mathbf{g}[G(\bar{\eta})] \rangle +$$

$$+ \sum_{j=1}^{s} \langle \bar{\theta}^{(j)}, \mathbf{h}^{(j)}[W(\mathbf{u}), G_j(\bar{\eta}^{(j)})] \rangle. \qquad (5.69)$$

To get optimality conditions for the Lagrange functional, we will again employ the Gateaux derivative-based method; see (5.60, 5.4.1). The corresponding first-order optimality conditions have the form

$$\ln \frac{W^*(\mathbf{u})}{W_0(\mathbf{u})} + 1 + \lambda +$$

$$+ \sum_{j=1}^{s} \langle \bar{\theta}^j, \mathbf{D}^{(j)}[\mathbf{x}((t_j, \tau_h), \mid h = \overline{1, (j-1)}), \mathbf{a}^- + A\,\mathbf{u}] \rangle = 0,$$

$$\ln \frac{G_j^*(\bar{\eta}^{(j)})}{G_j^0(\bar{\eta}^{(j)})} + 1 + \mu_j + \langle \bar{\theta}^{(j)}, \bar{\eta}^{(j)} \rangle = 0, \qquad (5.70)$$

$$j = \overline{1, s}.$$

The solution of these equations within the Lagrange multipliers λ, $\mu_1, \ldots, \mu_s$, and $\bar{\theta}^{(1)}, \ldots, \bar{\theta}^{(s)}$ is given by

$$W^*(\mathbf{u}) = W_0(\mathbf{u}) \exp[-1 - \lambda -$$

$$\sum_{j=1}^{s} \langle \bar{\theta}^j, \mathbf{D}^{(j)}[\mathbf{x}((t_j, \tau_h), \mid r = \overline{1, (j-1)}), \mathbf{a}^- + A\,\mathbf{u}] \rangle],$$

$$G_j^*(\bar{\eta}^{(j)}) = G_j^0(\bar{\eta}^{(j)}) \exp\left[-1 - \mu_j - \langle \bar{\theta}^{(j)}, \bar{\eta}^{(j)} \rangle\right], \qquad (5.71)$$

$$j = \overline{1, s}.$$

The Lagrange multipliers λ and $\mu_1, \ldots, \mu_s$ can be eliminated using the normalization conditions (5.28). Then formulas (5.71) yield the following entropy-optimal estimates for the PDFs of the RPM-Rel model parameters and measurement noises:

$$W^*(\mathbf{u}) = W_0(\mathbf{u}) \times$$

$$\times \frac{\exp\left[-\sum_{j=1}^s \langle \bar{\theta}^j, \mathbf{D}^{(j)}[\mathbf{x}((t_j, \tau_r), \,|\, r = \overline{1, (j-1)}), \mathbf{a}^- + \Delta\,\mathbf{u}]\rangle\right]}{\mathcal{W}(\bar{\theta}^{(1)}, \ldots, \bar{\theta}^{(s)})},$$

$$G_j^*(\bar{\eta}^{(j)}) = G_j^0(\bar{\eta}^{(j)}) \frac{\exp\left[-\sum_{j=1}^s \langle \bar{\theta}^{(j)}, \bar{\eta}^{(j)} \rangle\right]}{\mathcal{G}_j(\bar{\theta}^{(j)})}, \qquad (5.72)$$

$$j = \overline{1, s},$$

where

$$\mathcal{W}(\bar{\theta}^{(1)}, \ldots, \bar{\theta}^{(s)}) = \int_{\mathcal{U}} W_0(\mathbf{u}) \times$$

$$\times \exp\left[-\sum_{j=1}^s \langle \bar{\theta}^{(j)}, \mathbf{D}^{(j)}[\mathbf{x}((t_j, \tau_r), \,|\, r = \overline{1, (j-1)}), \mathbf{a}^- + \Delta\,\mathbf{u}]\rangle\right] d\mathbf{u},$$

$$\mathcal{G}_j(\bar{\theta}^{(j)}) = \int_{\mathcal{E}} G_j^0(\bar{\eta}^{(j)}) \exp\left[-\langle \bar{\theta}^{(j)}, \bar{\eta}^{(j)} \rangle\right] d\bar{\eta}^{(j)}, \qquad (5.73)$$

$$j = \overline{1, s}.$$

Like for the randomized models of the RPM-Orig class (see (5.4.1)), the entropy-optimal PDFs are parameterized by the Lagrange multipliers and represent exponential functions.

5.4.3 Randomized models of the RPM-Orig class with input and output errors

Consider the problem (5.2.3, 5.36, 5.2.3) and write out the Lagrange functional:

$$\mathcal{L}[F(\mathbf{a}, \mu), \bar{Q}(\bar{\xi})] = \mathcal{H}[F(\mathbf{a}, \mu), \bar{Q}(\bar{\xi})] + \lambda w[F(\mathbf{a}, \mu)] + \langle \bar{\eta}, \mathbf{q}[\bar{Q}(\bar{\xi})] \rangle +$$

$$+ \langle \bar{\theta}^{(j)}, \mathbf{c}^{(j)}[F(\mathbf{a}, \mu), \bar{Q}(\bar{\xi})] \rangle, \qquad (5.74)$$

where $\lambda; \bar{\eta}; \bar{\theta}$—Lagrange multipliers for corresponding constraints:

$$w[P(\mathbf{a})] = 1 - \int_{\mathcal{R}} F(\mathbf{a}, \mu) d\mathbf{a} d\mu = 0, \qquad (5.75)$$

$$\mathbf{q}[\bar{Q}(\bar{\xi})] = \{q_1[Q_1(\bar{\xi}^{(1)}], \ldots, q_s[Q_s(\bar{\xi}^{(s)}]\}, \qquad (5.76)$$

$$q_j[Q_j(\bar{\xi}^{(j)}] = 1 - \int_{\div_{\mid}} Q_j(\bar{\xi}^{(j)})d\bar{\xi}^{(j)} = 0.$$

$$\mathbf{c}^{(j)}[F(\mathbf{a},\mu),\bar{Q}(\bar{\xi})] = \mathbf{y}^{(j)} - \left[\int_{\mathcal{R}} F(\mathbf{a},\mu)\hat{\mathbf{y}}(t_j,\mathbf{a},\mu)\,d\mathbf{a}\,d\mu\right] +$$

$$+ \int_{\Xi_j} Q_j(\bar{\xi}^{(j)})\,\bar{\xi}^{(j)}\,d\bar{\xi}^{(j)} = 0, \qquad (5.77)$$

$$j = \overline{1,s}.$$

To get optimality conditions for the Lagrange functional, we will again employ the Gâteaux derivative-based method. The corresponding first-order optimality conditions have the form

$$F^*(\mathbf{a},\mu) = F_0(\mathbf{a},\mu)\times$$

$$\times \frac{\exp[-\sum_{j=1}^{s}\langle\bar{\theta}^{(j)},\mathbf{B}^{(j)}[\mathbf{x}((t_j,\tau_h)+\mu(t_j,\tau_h),\,|\,h = \overline{1,(j-1)}),\mathbf{a}]\rangle]}{\mathcal{F}(\bar{\theta}^{(1)},\ldots,\bar{\theta}^{(s)})},$$

$$Q_j^*(\bar{\xi}^{(j)}) = Q_j^0(\bar{\xi}^{(j)})\frac{\exp[-\langle\bar{\theta}^{(j)},\bar{\xi}^{(j)}\rangle]}{\mathcal{Q}_j(\bar{\theta}^{(j)})}, \quad j = \overline{1,s}. \qquad (5.78)$$

where

$$\mathcal{F}(\bar{\theta}^{(1)},\ldots,\bar{\theta}^{(s)}) = \int_{\mathcal{R}} F_0(\mathbf{a},\mu)\times$$

$$\exp\left[-\sum_{j=1}^{s}\langle\bar{\theta}^{(j)},\mathbf{B}^{(j)}[\mathbf{x}((t_j,\tau_h)+\mu(t_j,\tau_h),\,|\,h = \overline{1,(j-1)}),\mathbf{a}]\rangle\right]\,d\mathbf{a},$$

$$\mathcal{Q}_j(\bar{\theta}^{(j)}) = \int_{\Xi} Q_j^0(\bar{\xi}^{(j)})\exp\left[-\langle\bar{\theta}^{(j)},\bar{\xi}^{(j)}\rangle\right]d\bar{\xi}^{(j)}, \qquad (5.79)$$

$$j = \overline{1,s}.$$

The obtained PDF functions are parameterized by the Lagrange multipliers $\bar{\theta}$, which are determined from the balance equations.

5.5 EQUATIONS FOR LAGRANGE MULTIPLIERS

All entropy-optimal PDFs belong to the class of exponential functions parameterized by the Lagrange multipliers $\bar{\theta}^1,\ldots,\bar{\theta}^s$; see the formulas derived in the previous section. For calculating the Lagrange multipliers, we have to solve the equations that are obtained from the balance constraints (5.24, 5.31) with the entropy-optimal PDFs (5.64, 5.4.2).

1. RPM-Orig model with output errors

To form the corresponding equations, introduce the following auxiliary notations.

- the mean output of the model with the entropy-optimal randomized parameters

$$\mathbb{Y}_j(\bar{\theta}) = \mathcal{P}^{(-1)}(\bar{\theta}) \int_{\mathcal{A}} P_0(\mathbf{a}) \times$$

$$\times \exp\left[-\sum_{j=1}^{s} \langle \bar{\theta}^{(j)}, \mathcal{B}[\mathbf{x}((t_j, \tau_h) \,|\, h = \overline{1, (j-1)}), \mathbf{a}] \rangle\right] \times$$

$$\times \mathcal{B}[\mathbf{x}((t_j, \tau_h) \,|\, h = \overline{1, (j-1)}), \mathbf{a}] d\mathbf{a}; \qquad (5.80)$$

- the mean entropy-"worst" measurement noises

$$\mathbb{Q}_j(\bar{\theta}^{(j)}) = \mathcal{Q}_j^{(-1)}(\bar{\theta}^{(j)}) \int_{\Xi_j} Q_j^0(\bar{\xi}^{(j)}) \exp\left[-\langle \bar{\theta}^{(j)}, \bar{\xi}^{(j)} \rangle\right] d\bar{\xi}^{(j)}, \quad (5.81)$$

$$j = \overline{1, s}.$$

In these expressions, $\bar{\theta} = [\bar{\theta}^{(1)}, \ldots, \bar{\theta}^{(s)}]$.

Using the balance constraints (5.24), write the equations of the Lagrange multipliers in the form

$$\mathbb{Y}_j(\bar{\theta}) + \mathbb{Q}_j(\bar{\theta}^{(j)}) = \mathbf{y}^{(j)}, \qquad j = \overline{1, s}. \qquad (5.82)$$

2. RPM-Rel model with output errors

By analogy with the previous case, introduce the following auxiliary notations:

- the mean output of the model with the entropy-optimal randomized parameters

$$\mathbb{Z}_j(\bar{\theta}) = \mathcal{W}^{(-1)}(\bar{\theta}) \int_{\mathcal{U}} W_0(\mathbf{u}) \times$$

$$\times \exp[-\sum_{j=1}^{s} \langle \bar{\theta}^{(j)}, \mathcal{D}[\mathbf{x}((t_j, \tau_h) \,|\, h = \overline{1, (j-1)}), \mathbf{a}^- +$$

$$A\,\mathbf{u}] \rangle] \times$$

$$\times \mathcal{D}[\mathbf{x}((t_j, \tau_h) \,|\, h = \overline{1, (j-1)}), \mathbf{a}^- + A\,\mathbf{u}] d\mathbf{u}; \qquad (5.83)$$

- the mean entropy-"worst" measurement noises

$$\mathbb{G}_j(\bar{\theta}^{(j)}) = \mathcal{G}_j^{(-1)}(\bar{\theta}^{(j)}) \int_{\mathcal{E}} G_j^0(\bar{\eta}^{(j)}) \exp\left[-\langle\bar{\theta}^{(j)}, \bar{\eta}^{(j)}\rangle\right] d\bar{\eta}^{(j)}, \quad (5.84)$$

$$j = \overline{1, s}.$$

Using the balance constraints (5.31), write the equations of the Lagrange multipliers in the form

$$\mathbb{Z}_j(\bar{\theta}) + \mathbb{G}_j(\bar{\theta}^{(j)}) = \mathbf{y}^{(j)}, \qquad j = \overline{1, s}. \tag{5.85}$$

3. RPM-Orig model with input and output errors

To form the corresponding equations, introduce the following auxiliary notations.

- the mean output of the model with the entropy-optimal randomized parameters

$$\mathbb{Y}_j(\bar{\theta}) = \mathcal{F}^{(-1)}(\bar{\theta}) \int_{\mathcal{R}} F_0(\mathbf{a}, \mu) \times$$

$$\times \exp\left[-\sum_{j=1}^{s}\langle\bar{\theta}^{(j)}, \mathcal{B}[\mathbf{x}((t_j, \tau_h) + \mu(t_j, \tau_h) \,|\, h = \overline{1, (j-1)}), \mathbf{a}]\rangle\right] \times$$

$$\times \mathcal{B}[\mathbf{x}((t_j, \tau_h) + \mu(t_j, \tau_h) \,|\, h = \overline{1, (j-1)}), \mathbf{a}] d\mathbf{a} d\mu. \tag{5.86}$$

- the mean entropy-"worst" measurement noises

$$\mathbb{Q}_j(\bar{\theta}^{(j)}) = \mathcal{Q}_j^{(-1)}(\bar{\theta}^{(j)}) \int_{\Xi_j} Q_j^0(\bar{\xi}^{(j)}) \exp\left[-\langle\bar{\theta}^{(j)}, \bar{\xi}^{(j)}\rangle\right] d\bar{\xi}^{(j)}, \quad (5.87)$$

$$j = \overline{1, s}.$$

Then, write the equations of the Lagrange multipliers in the form

$$\mathbb{Y}_j(\bar{\theta}) + \mathbb{Q}_j(\bar{\theta}^{(j)}) = \mathbf{y}^{(j)}, \qquad j = \overline{1, s}. \tag{5.88}$$

As a matter of fact, equations (5.82, 5.85, 5.88) possess some specifics. They incorporate functions that depend on the Lagrange multipliers; in turn, the latter are calculated by integration over the multidimensional domains of definition of the parameters and measurement noises. To be fair, these domains have a simple structure, representing multidimensional parallelepipeds. In the sequel, they will be called *the parameterized integral components (PICs)* in order to underline their specifics. PICs calculation and solution of the balance equations are separate and nontrivial problems. Some approaches to solve these problems will be considered in detail in Chapter 6.

ENTROPY-ROBUST ESTIMATION METHODS FOR PROBABILITIES OF BELONGING IN MACHINE LEARNING PROCEDURES

The constrained maximization of entropy functionals represents an efficient tool for solving various problems, including the estimation of the probabilities of belonging to certain intervals for the model parameters and measurement noises. The original model parameters and measurement noises of the interval type are transformed into auxiliary nonnegative variables lying within the unit interval, which are treated as the probabilities of belonging [67]. Such a probabilistic interpretation of the auxiliary parameters and measurement noises justifies the applicability of entropy functionals to find best values of the auxiliary and original variables. The estimation algorithms for the probabilities of belonging are machine learning algorithms. The estimation problem is formulated

DOI: 10.1201/9781003306566-6

as an entropy-linear programming problem with an admissible set that is constructed using available learning data arrays.

6.1 ENTROPY-ROBUST ESTIMATION ALGORITHMS FOR PROBABILITIES OF BELONGING

An important class of the parameterized models used in ML procedures consists of the models with the interval-type *quasi randomized* parameters; their values are defined by the probabilities of belonging to corresponding intervals. Denote this class of models by *RPM-QuaRand* and their parameters by $\mathbf{c}$. These models have deterministic parameters with unknown values. Hence, the RPM-QuaRand models can be considered as randomized models with the PDFs $R(\mathbf{c}) = R\delta(\mathbf{c}^* - \mathbf{c})$.

The parameters $\mathbf{c}$ are of the interval type, i.e.,

$$\mathbf{c} \in \mathcal{C} = \bigotimes_{i=1}^{n} \mathcal{C}_i = [c_i^-, c_i^+]. \tag{6.1}$$

Introduce the probability p_i that the parameter c_i takes some value within the interval $\mathcal{C}_i$. The values p_i will be called *the probabilities of belonging (PO)*. Then a parameter c_i can be written as

$$c_i = c_i^- + (c_i^+ - c_i^-)\, p_i, \qquad \mathbf{c} = \mathbf{c}^- + C\,\mathbf{p}, \quad C = \mathrm{diag}[(c_i^+ - c_i^-) \,|\, i = \overline{1,n}]. \tag{6.2}$$

The probabilities of belonging are

$$\mathbf{p} \in \mathcal{I} = [\mathbf{0} \le \mathbf{p} \le \mathbf{1}]. \tag{6.3}$$

Some statements also involve an additional constraint known as the complete set condition (the normalization condition), i.e.,

$$\sum_{i=1}^{n} p_i = 1. \tag{6.4}$$

The definition of the probabilities of belonging and their properties (6.3, 6.4) formally match the definition of the relative parameters (4.96, 4.97). However, there exist some differences of principle between them. The relative parameters $\mathbf{u}$ are the randomized variables characterized by a given PDF. In contrast, the probabilities of belonging $\mathbf{p}$ are nonrandom variables taking values within the unit nonnegative cube.

A dynamic model of the RPM-QuaRand class generates a unique trajectory, that is, described by the following equality at discrete times $j = \overline{0, s} = \mathcal{T}_s$:

$$\begin{aligned} \mathbf{s}[j, \mathbf{p}] &= \mathcal{S}[\mathbf{x}(j, \tau), \mathbf{c}^- + C\mathbf{p}; (j, \tau) \in \mathcal{T}_s], \\ \mathbf{c} &= \mathbf{c}^- + C\mathbf{p}. \end{aligned} \qquad (6.5)$$

Here $\mathcal{S}$ means an integral functional.

This problem has an additive model of measurements: in each jth measurement, the noises are characterized by an interval-type vector

$$\bar{\zeta}^{(j)} \in \mathcal{Z}^{(j)} = \bigotimes_{i=1}^{m} \mathcal{Z}_i^{(j)}, \qquad \mathcal{Z}_i^{(j)} = [\zeta_-^{(j)}(i), \zeta_+^{(j)}(i)], \quad j = \overline{1, s}. \qquad (6.6)$$

Introduce the probability of belonging $q_i^{(j)}$ that the element $\zeta_i^{(j)}$ of the noise vector takes some value from the interval $\mathcal{Z}_i^{(j)}$. The probabilities $q_i^{(j)}$ lie within the unit nonnegative interval and the probability vector is

$$\mathbf{q}^{(j)} \in \mathcal{I} = [\mathbf{0} \le \mathbf{q}^{(j)} \le \mathbf{1}], \qquad j = \overline{1, s}. \qquad (6.7)$$

In some cases, like for the probabilities of belonging of the model parameters, it is necessary to consider the normalization conditions

$$\sum_{i=1}^{m} q_i^{(j)} = 1, \qquad j = \overline{1, s}. \qquad (6.8)$$

The noise vector can be written as

$$\bar{\zeta}^{(j)} = \bar{\zeta}^{(j)}(i) + L_j \mathbf{q}^{(j)}, \qquad L_j = \operatorname{diag}[\zeta_+^{(j)}(i) - \zeta_-^{(j)}(i), \, | \, i = \overline{1, m}]. \qquad (6.9)$$

The measurable output of the RPM-QuaRand model (6.5) is a unique trajectory $\mathbf{w}[j, \mathbf{p}, \mathbf{q}]$ composed of an additive mix of the vectors $\mathbf{s}[j, \mathbf{p}]$ (6.5) and $\bar{\zeta}^{(j)}$ (6.9), i.e.,

$$\mathbf{w}[j, (\mathbf{p}, \mathbf{q}^{(j)}] = \mathbf{s}[j, \mathbf{p}] + \bar{\zeta}^{(j)} - +L_j \mathbf{q}^{(j)}, \qquad j = \overline{1, s}. \qquad (6.10)$$

Therefore, as its variables the RPM-QuaRand model incorporates the probabilities of belonging $\mathbf{p}, \mathbf{q}^j, j = \overline{1, s}$.

6.1.1 ML algorithm for RPM-QuaRand model with normalized probabilities of belonging

Consider the case in which the vectors $\mathbf{p}$ and $\mathbf{q}^{(1)}, \ldots, \mathbf{q}^{(s)}$ are normalized and there exists information about the prior probabilities of belonging, $\mathbf{p}_0$ and $\mathbf{q}_0^{(1)}, \ldots, \mathbf{q}_0^{(s)}$.

Throughout the book, it has been repeatedly mentioned that the ML procedures operate under uncertainty measured in terms of entropy. Therefore, we will characterize the quality of estimation using the entropy approach. Since the measurement noises are additive, let us adopt the generalized Boltzmann information entropy [141]

$$H_B(\mathbf{p}, Q) = -\langle \mathbf{p}, \ln\frac{\mathbf{p}}{\mathbf{p}_0}\rangle - \sum_{j=1}^{s}\langle \mathbf{q}^{(j)}, \ln\frac{\mathbf{q}^{(j)}}{\mathbf{q}_0^{(j)}}\rangle, Q = [\mathbf{q}^{(1)}, \ldots, \mathbf{q}^{(s)}]. \quad (6.11)$$

Here $\ln(\bullet)$ denotes a vector with corresponding elements $\bullet$. Recall that the variables in this expression are nonnegative. Then *the entropy-robust estimation problem for the RPM-QuaRand model with the normalized probabilities of belonging* can be written as follows:

- maximize the entropy function (6.11)

$$H_B(\mathbf{p}, Q) \Rightarrow \max_{\mathbf{p}, Q} \quad (6.12)$$

- subject to

    –the normalization conditions

$$\langle \mathbf{p}, \mathbf{1}\rangle = 1, \quad \langle \mathbf{q}^{(j)}, \mathbf{1}\rangle = 1, \quad j = \overline{1, s}, \quad (6.13)$$

    and

    –the balance constraints

$$\mathcal{S}[\mathbf{x}[j, \tau], \mathbf{c}^- + C\,\mathbf{p}; (j, \tau) \in \mathcal{T}_s] + \bar{\zeta}^{(j)} + L_j\,\mathbf{q}^{(j)} = \mathbf{y}[j], \quad j = \overline{1, s}. \quad (6.14)$$

The constraints (6.13, 6.14) extract an admissible set Υ from the unit nonnegative $(m + r)$-dimensional cube $\mathcal{I}$.

In the general case, this problem is an entropy-nonlinear programming problem. Its nonlinear properties follow from the balance constraints (6.14), which contain a nonlinear functional $\mathcal{S}$ of the parameters. Note that the randomized approach to estimation leads to the optimization problems of entropy-linear functional programming.

Consequently, *the entropy-robust ML algorithm for the RPM-QuaRand model with the normalized probabilities of belonging* has the form

$$(\mathbf{p}, Q)^* = \arg\max_{(\mathbf{p}, Q)}[H_B, \mid (\mathbf{p}, Q) \in \Upsilon]. \quad (6.15)$$

6.1.2 ML algorithm for RPM-QuaRand model with interval-type probabilities of belonging

For the models of this class, the probabilities of belonging take values within the nonnegative r-dimensional cube—the normalization condition is not imposed. The estimation problems arising for these RPM-QuaRand models have two modifications depending on entropy choice, namely, the statements with the Boltzmann or Fermi entropies [149].

When using the generalized Boltzmann information entropy (6.11), *the Boltzmann entropy-robust estimation problem for the RPM-QuaRand model with the interval-type probabilities of belonging* can be written as follows:

- maximize the entropy function (6.11)

$$H_B(\mathbf{p}, Q) \Rightarrow \max_{\mathbf{p}, Q} \tag{6.16}$$

- subject to:

  –the interval-type conditions

$$\mathbf{p} \in \mathcal{I}, \qquad \mathbf{q}^{(j)} \in \mathcal{I}, \quad j = \overline{1, s}, \tag{6.17}$$

  and

  –the balance constraints

$$\mathcal{S}[\mathbf{x}[j, \tau], \mathbf{c}^- + C\,\mathbf{p}; (j, \tau) \in \mathcal{T}_s] + \bar{\zeta}_-^{(j)} + L_j\,\mathbf{q}^{(j)} = \mathbf{y}^{(j)}, \qquad j = \overline{1, s}. \tag{6.18}$$

The constraints (6.17, 6.18) extract an admissible set Γ from the unit nonnegative r-dimensional cube $\mathcal{I}$.

The Boltzmann entropy-robust ML algorithm for the RPM-QuaRand model with the interval-type probabilities of belonging has the form

$$(\mathbf{p}, Q)^* = \arg\max_{(\mathbf{p}, Q)}[H_B, \,|\,(\mathbf{p}, Q) \in \Gamma]. \tag{6.19}$$

In some applications, it is necessary to choose the generalized Fermi information entropy in order to eliminate the interval-type conditions:

$$H_F(\mathbf{p}, Q) = -\langle \mathbf{p}, \ln\frac{\mathbf{p}}{\mathbf{p}_0}\rangle - \langle (1 - \mathbf{p}), \ln(1 - \mathbf{p})\rangle -$$

$$- \sum_{j=1}^{s}\left(\langle \mathbf{q}^{(j)}, \ln\frac{\mathbf{q}^{(j)}}{\mathbf{q}_0^{(j)}}\rangle + \langle (1 - \mathbf{q}^{(j)}), \ln(1 - \mathbf{q}^{(j)})\rangle\right), \tag{6.20}$$

where $Q = [\mathbf{q}^{(1)}, \ldots, \mathbf{q}^{(s)}]$.

Clearly, the definition of this function explicitly considers the interval property of the variables. The admissible set is characterized by the balance constraints only.

The Fermi entropy-robust estimation problem for the RPM-QuaRand model with the interval-type probabilities of belonging can be written as follows:

- maximize the entropy function (6.1.2)

$$H_F(\mathbf{p}, Q) \Rightarrow \max_{\mathbf{p}, Q} \qquad (6.21)$$

- subject to the balance constraints

$$\mathcal{S}[\mathbf{x}[j, \tau], \mathbf{c}^- + C\mathbf{p}; (j, \tau) \in \mathcal{T}_s] + \bar{\zeta}_-^{(j)} + L_j \mathbf{q}^{(j)} = \mathbf{y}[j], \qquad j = \overline{1, s}. \qquad (6.22)$$

The constraints (6.22) extract an admissible set Θ from the unit non-negative n-dimensional cube $\mathcal{I}$.

The Fermi entropy-robust ML algorithm for the RPM-QuaRand model with the interval-type probabilities of belonging has the form

$$(\mathbf{p}, Q)^* = \arg \max_{(\mathbf{p}, Q)} [H_F, \,|\, (\mathbf{p}, Q) \in \Theta], \qquad (6.23)$$

where Θ is the set defined by the balance constraints. Note that the problem (6.21, 6.22) contains $(r + ms)$ less constraints than the problem (6.16–6.18). Due to this feature, the computational complexity of the estimation problem (6.21, 6.22) is smaller in comparison with its counterpart (6.16–6.18).

6.2 FUNCTIONAL DESCRIPTION OF DYNAMIC RPM-QUARAND MODELS

As established in the previous section, the estimation problems for the probabilities of belonging generally represent finite-dimensional nonlinear programming problems with entropy objective functions. The nonlinearity of these problems is caused by the nonlinearity of the functional $\mathcal{S}$ that characterizes the model (6.5). Consider the models with one input and one output that are measured at discrete times from the interval $\mathcal{T}$.

1. Dynamic models with unstructured power nonlinearities
A functional description of such a model has the following form

$$s[j] = \sum_{h=1}^{R} \sum_{(n_1,\ldots,n_h)=0}^{K_h} w_h[n_1,\ldots,n_h]x[j-n_1]\cdots x[j-n_h], \qquad (6.24)$$

$$j \geq maxK_h.$$

(Also, see Fig. 4.8.)

Its dynamic and nonlinear properties are reflected by the multidimensional pulse characteristics $w_h[n_1,\ldots,n_h]$, $h = \overline{1,R}$, with a finite "memory" K_h, i.e.,

$$w_h[n_1,\ldots,n_h] \neq 0, \quad \text{for } 0 \leq (n_1,\ldots,n_h) \leq K_h;$$
$$w_h[n_1,\ldots,n_h] = 0, \quad \text{for } (n_1,\ldots,n_h) > K_h. \qquad (6.25)$$

As the model parameters consider the ordinates of its pulse characteristics. A lexicographical ordering of the variables yields

$$(n_1,\ldots,n_h) \to v, \qquad v = \overline{1,V_h}, \ V_h = (K_h+1)\,h, \qquad (6.26)$$

Define a vector $\mathbf{c}^{(h)} = \{c_1^{(h)},\ldots,c_{V_h}^{(h)}\}$ composed of the ordinates of the pulse characteristic $w_h[n_1,\ldots,n_h]$. Perform a similar operation with the products $x[j-n_1]\cdots x[j-n_h]$. Introduce another vector

$$\mathbf{X}^{(h)}[j] = \{x_1^{(h)}[j],\ldots,x_{V_h}^{(h)}[j]\} \qquad (6.27)$$

composed of the lexicographically ordered products $x[j-n_1]\cdots x[j-n_h]$. Then the model equation (6.1.2) can be transformed into

$$s[j] = \sum_{h=1}^{R} \langle \mathbf{c}^{(h)}, \mathbf{X}^{(h)}[j]\rangle. \qquad (6.28)$$

As the result of all these manipulations, the original nonlinear model (6.1.2) has been reduced to its linear analog (6.28) of considerably higher dimension.

Using the probabilities of belonging $\mathbf{p}$ from Section 6.3, let us write the parameter vector $\mathbf{c}^{(h)}$ as

$$\mathbf{c}^{(h)} = \mathbf{c}_{-}^{(h)} + C^{(h)}\mathbf{p}^{(h)}, \ C_h = \text{diag}\,[(c_i^{(h,+)} - c_i^{(h,-)})\,|\,i = \overline{1,V_h}], \qquad (6.29)$$
$$h = \overline{1,R}.$$

Substituting (6.29) into (6.28) gives the equation

$$s[j] = \hat{s}[j] + \sum_{h=1}^{R} \langle C_h \, \mathbf{p}^{(h)}, \mathbf{X}^{(h)}[j] \rangle,$$

$$.\hat{s}[j] = \sum_{h=1}^{R} \langle \mathbf{c}_{-}^{(h)}, \mathbf{X}^{(h)}[j] \rangle. \tag{6.30}$$

2. Dynamic models with structured power nonlinearities

For this class of dynamic models, a functional description has the following form:

$$s[j] = \sum_{h=1}^{R} \sum_{n=0}^{K_h} (w_h[n] x[j-n])^h, \qquad j > K. \tag{6.31}$$

(Also, see Fig. 4.11a.) Here the dynamic properties are reflected by the pulse characteristics $w_h[n]$, $h = \overline{1, R}$, with a finite "memory" K_h, i.e.,

$$w_h[n] \neq 0, \quad \text{for } 0 \leq n \leq K_h; \qquad r_h[n] = 0, \quad \text{for } n > K_h. \tag{6.32}$$

As the model parameters consider the ordinates of the pulse characteristics. In other words, the parameters make the vector

$$\mathbf{c}^{(h)} = \{c_1^{(h)}, \ldots, c_{K_h}^{(h)}\}, \tag{6.33}$$

where

$$c_n^{(h)} = w_h[n], \qquad n = \overline{0, K_h}. \tag{6.34}$$

Using the probabilities of belonging $\mathbf{p}$, write the parameter vector $\mathbf{c}^{(h)}$ in the form (6.29). Substituting these equalities into (6.31) yields the model equation

$$s[j] = \sum_{h=1}^{R} \left(\langle \mathbf{c}_{-}^{(h)}, \mathbf{x}[j] \rangle + \langle C_h \, \mathbf{p}^{(h)}, \mathbf{x}[j] \rangle \right)^h, \tag{6.35}$$

where $\mathbf{x}[j] = \{x[j], \ldots, x[j - K_h]\}$.

Thus, the relation between the model's output and parameters is essentially nonlinear (more specifically, a power function).

3. Dynamic models with structured power nonlinearities

For this class of dynamic models, a functional description has the following form:

$$s[j] = \sum_{h=1}^{R} \sum_{n=0}^{K_h} w_h[n] x^h[j-n], \qquad j > K. \tag{6.36}$$

(Also, see Fig. 4.11b.) Its dynamic properties are described by the pulse characteristics $w_h[n]$, $h = \overline{1, R}$, with a finite "memory" K_h (6.32). As the parameters of this model adopt the ordinates of its pulse characteristics, i.e., the parameter vector is $\mathbf{c}^{(h)}$ (6.34). Define a vector $\tilde{\mathbf{X}}^h[j]$ composed of $x^h[j], \ldots, x^h[j - K_h]$. Then the model equation (6.36) can be transformed into

$$s[j] = \sum_{h=1}^{R} \langle \mathbf{c}, \tilde{\mathbf{X}}^{(h)}[j] \rangle, \qquad j > K. \tag{6.37}$$

Using the expression (6.29) in this equality, write

$$
\begin{aligned}
s[j] &= \hat{s}[j] + \sum_{h=1}^{R} \langle C_h \, \mathbf{p}^{(h)}, \tilde{\mathbf{X}}^{(h)}[j] \rangle, \\
\hat{s}[j] &= \sum_{h=1}^{R} \langle \mathbf{c}_{-}^{(h)}, \tilde{\mathbf{X}}^{(h)}[j] \rangle.
\end{aligned}
\tag{6.38}
$$

Thus, the models of this structural class are linear in their parameters.

6.3 OPTIMALITY CONDITIONS AND STRUCTURE OF ENTROPY-OPTIMAL PROBABILITIES OF BELONGING

The entropy-robust estimation problems for the RPM-QuaRand models are classified as finite-dimensional entropy-nonlinear programming problems. Their nonlinear properties fully depend on the nonlinearity of the RPM-QuaRand model. In particular, if the dynamic model is represented by a functional Volterra series with the structure shown in Fig. 4.11a, then the admissible set in the estimation problems is described by a system of equalities with power nonlinearities. In this case, the admissible set often has a nonconvex configuration, and the optimality conditions turn out to be local and therefore can be used only numerically. As a result, we fail in extracting common structural properties of the solutions (the probabilities of belonging) from these conditions. On the other hand, for the one-dimensional RPM-QuaRand models that are reducible to the linear form (6.30, 6.38), we can establish global optimality conditions and analyze the structural properties of the solutions.

1. Normalized probabilities of belonging

The general problem within the functional model framework is characterized by equalities (6.12–6.14). It will be adapted to the case in which the functional model is one-dimensional and reducible to the linear form (6.30):

$$s[j] = \hat{s}[j] + \sum_{h=1}^{R} \sum_{i=1}^{V_h} p_i^{(h)} z_i^{(h)}[j] + \zeta_j^- + Z_j q_j, \tag{6.39}$$

$$\hat{s}[j] = \sum_{h=1}^{R} \sum_{i=1}^{V_h} c_-^{(h)}(i) z_i^{(h)}[j],$$

$$z_i^{(h)}[j] = \delta_i^{(h)} x_i^{(h)}[j], \qquad \delta_i^{(h)} = c_i^{(h,+)} - c_i^{(h,-)}, \quad i = \overline{1, V_h},$$

$$Z_j = \zeta_j^+ - \zeta_j^-, \qquad j = \overline{1, s}.$$
$$\tag{6.40}$$

Then the constraints (6.13, 6.14) in the estimation problem (6.12) take the following form.

- The normalization conditions for the probabilities of belonging of the model parameters in the power blocks of the functional model are

$$\sum_{i=1}^{V_h} p_i^{(h)} = 1, \quad h = \overline{1, R}; \tag{6.41}$$

- The normalization conditions for the probabilities of belonging of the measurement noises are

$$\sum_{j=1}^{s} q_j = 1; \tag{6.42}$$

- The empirical balance conditions are

$$\tilde{s}[j] + \sum_{h=1}^{R} \sum_{i=1}^{V_h} p_i^{(h)} z_i^{(h)}[j] + Z_j q_j - y[j] = 0, \quad j = \overline{1, s},$$

$$\tilde{s}[j] = \hat{s}[j] + \zeta_j^-. \tag{6.43}$$

The entropy function is given by

$$H_B(P, Q) = -\sum_{h=1}^{R} \sum_{i=1}^{V_h} p_i^{(h)} \ln \frac{p_i^{(h)}}{\tilde{p}_i^{(h)}} - \sum_{j=1}^{s} q_j \ln \frac{q_j}{\tilde{q}_j}. \tag{6.44}$$

Thus, the estimation problem can be written as

$$H_B(P, Q) \Rightarrow \max \tag{6.45}$$

subject to the constraints (6.41, 6.42, 6.43).

In this problem, the admissible set Υ is described by a system of equalities and hence it represents a constrained optimization problem. To obtain optimality conditions, let us define the Lagrange function

$$
\begin{aligned}
L_B &= H_B(P,Q) + \sum_{h=1}^{R} \lambda_h \left(1 - \sum_{i=1}^{V_h} p_i^{(h)} \right) + \mu \left(1 - \sum_{j=1}^{s} q_j \right) \\
&+ \sum_{j=1}^{s} \theta^{(j)} \left(y[j] - \tilde{s}[j] - \sum_{h=1}^{R}\sum_{i=1}^{V_h} p_i^{(h)} \delta_i^{(h)} x_i^{(h)}[j] + Z_j q_j \right) \quad (6.46)
\end{aligned}
$$

Here $\lambda_1, \ldots, \lambda_R; \mu; \theta_1, \ldots, \theta_s$ denote the Lagrange multipliers for the corresponding constraints (6.41, 6.42, 6.43). The first-order optimality conditions in the direct variables (P,Q) yield the following system of equations:

$$
\frac{\partial L}{\partial p_i^{(h)}} = -\ln \frac{p_i^{(h)}}{\tilde{p}_i^{(h)}} - 1 - \lambda_h - \sum_{j=1}^{s} \theta_j z_i^{(h)}[j] = 0, \quad (6.47)
$$
$$
h = \overline{1,s}; \quad i = \overline{1,V_h}.
$$

$$
\frac{\partial L}{\partial q_j} = \ln \frac{(1-q_j)\tilde{q}_j}{q_j} - \mu - Z_j = 0, \quad j = \overline{1,s}. \quad (6.48)
$$

In combination with the normalization conditions (6.41, 6.42), equations (6.47, 6.48) lead to the following expressions for the entropy-optimal probabilities of belonging of the model parameters and measurement noises that are parameterized by the Lagrange multipliers $\theta_1, \ldots, \theta_s$:

$$
[p_i^{(h)}]^* = \frac{\tilde{p}_i^{(h)} \exp\left(-\sum_{j=1}^{s} \theta_j z_i^{(h)}[j] \right)}{\sum_{k=1}^{V_h} \tilde{p}_k^{(h)} \exp\left(-\sum_{m=1}^{s} \theta_j z_k^{(h)}[m] \right)}, \quad (6.49)
$$

$$
q_j^* = \frac{\tilde{q}_j \exp(-\theta_j Z_j)}{\sum_{m=1}^{s} \tilde{q}_m \exp(-\theta_m Z_m)}.
$$

$$
h = \overline{1,R}, \quad i = \overline{1,V_h}, \quad j = \overline{1,s}
$$

Introduce V_h functions of integer argument $P^{(h)}[i]$, $i = 1, V_h$, whose values are defined by the entropy-optimal probabilities of belonging. By this definition, $P^{(h)}[i]$, $i = 1, V_h$, represent distribution functions for the probabilities of belonging. According to (6.49), they are exponential functions.

The Lagrange multipliers $\theta_1, \ldots, \theta_s$ are calculated from the empirical balance constraints (6.43) after substitution of the entropy-optimal probabilities of belonging (6.49). For notational convenience, introduce

$$S[j] = \hat{s}[j] + \zeta_j^- - y[j],$$

$$\varphi^{(h)}(\theta_1, \ldots, \theta_s) = \sum_{k=1}^{(V_h)} \tilde{p}_k^{(h)} \exp\left(-\sum_{l=1}^{s} \theta_l z_k^{(h)}[l]\right), \qquad (6.50)$$

$$\psi_j^{(h)}(\theta_1, \ldots, \theta_s) = \sum_{k=1}^{(V_h)} \tilde{p}_k^{(h)} z_k^{(h)}[j] \exp\left(-\sum_{l=1}^{s} \theta_l z_k^{(h)}[l]\right),$$

and

$$\mu(\theta_1, \ldots, \theta_s) = \sum_{l=1}^{s} \tilde{q}_l \exp(-\theta_l Z_l), \quad \nu_j(\theta_j) = \tilde{q}_j Z_j \exp(-\theta_j Z_j), \quad (6.51)$$

$$j = \overline{1, s}.$$

With these notations, the empirical balance constraints take the form

$$\sum_{h=1}^{R} \frac{\psi_j^{(h)}(\theta_1, \ldots, \theta_s)}{\varphi^{(h)}(\theta_1, \ldots, \theta_s)} + \frac{\nu_j(\theta_j)}{\mu(\theta_1, \ldots, \theta_s)} + S[j] = 0, \qquad j = \overline{1, s}. \quad (6.52)$$

2. Interval-type probabilities of belonging

The entropy-optimal estimation problem for the RPM-QuaRand model can be stated using the generalized Boltzmann information entropy (6.16–6.18). This form of the estimation problem involves a considerably larger number of constraints than the previous one (some of them are inequalities). Therefore, it moves to the class of finite-dimensional mathematical programming problem, and optimality conditions are defined by the Kuhn–Tucker theorem [134].

The solution procedure of the entropy-optimal estimation problem can be significantly simplified if we use the generalized Fermi information entropy (6.1.2) as the objective function. The interval properties of the probabilities of belonging for the model parameters and measurement noises are embedded in the entropy function, which makes the estimation problem a constrained optimization problem. Like before, optimality conditions can be derived using Lagrange's method of multipliers.

The Lagrange function for the problem (6.21, 6.22) takes the form

$$L_F = H_F(P, Q) + \sum_{j=1}^{s} \theta^{(j)} \left(y[j] - \hat{s}[j] - \sum_{h=1}^{R} \sum_{i=1}^{V_h} p_i^{(h)} z_i^{(h)}[j] + \zeta_j^- + Z_j q_j \right).$$

$$(6.53)$$

The first-order optimality conditions are given by the system of inequalities

$$\frac{\partial L}{\partial p_i^{(h)}} = \ln p_i^{(0,h)} \frac{1 - p_i^{(h)}}{p_i^{(h)}} - \sum_{j=1}^{s} \theta_j z_i^{(h)}[j], \tag{6.54}$$

$$h = \overline{1, R}, \ i = \overline{1, V_h};$$

$$\frac{\partial L}{\partial q_j} = \ln q_j^0 \frac{1 - q_j}{q_j} - \theta_j Z_j = 0, \qquad j = \overline{1, s}. \tag{6.55}$$

Then the entropy-optimal probabilities of belonging of the model parameters and measurement noises that are parameterized by the Lagrange multipliers have the form

$$p_i^{(h),*} = \frac{p_i^{(h,0)}}{p_i^{(h,0)} + \exp\left(\sum_{j=1}^{s} \theta_j z_i^{(h)}[j]\right)}, \tag{6.56}$$

$$h = \overline{1, R}, \quad i = \overline{1, V_h},$$

$$q_j^* = \frac{q_j^0}{q_j^0 + \exp(\theta_j Z_j)}, \qquad j = \overline{1, s}.$$

Clearly, the entropy-optimal probabilities of belonging lie within the unit intervals. To find the Lagrange multipliers, substitute the expressions (6.56) into the empirical balance constraints (6.22). By analogy with the previous case, introduce the following notations:

$$S[j] = \hat{s}[j] + \zeta_j^- - y[j],$$

$$\varphi_j(\theta_j) = Z_j \frac{q_j^0}{q_j^0 + \exp(\theta_j Z_j)}, \tag{6.57}$$

$$\psi_j^{(h)}(\theta_1, \ldots, \theta_s) = \sum_{i=1}^{V_h} \frac{p_i^{(h,0)}}{p_i^{(h,0)} + \exp\left(\sum_{l=1}^{s} \theta_l z_i^{(h)}[l]\right)}.$$

Then the empirical balance equations can be written as

$$\sum_{h=1}^{R} \psi_j^{(h)}(\theta_1, \ldots, \theta_s) + \varphi_j(\theta_j) + S[j] = 0, \qquad j = \overline{1, s}. \tag{6.58}$$

The left-hand sides of equations (6.52, 6.58) incorporate strictly monotonically decreasing functions, which guarantees the existence of a unique solution. (The proof of this fact is left to the interested reader). Chapter 7 below describes an algorithm for solving such multiplicative-type equations.

COMPUTATIONAL METHODS OF RANDOMIZED MACHINE LEARNING

The general randomized machine learning problem is formulated in terms of functional entropy-linear programming, which allows us to obtain the optimal solution structure parameterized by the Lagrange multipliers. To calculate their values, we have to solve the nonstandard nonlinear equations with the so-called parameterized integral components, which can be found only numerically. Therefore, it is impossible to analytically estimate such important parameters of computational procedures as the Lipschitz constant, the Hölder constant, the degree of smoothness, and others. For solving these equations, our idea is to extend the randomization approach to the computational algorithms intended for solving the RML problems. The randomization approach itself was declared and repeatedly used before; see [135]. However, the major properties of the randomized problem-oriented algorithms arise in the course of their implementation. In what follows, we will develop the algorithms for solving the empirical balance equations and calculating the multidimensional integrals on the basis of "packet" Monte Carlo simulations, which are implemented using heterogeneous-structure computing aids.

An important special class of the RML problems includes the machine learning problems with the probabilities of belonging and linear

DOI: 10.1201/9781003306566-7

models. For this class, it is possible to obtain a structural representation of the optimal solutions parameterized by the Lagrange multipliers. The resulting balance constraints have a nonlinear form but without the integral components. For solving these equations, the multiplicative algorithms [140, 142] will be employed.

7.1 CLASSES OF BALANCE EQUATIONS IN RML AND ML PROCEDURES

In Chapters 4 and 5, we have derived the empirical balance equations for the RPM-Orig, RPM-Rel and RPM-CauRand models, in which the role of variables is played by the Lagrange multipliers. An analysis of these equations (more specifically, of the functions that form their components) allows us to separate out two classes of the nonlinear equations for further study.

1. Equations with integral components

The equations of this class arise in the RML procedures with the RPM-Orig and RPM-Rel models (5.74–5.76). Recall that the equations under consideration contain the definite multidimensional integrals that are parameterized by the Lagrange multipliers. For example, for the RPMs with the original parameters (RPM-Orig), the integral components have the form

$$\mathbb{M}_j(\theta) - \mathbf{y}^{(j)} = 0, \qquad j = \overline{1, s}, \tag{7.1}$$

where

$$\mathbb{M}_j(\theta) = \mathcal{P}^{(-1)}(\theta) \int_{\mathcal{A}} P_0(\mathbf{a}) \times$$

$$\times \exp\left[-\sum_{j=1}^{s} \langle \bar{\theta}^{(j)}, \mathcal{B}[\mathbf{x}((t_j, \tau_h) \mid h = \overline{1, (j-1)}), \mathbf{a}] \rangle \right] \times$$

$$\times \mathcal{B}[\mathbf{x}((t_j, \tau_h) \mid h = \overline{1, (j-1)}), \mathbf{a}] d\mathbf{a} + \mathbb{Q}_j(\bar{\theta}^{(j)}), \tag{7.2}$$

$$j = \overline{1, s}.$$

In these equalities, the matrix $\theta = [\bar{\theta}^{(1)}, \dots, \bar{\theta}^{(s)}]$ and the functions $\mathcal{P}(\bar{\theta})$ and $\mathbb{Q}_j(\bar{\theta}^{(j)})$ are given by (5.4.1) and (5.75), respectively.

A similar situation also occurs for the randomized parametric models with the relative parameters (RPM-Rel). The balance equations for this class of models can be written as

$$\mathbb{N}_j(\bar{\theta}) - \mathbf{y}^{(j)} = 0, \qquad j = \overline{1, s}, \tag{7.3}$$

where

$$\mathbb{N}_j(\theta) = \mathcal{W}^{(-1)}(\theta) \int_{\mathcal{U}} W_0(\mathbf{u}) \times$$

$$\times \exp\left[-\sum_{j=1}^{s}\langle\bar{\theta}^{(j)}, \mathcal{D}[\mathbf{x}((t_j, \tau_h) \mid h = \overline{1, (j-1)}), \mathbf{a}^- + \Delta\mathbf{u}]\rangle\right] \times$$

$$\times \mathcal{D}[\mathbf{x}((t_j, \tau_h) \mid h = \overline{1, (j-1)}), \mathbf{a}^- + \Delta\mathbf{u}]d\mathbf{u} + \mathbb{G}_j(\bar{\theta}^{(j)}), \quad (7.4)$$

$$j = \overline{1, s}.$$

These expressions include $\mathcal{W}(\theta)$ (5.4.2) and $\mathbb{G}_j(\bar{\theta}^{(j)})$ (5.4.3). Note that the functionals $\mathcal{B}$ and $\mathcal{D}$ are of the power type.

Equations (7.1, 7.3) have a similar structure; using them, construct a class of nonlinear equations of the following form:

$$\mathbf{B}(\mathbf{u}) = \mathbf{B}\left[\mathbb{C}(\mathbf{f})\right] = \mathbf{0}, \quad (7.5)$$

with the integral components

$$\mathbb{C}(\mathbf{f}) = \int_{\mathcal{E}} \exp\left(-E(\mathbf{f}, \mathbf{e})\right) d\mathbf{e}, \quad (7.6)$$

or

$$\mathbb{C}(\mathbf{f}) = \int_{\mathcal{E}} E(\mathbf{f}, \mathbf{e}) \exp\left(-\mathcal{E}(\mathbf{f}, \mathbf{e})\right) d\mathbf{e}, \quad (7.7)$$

where

$$E(\mathbf{f}, \mathbf{e}) = \sum_{h=1}^{R} \sum_{(n_1,\dots,n_h)=1}^{m} A_{(n_1,\dots,n_h)}(\mathbf{f}) \prod_{k=1}^{h} e_{n_k}. \quad (7.8)$$

The notations are as follows: $\mathbf{f} \in R^s$ as the vector of the desired variables, i.e., the parameters of the integral components; $\mathbf{B}(\mathbf{u})$ as a vector function that contains the vectors of the integral components (7.6) or (7.7); $\mathbf{e} \in E \subset R^r$ as the vector of the integration variables in (7.6, 7.7); finally, E as the r-dimensional parallelepiped.

A numerical solution procedure for equation (7.5) consists of the following steps: (a) calculating the integral components $\mathbb{C}(\mathbf{f})$ under fixed variables $\mathbf{f}$ and (b) calculating the new vector of the variables that is the solution of the corresponding balance equations. This procedure corresponds to an iterative process with the structure illustrated in Fig. 7.1. It has the main iterative loop (see double lines) intended for solving equation (7.5) using the algorithm

$$\mathbf{f}^{(k+1)} = \Omega\left(\mathbf{f}^k, \mathbf{B}\left[\mathbb{C}(\mathbf{f}^k)\right]\right), \quad (7.9)$$

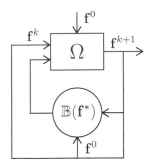

Figure 7.1

and also the secondary loop (see single lines) intended for calculating the function $\mathbf{B}$ with the integral components (7.6, 7.7).

Get back to equation (7.5), reducing it to a more convenient form for calculations. Introduce a change of variables that maps the space R^s and the domain $\mathcal{E} \in R^r$ into the unit nonnegative cubes of proper dimensions. Since the domain $\mathcal{E}$ is the r-dimensional parallelepiped, this transformation can be performed with the linear change of variables

$$\mathbf{w} = L\,\mathbf{e} - \mathbf{e}^-, \quad L = \operatorname{diag}\left[1/(e_i^+ - e_i^-), \,|\, i = \overline{1, r}\right],$$
$$\mathbf{w} \in I^r \subset R_+^r = [\mathbf{0}, \mathbf{1}]. \tag{7.10}$$

The space R^s can be mapped into the unit nonnegative s-dimensional cube by nonlinear transformations, e.g.,

$$\mathbf{z} = \{\frac{1}{1 + \exp f_i}, \,|\, i = \overline{1, s}\}, \qquad \mathbf{z} \in I^s. \tag{7.11}$$

Using these transformations in (7.5), write

$$\mathbf{W}(\mathbf{z}) = \mathbf{0}, \qquad \mathbf{z} \in I^s. \tag{7.12}$$

The integral components in this equation are

$$\mathbf{C}(\mathbf{z}) = \int_{I^r} \exp\left(-E(\mathbf{z}, \mathbf{w})\right) d\mathbf{w} = \mathbf{0}, \qquad \mathbf{z} \in I^s, \quad \mathbf{w} \in I^r, \tag{7.13}$$

or

$$\mathbf{G}(\mathbf{z}) = \int_{I^r} \exp\left(-E(\mathbf{z}, \mathbf{w})\right) \exp\left(-E(\mathbf{z}, \mathbf{w})\right) d\mathbf{w} = \mathbf{0}, \tag{7.14}$$
$$\mathbf{z} \in I^s, \quad \mathbf{w} \in I^r,$$

where

$$E(\mathbf{z}, \mathbf{w}) = \sum_{h=1}^{R} \sum_{(n_1,\ldots,n_h)=1}^{m} A_{(n_1,\ldots,n_h)}(\mathbf{z}) \prod_{k=1}^{h} w_{n_k}. \qquad (7.15)$$

Equation (7.12) will be solved by minimizing the residual function

$$F(\mathbf{z}) = \|\mathbf{W}(\mathbf{z})\| \Rightarrow \min_{\mathbf{z}}, \qquad \mathbf{z} \in I^s. \qquad (7.16)$$

Note that it is impossible to establish any characteristics of the residual function analytically due to its integral components. Thus, we have to hypothesize several local minima of the function $F(\mathbf{z})$ on the admissible set I^s. Hence, the problem (7.16) comes to a global minimization of this function.

2. Equations with analytical functions

The balance equation of this class arise in the ML problems with the RPM-QuaRand models. The probabilities of belonging that characterize the parameters of these models can be normalized or interval-type. In the former case, the ML-based estimation algorithm is formulated as a constrained maximization problem for the generalized Boltzmann information entropy. In the latter case, it seems more convenient to choose the generalized Fermi information entropy.

Write the empirical balance equations for the normalized probabilities of belonging (6.52) in the form

$$m_j(\mathbf{v}) = \sum_{h=1}^{R} \frac{\beta^{(h)}(v_1,\ldots,v_s; j)}{\alpha^{(h)}(v_1,\ldots,v_s)} + \gamma_j(v_1,\ldots,v_s) = 0, \qquad j = \overline{1,s}, \qquad (7.17)$$

where

$$\alpha^{(h)}(v_1,\ldots,v_s) = \sum_{k=1}^{V_h} p_k^{(h,0)} \prod_{l=1}^{V_h} v_l^{z_k^{(h)}[l]},$$

$$\beta_k^{(h)}(v_1,\ldots,v_s; j) = \sum_{k=1}^{V_h} z_k^{(h)}[j] p_k^{(h,0)} \prod_{l=1}^{V_h} v_l^{z_k^{(h)}[l]}, \qquad (7.18)$$

$$\gamma_j(v_1,\ldots,v_s) = S[j] + \frac{q_j^0 Z_j v_j^{Z_j}}{\sum_{l=1}^{s} q_j^0 v_j^{Z_l}} - y[j],$$

$$v_l = \exp(-\theta_l), \qquad l = \overline{1,s}.$$

The variables in these equations are defined on the nonnegative orthant R_+^s. Sometimes, they are called the exponential Lagrange multipliers.

Consider the empirical balance equations for the interval probabilities of belonging (6.58) and represent them in terms of the variables $(v_1, \ldots, v_s) \geq 0$ (7.18). Introduce the following notations:

$$\varrho_j^{(h)}(v_1, \ldots, v_s) = \sum_{i=1}^{V_h} \frac{p_i^{(h,0)}}{p_i^{(h,0)} + \prod_{l=1}^{s} v_i^{z_i^{(h)}[l]}},$$

$$\vartheta_j(v_j) = Z_j \frac{q_j^0}{q_j^0 + v_j^{Z_j}}, \qquad (7.19)$$

$$\omega[j] = \vartheta_j(v_j) + \hat{s}[j] + \zeta_j^- - y[j].$$

Write equations (6.58) as

$$n_j(\mathbf{v}) = \sum_{h=1}^{R} \varrho_j^{(h)}(v_1, \ldots, v_s) + \omega[j] = 0, \qquad j = \overline{1, s}. \qquad (7.20)$$

The final forms of the empirical balance equations allow us to identify some general properties of these equations, which can be fruitful for choosing appropriate computational methods.

7.2 MONTE CARLO PACKET ITERATIONS FOR GLOBAL OPTIMIZATION

The RML algorithms for the RPM-Orig and RPM-Rel models are intended for solving the empirical balance equations, in which the role of variables is played by the Lagrange multipliers or by the auxiliary variables taking their values within the unit nonnegative cube. An approach to design numerical methods proceeds from minimization of a residual function (generally, a quadratic norm). In an RML problem, there exists an algorithmically defined relation between the residual function and variables, and this feature of the problem finally reduces it to a global minimization problem. Under such conditions, the Monte Carlo method seems to be an suitable design tool for RML algorithms.

Vast amounts of literature have been published dealing with applications of the Monte Carlo method in different global optimization problems. Some insight into these problems and methods to solve them can be gained from [135, 158, 171–173, 196, 199].

In contrast to the existing methods, the idea put forward here is to use *simple Monte Carlo simulations*, i.e., the generation of independent random vectors with the uniform distribution on the nonnegative unit cube.

7.2.1 Canonical form of global optimization problem

Consider the global minimization problem

$$f(\mathbf{x}) \Rightarrow \text{globmin}$$

subject to the following constraints:

$$\mathbf{x} \in \mathcal{G}_{\mathbf{x}} = \{\mathbf{x} : \mathbf{g}(\mathbf{x}) \le \mathbf{0}, \ \mathbf{x} \in R^d\}, \quad \mathbf{g} \in R^r, r < d. \quad (7.21)$$

The objective function $f(\mathbf{x})$ satisfies the Hölder condition with constants $H > 0$ and $w > 0$, i.e., its modulus of continuity has the form

$$\omega(t) = \max_{(\tilde{\mathbf{v}}, \tilde{\mathbf{y}}) \in R^d : \|\tilde{\mathbf{v}} - \tilde{\mathbf{y}}\| \le t} |f(\tilde{\mathbf{v}}) - f(\tilde{\mathbf{y}})| \le H t^w. \quad (7.22)$$

The functions $\mathbf{g}(\mathbf{x})$ in (7.22) are assumed to be continuous.
Write *the canonical form* of the problem (7.21) as

$$F(\mathbf{z}) \Rightarrow \text{globmin} \quad (7.23)$$

subject to the following constraints:

$$\mathbf{z} \in \mathcal{K}_{\mathbf{z}} = \mathcal{G}_{\mathbf{z}} \bigcap Z_+^d,$$

$$Z_+^d = \{\mathbf{z} : \mathbf{0} \le \mathbf{z} \le \mathbf{1}\} \subset R^d,$$

$$\mathcal{G}_{\mathbf{z}} = \{\mathbf{z} : \mathbf{w}(\mathbf{z}) \le \mathbf{0}, \ \mathbf{z} \in Z_+^d\}, \quad \mathbf{w} \in R^r, r < d. \quad (7.24)$$

A relation between the original problem and its canonical form is defined by a continuously differentiable bijective transformation of the variables, i.e.,

$$\mathbf{z} = \mathcal{T}(\mathbf{x}), \qquad \mathbf{x} = \mathcal{T}^{(-1)}(\mathbf{z}). \quad (7.25)$$

In fact, quite a few transformations answer these conditions, e.g.,

$$\mathbf{z} = \{\frac{1}{1 + exp(-a_i x_i)}, \ | \, i = \overline{1, d}\},$$

$$\mathbf{x} = \{\frac{1}{a_i} \ln \frac{z_i}{1 - z_i}, \ | \, i = \overline{1, d}\}. \quad (7.26)$$

In many applications, the domain of definition of the variables $\mathbf{x}$ is not the entire space R^d but its subset—the d-dimensional parallelepiped $\Pi_{\mathbf{x}}$. Then the problem (7.21) is stated as follows:

$$f(\mathbf{x}) \Rightarrow \text{globmin}$$

subject to the constraints

$$\mathbf{x} \in \Pi_{\mathbf{x}} \bigcap \mathcal{G}_{\mathbf{x}}, \quad (7.27)$$

where
$$\Pi_{\mathbf{x}} = \{\mathbf{x} : \mathbf{a} \le \mathbf{x} \le \mathbf{b}\}, \qquad \mathbf{x} \in R^d. \tag{7.28}$$

In these expressions, the d-dimensional vectors $\mathbf{a}$ and $\mathbf{b}$ have fixed elements a_i and b_i, $i = \overline{1, d}$. The transformation (7.26) is simplified to

$$\mathbf{z} = \{\frac{x_i - a_i^-}{a_i^+ - a_i^-}, \, | \, i = \overline{1, d}\}. \tag{7.29}$$

Due to the continuity of the transformation (7.26), the function $F(\mathbf{z})$ is bounded below on $\mathcal{K}_{\mathbf{z}}$, i.e., there exists a constant c such that

$$F^* \ge c. \tag{7.30}$$

The constant c is generally unknown but equals 0 in some special global minimization problems. For example, this takes place for the problems in which the function $F(\mathbf{z})$ is defined as the residual between the left- and right-hand sides of a certain system of equations.

The function $F(\mathbf{z})$ also satisfies the Hölder condition with other constants L and p, i.e.,

$$\omega(h) = \max_{(\mathbf{v},\mathbf{y}) \in \mathcal{K}_{\mathbf{x}} : \|\mathbf{v}-\mathbf{y}\| \le h} |F(\mathbf{v}) - F(\mathbf{y})| \le Lh^s. \tag{7.31}$$

Denote by $\mathbb{Z}^*$ the set of all global minimum points $\mathbf{z}^*$ and by $F(\mathbf{z}^*) = F^*$ the global minimum value in the problem (7.23). In other words, assume that a global minimum is achieved not at a unique point but on a finite set $\mathbb{Z}^*$.

How should we comprehend the *solution* of the problem (7.23)? This issue is of crucial importance, as its obviousness disappears for real computational processes. Actually, it is impossible to achieve an exact value of the global minimum F^* even with an arbitrarily high accuracy of calculations. We can speak about getting into some r-neighborhood of the exact global minimum value. In other words, the matter concerns an *approximate* global minimum value $\tilde{F}^*$ within the r-neighborhood. At the very beginning, we have declared to employ the Monte Carlo method for global minimization. Hence, the value $\tilde{F}^*$ gets into the r-neighborhood with a certain probability, less or equal to 1.

Also note that computations are implemented through iterative procedures that represent a sequence of steps with numbers k. Each step yields an intermediate result. In the problem (7.23), the intermediate results are the values $\tilde{F}_1^*, \ldots, \tilde{F}_k^*, \ldots$. Assume that at some step k^* an approximate global minimum value $\tilde{F}_{k^*}^*$ has entered an r-neighborhood. Nevertheless, this event is random and it may happen that at step

$(k^* + 1)$, the value leaves the neighborhood. For increasing the reliability of approximate global minimization, introduce a positive number ϱ that characterizes the number of steps during which the values $\tilde{F}^*_{k^*}, \tilde{F}^*_{(k^*+k_1)}, \ldots, \tilde{F}^*_{(k^*+k_\varrho)}$ stay within the r-neighborhood after first getting into it at step k^*.

7.2.2 Idea of method and concept of solution

Consider a simplified statement of the problem (7.23) in which the admissible set is the d-dimensional cube Z^d_+. This problem is solved by an algorithm that represents an iterative procedure to construct a sequence of records[1] and a sequence of their decrements.

At step k, the algorithm generates a batch $\mathcal{Z}_k$ of the $N_k = M^d_k$ independent random vectors $\mathbf{z}^1, \ldots, \mathbf{z}^{N_k}$ obeying the uniform distribution on Z^d_+. Here M_k denotes the number of independent uniformly distributed random variables on the interval $[0, 1]$ that are produced at step k by a random generator for each of the d coordinates.

For the kth batch, the algorithm calculates the values of the function $F(\mathbf{z}^j)$, $j = \overline{1, N_k}$, as well as its minimum value F^*_k (the kth record) and argument $\mathbf{z}^*_k$. Due to a random character of the Monte Carlo simulations, the value F^*_k is also random.

Transition to step $(k+1)$ consists in increasing the number of random vectors by the rule

$$M_{k+1} = \alpha M_k, \qquad \alpha > 1, \tag{7.32}$$

and calculating a new record F^*_{k+1} for the $(k + 1)$th packet. This yields a monotonically decreasing sequence

$$\mathcal{F} = \{F^*_0 \geq \ldots, \geq F^*_k, \ldots\}. \tag{7.33}$$

In addition to the records sequence (7.33), the algorithm constructs a sequence of their decrements

$$\mathcal{U} = \{u_1, \ldots, u_k, \ldots\}, \tag{7.34}$$

where $u_k = |F^*_k - F^*_{k+1}|$.

The algorithm stops when the sequence of random variables u_k stays within a given δ-neighborhood of the origin during ρ successive steps, i.e., the number of steps till its termination is

$$\tau = \min\{k : \max_{k-\rho \leq s \leq k} u_s \leq \delta\}. \tag{7.35}$$

---

[1] A record is a minimum value of an objective function at step k of the procedure.

In the sequel, this algorithm will be called the *GFS algorithm (Generation, Filtration, Selection)*.

The solution of the problem (7.23) is represented by the quadruple

$$S = \{F_\tau^*, \mathbf{z}_\tau^*, r_\tau^*, \mathbf{P}(\mathcal{E}_{r_\tau})\}, \tag{7.36}$$

where

- $F_\tau^*, \mathbf{z}_\tau^*$ indicate an approximate global minimum value and its argument, respectively, at the stop time τ;

- r_τ^* is the estimated size of the neighborhood of the exact global minimum value at the stop time τ;

- the event

$$\mathcal{E}_{r_\tau^*} = \{F_\tau^* - F^* \le r_\tau\}, \tag{7.37}$$

  where F^* gives the exact global minimum value on the set Z_+^d;

- $\mathbf{P}(\mathcal{E}_{r_\tau^*})$ denotes an estimated probability of the event (7.37).

A real algorithm involves a strictly monotonically decreasing subsequence

$$\mathbb{F}^* = \{F_1^* > \dots, > F_k^*, \dots\} \tag{7.38}$$

of the sequence $\mathcal{F}^*$ (7.33). A theoretical analysis below is based on exploring the properties of the sequence $\mathcal{F}^*$. As $\mathbb{F}^*$ forms a subsequence of $\mathcal{F}^*$, in qualitative terms the analysis results remain the same for the subsequence $\mathbb{F}^*$.

7.2.3   Probabilistic characteristics of random sequences $\mathcal{F}$ and $\mathcal{U}$

A study of the random sequences generated by the packets of simple Monte Carlo simulations is an important stage to investigate the fundamental properties of the GFS algorithm in terms of computations. These properties include convergence analysis (in particular, convergence conditions) and estimation of the numerical characteristics of a resulting solution [35].

1. Probabilistic characteristics of packets $\mathcal{Z}_k$

In each stage k, the algorithm generates a packet $\mathcal{Z}_k$ of N_k independent random vectors with the uniform distribution on the d-dimensional cube. The source of the packet $\mathcal{Z}_k$ is a random generator: for each of the d coordinates at step k it produces an array of M_k independent uniformly distributed random variables on the interval $[0, 1]$ (simple Monte

Carlo simulations). As a result, at step k we have a packet of $(M_k)^d$ random vectors obeying the uniform distribution on the unit cube.

Consider the unit cube Z_+^d (7.23) and its partition into elementary cubes using a *virtual* uniform lattice with the step

$$\eta_k = M_k^{-q} \tag{7.39}$$

for each of the d coordinates. Here q specifies a lattice parameter such that

$$0 < \gamma \le q \le \beta < 1. \tag{7.40}$$

Denote by $P(M_k, d, q)$ the probability that there exists an elementary *virtual* cube containing none of the random $(M_k)^d$ vectors generated at step k. This probability can be estimated as follows.

Lemma 1 The probability $P(M_k, d, q)$ has the upper bound

$$P(M_k, d, q) \le (M_k^q + 1)^d \left(1 - \frac{1}{M_k^{dq}} \right)^{M_k^d} = Q(M_k, d, q). \tag{7.41}$$

As $k \to \infty$,

$$Q(M_k, d, q) \sim \mathbf{Q}(M_k, d, q) = M_k^{dq} \exp\left(-(M_k)^{(1-q)d} \right). \tag{7.42}$$

Proof. Consider a certain *virtual* elementary cube of the volume $(\eta_k)^d$. The generated random vectors are independent and obey the uniform distribution on the unit cube; at step k, their number is $(M_k)^d$. Hence, the probability that this elementary virtual cube *contains none of these vectors* makes up

$$\left(1 - (\eta_k)^d \right)^{M_k^d} = \left(1 - \frac{1}{M_k^{dq}} \right)^{M_k^d}. \tag{7.43}$$

Let B be the event that such an elementary *virtual* cube exists. According to (7.41),

$$P(B) \le Q(M_k, d, q), \tag{7.44}$$

because $(M_k^q + 1)^d$ is an upper bound for the number of all elementary virtual cubes with the side $\eta_k = M_k^{-q}$ for each coordinate.

Due to

$$\lim_{x \to \infty} \frac{(1 - x^{-q})^x}{\exp(-x^{(1-q)})} = 1,$$

letting $k \to \infty$ in (7.44) gives the inequality

$$P(B) \leq Q(M_k, d, q) \sim M_k^{dq} \exp\left(-M_k^{(1-q)d}\right). \qquad (7.45)$$

■

Corollary 1 *With a probability not smaller than* $(1 - Q(M_k, d, q))$, *each elementary cube with the side* M_k^{-q} *contains at least one of the random vectors from the packet generated at step* k.

2. Probabilistic properties of sequence $\mathcal{F}^*$

Consider a sequence of "quasi global" minimum values $\mathcal{F}^* = \{F_0^*, \ldots, F_k^*, \ldots\}$ and a sequence of the corresponding arguments $\mathcal{Z}^* = \{\mathbf{z}_0^*, \ldots, \mathbf{z}_k^*, \ldots\}$.

Denote by $\mathbb{Z}^*$ the set of all global minimum points of the function $F(\cdot)$ on the unit cube. Introduce the distance $l(\cdot)$ between an arbitrary point $\mathbf{z}$ and the compact set $\mathbb{Z}^*$ in the form

$$l(\mathbf{z}, \mathbb{Z}^*) = \min_{\mathbf{y} \in \mathbb{Z}^*} \|\mathbf{z} - \mathbf{y}\|. \qquad (7.46)$$

Lemma 2 *With a probability not smaller than* $(1 - \mathbf{Q}(M_k, d, q))$,

$$0 \leq F_k^* - F^* \leq \omega(h_k), \qquad (7.47)$$

where $\omega(\cdot)$ *is the modulus of continuity of the function* $F(\mathbf{z})$, $h_k = \eta_k \sqrt{d}/2$, *and* F^* *gives an unknown value of the global minimum of the function* $F(\cdot)$ *on the set* Z_+^d.

Proof. Let $\hat{\mathbf{z}}$ be the closest point (in terms of the distance (7.47)) to the set $\mathbb{Z}^*$ from the packet of the random points generated at step k. Recall that, with a probability not smaller than $(1 - Q(M_k, d, q))$, at least one of the generated points gets into each *virtual* elementary cube with the side η_k. Therefore, the distance is

$$h_k = \|\mathbf{z}^* - \hat{\mathbf{z}}\| \leq \eta_k \frac{\sqrt{d}}{2}. \qquad (7.48)$$

This happens if the point $\mathbf{z}^* \in \mathbb{Z}^*$ implementing the distance to the set $\mathbb{Z}^*$ is at the center of the cube with the side η_k. In addition, the closest random points are located at the corners of this cube and also each cube contains at least one random point.

Then, by the Hölder condition,

$$0 \leq F(\hat{\mathbf{z}}) - F^* \leq \omega(h_k). \qquad (7.49)$$

On the other hand,

$$F^* = F(\mathbf{z}^*) = \min_{\mathbf{z} \in Z_+^d} F(\mathbf{z}) \leq \min_{1 \leq s \leq k} F_s^* \leq F_k^* \leq F(\hat{\mathbf{z}}). \tag{7.50}$$

It follows from (7.50) that

$$0 \leq F_k^* - F^* \leq F(\hat{\mathbf{z}}) - F^*. \tag{7.51}$$

Finally, the expressions (7.51) and (7.49) yield the relation

$$0 \leq F_k^* - F^* \leq \omega(h_k), \tag{7.52}$$

which holds with a probability not smaller than $(1 - Q(M_k, d, q))$. ■

Corollary 2 *For the functions satisfying the Hölder condition (7.31), there exist constants*

$$D = L\left(\frac{\sqrt{d}}{2}\right)^s, \qquad p = qs$$

such that the upper bound

$$0 \leq F_k^* - F^* \leq DM_k^{-p} = r_k(q) \tag{7.53}$$

is valid with a probability not smaller than $1 - Q(M_k, d, q)$, *i.e.,*

$$P\{0 \leq F_k^* - F^* \leq r_{(q)}\} \geq 1 - Q(M_k, d, q). \tag{7.54}$$

Note that the upper bounds (7.53)–(7.54) depend on the parameter q, which affects the volume of the *virtual* elementary cube with the side (7.39) for each coordinate. The constants β and γ in (7.40) should be adjusted for reaching an acceptable result in terms of the size of the desired neighborhood and the corresponding probability.

7.2.4   Convergence of GFS algorithm

Using the probabilistic characteristics of the records sequence $\mathcal{F}_k^*$, we can establish the following result on its convergence.

Theorem 1 *For the functions that satisfy the Hölder condition,*

(i) the sequence $\mathcal{F}_k^*$ of the records F_k^* converges with probability 1 to the true global minimum F^* with the exponential rate;

(ii) the sequence of the record arguments $\mathbf{z}_k^*$ converges to the set $\mathbb{Z}^*$ (in terms of the distance (7.48)) with probability 1.

Proof. Use the bound $\mathbf{Q}(M_k, d, q)$ (7.42), which is asymptotically equivalent to $Q(M_k, d, q)$. Therefore, further considerations will rest on the assumption that k is sufficiently large. The parameter q has a fixed value within the interval (7.40).

(**i**). Choose a sufficiently small value $\varepsilon > 0$ and find $K(\varepsilon)$ from the condition $K(\varepsilon) = \min\{k : r_k(q) \le \varepsilon\}$, where $r_k(q)$ is defined by (7.53).

Then, for $k > K(\varepsilon)$,

$$\{F_k^* - F^* > \varepsilon\} \subset \{F_k^* - F^* > r_k\}. \tag{7.55}$$

It follows from (7.54) and (7.42) that the probability of the latter event can be estimated as

$$\mathbf{P}\{F_k^* - F^* > \varepsilon\} \le \mathbf{Q}(M_k, d, q) \tag{7.56}$$

for sufficiently large k. According to (7.32), the obvious inequality is

$$\alpha^s > s \text{ for } \alpha > 1,\ s > 0. \tag{7.57}$$

In this case, write the estimate (7.42) as

$$\mathbf{Q}(M_k, d, q) = \alpha^{qkd} M_0^{qd} \exp\left(-\frac{1}{2}\alpha^{(1-q)kd} M_0^{(1-q)d}\right) \times$$
$$\times \exp\left(-\frac{1}{2}\alpha^{(1-q)kd} M_0^{(1-q)d}\right). \tag{7.58}$$

Using (7.57), one of the exponents can be assigned an upper bound of the form

$$\exp\left(-\frac{1}{2}\alpha^{(1-q)kd} M_0^{(1-q)d}\right) < \exp\left(-\frac{kd(1-q)}{2} M_0^{(1-q)d}\right). \tag{7.59}$$

Next, denoting $x = \alpha^{kd} M_0^d$, from (7.2.4) and (7.59) write

$$\mathbb{Q}(M_k, d, q) \le \tilde{Q} \exp\left(-\frac{kd(1-q)}{2} M_0^{(1-q)d}\right), \tag{7.60}$$

where

$$\tilde{Q} = \max_{x \ge 0} x^q \exp\frac{-x^{(1-q)}}{2}. \tag{7.61}$$

Let

$$C = \frac{(1-q)d}{2} M_0^{(1-q)d}. \tag{7.62}$$

Then the probability (7.56) can be estimated in the following way:

$$\mathbf{P}\{F_k^* - F^* > \varepsilon\} \le \tilde{Q} \exp(-Ck) \quad \text{for } k > K(\varepsilon). \tag{7.63}$$

Due to (7.63),

$$\mathbf{P}\{\sup_{s\geq 0}|F_k^* - F_{k+s}^*| > \varepsilon\} \leq 2\mathbf{P}\{\sup_{s\geq k}(F_s^* - F^*) > \varepsilon/2\} \leq$$

$$\leq 2\sum_{s=k}^{\infty}\mathbf{P}\{(F_s^* - F^*) > \varepsilon/2\} \leq\leq 2\tilde{Q}\sum_{s=k}^{\infty}\exp\left(-Cs\right) \qquad (7.64)$$

$$\text{for} \quad k > K(\varepsilon/2).$$

For sufficiently small $\varepsilon > 0$, the latter inequality implies

$$\lim_{k\to\infty}\mathbf{P}\{\sup_{s\geq 0}|F_k^* - F_{(k+s)}^*| \geq \varepsilon\} = 0, \qquad (7.65)$$

which is a necessary and sufficient condition for the almost sure convergence of the sequence $\{F_k^*\}$. In addition, the derived estimates show that this convergence has the exponential rate.

ii. This item of Theorem 1 is proved *by contradiction*. Assume on the contrary that there exists a subsequence $\{\hat{\mathbf{z}}_k^*\}$ of the sequence $\{\mathbf{z}_k^*\}$ (without loss of generality, the sequence $\{\mathbf{z}_k^*\}$ itself) that converges to a point $\tilde{\mathbf{z}} \notin \mathbb{Z}^*$ with a positive probability. Since the function $F(\cdot)$ is continuous, it follows that $\lim_{k\to\infty} F(\mathbf{z}_k^*) = F(\tilde{\mathbf{z}}) > F^*$ with a positive probability (on the one part) and also $\lim_{k\to\infty} F(\mathbf{z}_k^*) = \lim_{k\to\infty} F_k^* = F^* = F(\tilde{\mathbf{z}})$ with probability 1 by item **i** (on the other part). This contradiction concludes the proof.■

Corollary 3 *A strictly monotonic subsequence $\mathbb{F}^*$ of the records sequence $\mathcal{F}^*$ converges with probability 1 to the exact global minimum value with the exponential rate.*

Corollary 4 *The decrements sequence $\mathcal{U}_k$ (7.34) vanishes with probability 1 with the exponential rate.*

7.2.5 Study of decrements sequence $\mathcal{U}$

For the Hölder-type functions, the upper bound (7.53) holds with the probability $1 - Q(M_k, d, q)$; see Corollary 2. Because $r_{k+1}(q) < r_k(q)$,

$$u_k = |F_{k+1}^* - F_k^*| \leq DM_k^{-p}. \qquad (7.66)$$

Then it follows that, for sufficiently large k,

$$u_k \approx DM_k^{-p}. \qquad (7.67)$$

Let us take the logarithm of both sides of this approximate expression:

$$v_k = \log u_k \approx \log D - p\log M_k. \qquad (7.68)$$

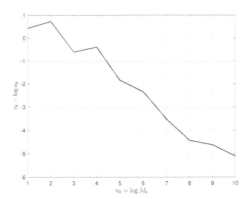

Figure 7.2

An example of the relation $v_k(n_k)$ for a computational process is illustrated in Fig. 7.2.

Consider a sequence of independent random variables ζ_k with zero mean and a finite variance σ^2. Add this sequence to the right-hand side of the approximate expression to simulate a random character of the sequence of v_k's, i.e.,

$$v_k = A - Bn_k + \zeta_k, \qquad (7.69)$$

where

$$A = \log D, \qquad B = p, \qquad n_k = \log M_k. \qquad (7.70)$$

The parameters A and B related to the Hölder constants and σ^2 are unknown. Their estimates $\hat{A}(k)$, $\hat{B}(k)$ and $\hat{\sigma}^2(k)$ will be constructed using the least-squares method based on the available information about the elements of the sequence $\{F_k^*\}$. This is easy to implement, since (7.69) represents a *linear* model (in this context, note once again that the linear regression (7.69) in *the log-log scale* directly follows from the upper bound (7.53)).

With these estimates, we can also estimate the size $r_k(q)$ of the corresponding neighborhood of the exact global minimum. According to (7.53),

$$r_k(q) = \log \hat{A}^{(k)} M_k^{\left(\frac{\hat{B}(k)}{q}\right)}. \qquad (7.71)$$

Clearly, the relation $r_k(q)$ is monotonically decreasing on the interval $[\gamma, \beta]$ (7.40). Hence, it achieves minimum at $q = \beta$. The estimated size of the neighborhood at the stop time τ is

$$r_\tau(\beta) = \log \hat{A}^{(\tau)} M_\tau^{\left(\frac{\hat{B}(\tau)}{\beta}\right)}, \tag{7.72}$$

and a lower bound for its probability (7.54) (see Corollary 2) has the form

$$P\{0 \le F_\tau^* - F^* \le r_\tau\} \ge M_\tau^{\beta d} \exp\left(-M_\tau^{(1-\beta)d}\right). \tag{7.73}$$

7.2.6   Admissible set $\mathcal{K}(\mathbf{z})$ of general form

Up to this point, we have considered the global minimization problem in which the domain of definition of variables $\mathbf{z}$ and the admissible set coincide. Now let us address the general case, where the admissible set $\mathcal{K}(\mathbf{z})$ (7.23) represents a part of the unit cube $\mathcal{Z}_+^d$. More specifically, it forms an intersection of the unit cube and a subset $\mathcal{G}$ described by a system of inequalities with continuous functions.

The admissible set under study has a smaller volume, and hence the number of elementary cubes in its virtual partition is decreased. However, this number necessarily does not exceed the number of such cubes in the unit cube partition. Therefore, the derived probability estimates (ergo, the propositions on the algorithm convergence) remain valid for a compact set that is defined by a system of continuous functions and has a positive Lebesgue measure.

Consider the admissible domain $\mathcal{K}_\mathbf{z}$ (7.23) in the unit cube. Assume that it contains no isolated points and its d-dimensional Lebesgue measure ν is nonzero, i.e., $0 < \nu_0 \le \nu \le 1$.

First, analyze the case in which the value ν is known. Then the probability that a specific vector from a generated sequence of independent random vectors with the uniform distribution on the unit cube gets into the admissible domain is ν. Let N_k independent random vectors be generated at step k with the uniform distribution on Z_+^d, including $\hat{N}_k$ vectors within the admissible domain. Denote by κ_{N_k} the frequency with which the generated vectors occur within the admissible set $\mathcal{K}_\mathbf{z}$, i.e.,

$$\kappa_{N_k} = \frac{\hat{N}_k}{N_k}. \tag{7.74}$$

By the law of large numbers, for any (sufficiently small) $\epsilon > 0$, $\eta > 0$ and the given value ν, there exists a number $N_0(\epsilon, \eta, \nu)$ such that, for all $N_k > N_0(\epsilon, \eta, \nu)$, the deviation $|\kappa_{N_k} - \nu|$ is less than $\epsilon \nu$ with a probability not smaller than $(1 - \eta)$.

Thus, for $N_k > N_0(\epsilon, \eta, \nu)$, the inequality

$$\kappa_{N_k} \geq (1 - \epsilon)\nu \tag{7.75}$$

holds with a probability not smaller than $(1 - \eta)$. Hence, the number $\hat{N}_k$ of admissible vectors in a packet composed of N_k vectors satisfies the inequality

$$\hat{N}_k \geq N_k(1 - \epsilon)\nu \tag{7.76}$$

with a probability not smaller than $(1 - \eta)$.

Consequently, in all the formulas incorporating the number N_k of generated vectors, we should replace it (for $N_k > N_0(\cdot)$) by the number $\hat{N}_k = (1 - \epsilon)\nu N_k$.

Now let us estimate the number $N_0(\epsilon, \eta, \nu)$. For $\eta > 0$, define a function $c(\eta)$ using the condition

$$\frac{1}{\sqrt{2\pi}} \int_{-c(\eta)}^{c(\eta)} \exp(-x^2/2)dx = 1 - \eta. \tag{7.77}$$

Then the number $N_0(\epsilon, \eta, \nu)$ can be found from the relation

$$N_0(\epsilon, \eta, \nu) \geq \frac{c^2(\eta)(1 - \nu)}{\epsilon^2 \nu}. \tag{7.78}$$

Second, consider the case in which the value ν is unknown but we know a value ν_0 such that $0 < \nu_0 \leq \nu$. Replacing ν by ν_0 in (7.78) yields an estimate for the minimum number of generated vectors under which the frequency of admissible vectors differs from the unknown value ν at most by $\epsilon\nu$ with a probability not smaller than $(1 - \eta)$. In this case, for $N > N_0(\epsilon, \eta, \nu)$, the value ν can be estimated as

$$\hat{\nu} = \kappa_N, \tag{7.79}$$

and, with a probability not smaller than $(1 - \eta)$, $|\kappa_N - \nu| \leq \epsilon\nu$, i.e.,

$$\frac{\kappa_N}{1 - \epsilon} \geq \nu \geq \frac{\kappa_N}{1 + \epsilon}. \tag{7.80}$$

The lower bound from this estimate can be used instead of ν in formula (7.78).

7.2.7 Logical structure of GFS algorithm

The GFS algorithm is intended for calculating a global minimum in the canonical form of the problem (7.23). It comprises two procedures, SeqLocMin and EstProb. The SeqLocMin procedure yields a sequence of records and their decrements and includes a stopping rule. The EstProb procedure estimates a minimum size for the neighborhood of the exact global minimum and also calculates a lower bound for the probability of getting into this neighborhood.

The procedures consist of a sequence of k steps, each performing t operations. Therefore, paired indexes (k, t) will be associated with the stages of each procedure.

SeqLocMin procedure

1. *Generation*—creation of a packet of random vectors:

    $(k, 1)$ generate a sequence of N_k independent random vectors with the uniform distribution on the unit cube;

    $(k, 2)$ form a packet $\mathcal{Z}_k = \{\mathbf{z}^1, \ldots, \mathbf{z}^{N_k}\}$ of these random vectors.

2. *Filtration*—choice of admissible vectors:

    $(k, 3)$ choose vectors from the packet $\mathcal{Z}_k$ that belong to the admissible set $\mathcal{K}_{\mathbf{z}}$ and form an admissible packet $\tilde{\mathcal{Z}}_k$ of size $\tilde{N}_k$.

3. *Selection*—definition of the quasi global minimum for the packet and calculation of the decrements:

    $(k, 4)$ calculate the values $F(\tilde{\mathbf{z}}^i)$, $i = \overline{1, \tilde{N}_k}$, of the objective function for the vectors from the packet $\tilde{\mathcal{Z}}_k$;

    $(k, 5)$ choose the minimum value $F_k^* = \min F(\tilde{\mathbf{z}}^i)$, $1 \leq i \leq \tilde{N}_k$;

    $(k, 6)$ calculate the decrements $u_k = |F_{k+1}^* - F_k^*|$ and the decrement logarithms $v_k = \log u_k$ and $n_k = \log N_k$; form the sequences $\mathcal{U} = \{u_k\}, \mathcal{V} = \{v_k\}, \mathcal{N} = \{n_k\}$.

If $v_k \leq \log \delta$, the value v_k is included in the sequence $\tilde{\mathcal{V}}$. As soon as the sequence $\tilde{\mathcal{V}}_k$ accumulates ρ elements, $k = \tau$ and the algorithm stops. Then a transition to the EstProb procedure is performed.

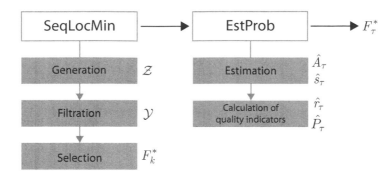

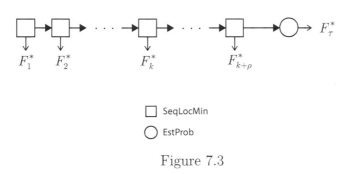

Figure 7.3

EstProb procedure

The computational scheme of this procedure consists in the following.

1. *Parameter estimation for the regression model of decrements*:

   $(\tau, 1)$ using sequences $\mathcal{N}$ and $\mathcal{V}$, estimate parameters $\hat{A}_\tau$ and $\hat{B}_\tau$ by the least-squares method.

2. *Estimation of solution characteristics (quality indicators)*:

   $(\tau, 2)$ estimate the size r_τ (7.72) for the neighborhood of the global minimum value;

   $(\tau, 3)$ calculate the lower bound for the probability $\hat{P}_\tau$ (7.73).

The logical structure of the GFS algorithm is presented in Fig. 7.3.

7.2.8 Experimental study of GFS algorithm

According to the previous subsections, the probabilistic properties of the GFS algorithm established by analytical study possess an asymptotical character. In other words, they hold for a sufficiently large number of iterations. Even the estimates derived under a finite number of steps assume that it is great enough. The aim of an experimental study, which always has a finite number of steps, is to test the operability of this algorithm in typical problems. More specifically, we will check the estimates for the size of the exact global minimum neighborhood and the probability of getting into it. Recall that these estimates have been obtained using the least-squares-based estimation of the Hölder constants for the objective function. Therefore, in the course of a computational experiment, we will confirm or doubt the hypothesis about the linear regression in the log-log scale.

In the general statement, the GFS algorithm is intended for solving the global minimization problems in which it is possible to calculate the values of the associated functions (the objective function and constraints). Below we will consider the entropy estimation problem for the PDFs of the parameters of a randomized model that possesses these properties. Let us begin the experimental study of the GFS algorithm with test problems using analytical functions.

1. Analytical tests

Our testing employs some minimization problems from [4,5,58,173], in which the admissible domain is the d-dimensional parallelepiped $\Pi_{\mathbf{z}}$.

A1. The Shekel function

$$F(\mathbf{x}) = -\sum_{i=1}^{m} \left(c_i + \sum_{j=1}^{d} (x_j - a_{ji})^2 \right)^{-1}, \quad m = 10, \qquad (7.81)$$

$$\mathbf{x}^* = 4.0,$$
$$F^* = -10.5364, \quad d = 4,$$
$$F^* = -11.0216, \quad d = 2,$$

where

$$A = \begin{pmatrix} 4 & 1 & 8 & 6 & 3 & 2 & 5 & 8 & 6 & 7 \\ 4 & 1 & 8 & 6 & 7 & 9 & 3 & 1 & 2 & 3 \\ 4 & 1 & 8 & 6 & 3 & 2 & 5 & 8 & 6 & 7 \\ 4 & 1 & 8 & 6 & 7 & 9 & 3 & 1 & 2 & 3 \end{pmatrix},$$

and

$$c = \frac{1}{10}(1, 2, 2, 4, 4, 6, 3, 7, 5, 5).$$

A2. The Rosenbrock function

$$F(\mathbf{x}) = \sum_{i=1}^{d} 100(x_{i+1} - x_i)^2 + (x_i - 1)^2, \qquad (7.82)$$

$$\mathbf{x}^* = \mathbf{1}, \ F^* = 0 \text{ for any } d.$$

A3. The Powell function

$$F(\mathbf{x}) = \sum_{i=1}^{d-2} (x_{i-1} + 10x_i)^2 + 5(x_{i+1} - x_{i+2})^2 +$$

$$+ (x_i - 2x_{i+1})^4 + 10(x_{i-1} - x_{i+2})^4, \qquad (7.83)$$

$$\mathbf{x}^* = \mathbf{0}, \ F^* = 0 \text{ for any } d.$$

A4. The trigonometric function

$$F(\mathbf{x}) = 1 + \sum_{i=1}^{d} 8 \sin^2 7(x_i - 0.9)^2 + 6 \sin^2 14(x_i - 0.9)^2 + x_i - 0.9,$$

$$\mathbf{x}^* = \mathbf{0.9}, \ F^* = 1 \text{ for any } d. \qquad (7.84)$$

A5. The Griewank function

$$F(\mathbf{x}) = 1 + \sum_{i=1}^{d} \frac{x_i^2}{4000} - \prod_{i=1}^{d} \cos\left(\frac{x_i}{\sqrt{i+1}}\right), \qquad (7.85)$$

$$\mathbf{x}^* = \mathbf{0}, \ F^* = 0 \text{ for any } d.$$

For all computations, $N_0 = 10$, $\alpha = 10$, $\delta = 0.01$, $a = 10$, $\rho = 2$, $\tau = 13$, and $\varepsilon_\tau = |F^* - F_\tau^*|$. The results of computations are illustrated in Table 7.1, where t denotes the solution time (in s), r_τ and P_τ give the estimated size of the exact global minimum neighborhood and the corresponding estimated probability, respectively. Clearly, the approximate solutions in all the tests belong to the estimated neighborhoods of the exact global minimum with a probability close to 1.

A feature of this algorithm consists in the estimation method for the Hölder constants based on the records sequence. The results of computations using this method can be observed in the graphs of Fig. 7.4, including the sequences of the records F_k^* and the estimates of the Hölder constants $\hat{L}$ and $\hat{s}$ that are yielded by approximating the decrements sequences.

For this collection of tests, the GFS algorithm was implemented in C++ on hybrid architecture computers with NVIDIA accelerators. Our computing system consists of two nodes with CPUs Intel Xeon X5650 and E5-2650 and two accelerators Tesla M2050 and K20Xm.

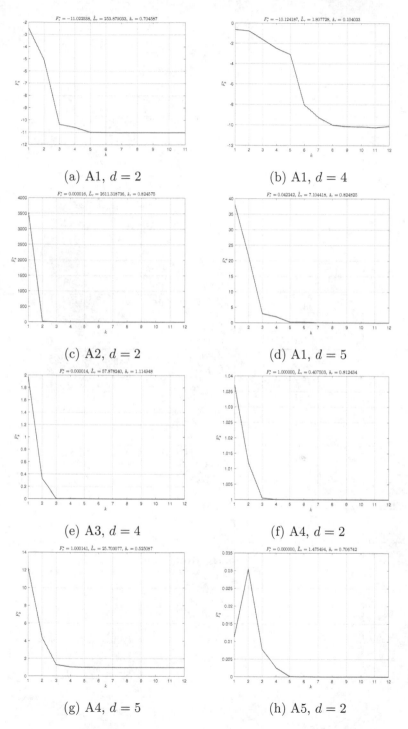

(a) A1, $d = 2$

(b) A1, $d = 4$

(c) A2, $d = 2$

(d) A1, $d = 5$

(e) A3, $d = 4$

(f) A4, $d = 2$

(g) A4, $d = 5$

(h) A5, $d = 2$

Figure 7.4: *(Continued)*

Table 7.1: Results of computations.

Function	$\Pi_{\mathbf{z}}$	F_τ^*	$\mathbf{z}_\tau^*$	ε_τ	$\hat{r}_\tau$	$\hat{P}_\tau$	t, s
A1 $d=2$	$[0, 10]$	-11.0225	4.0031 3.9983	0.0010	0.0265	≈ 1.0	1286
A1 $d=4$	$[0, 10]$	-10.1242	4.0475 4.0160 4.0135 3.9590	0.4122	0.8810	≈ 1.0	2262
A2 $d=2$	$[-10, 10]$	0.000016	0.9966 0.9930	0.000016	0.0221	≈ 1.0	2656
A2 $d=5$	$[-0.5, 1.5]$	0.0288	0.9852 0.9843 0.9706 0.9357 0.8773	0.0288	3.3406	≈ 1.0	3186
A3 $d=4$	$[-0.5, 0.5]$	0.000013	0.0405 -0.0043 0.0183 0.0181	0.000013	0.0479	≈ 1.0	1715
A4 $d=2$	$[0, 1]$	1.0000	0.9000 0.9000	$2 \cdot 10^{-9}$	0.0000041	≈ 1.0	222
A4 $d=5$	$[0, 1]$	1.0001	0.9002 0.8918 0.9007 0.8989 0.8923	0.0001	1.4970	≈ 1.0	599
A5 $d=2$	$[-5, 5]$	0.0000	-0.0004 0.0000	$8 \cdot 10^{-8}$	0.00007	≈ 1.0	281
A5 $d=5$	$[-5, 5]$	0.0018	-0.0294 -0.0140 -0.0206 -0.0036 -0.1094	0.0018	0.2533	≈ 1.0	584
A5 $d=10$	$[-5, 5]$	0.1245	3.0566 -4.4010 -0.3254 -0.0586 0.2304 0.3659 0.1528 -0.9495 0.6693 0.1680	0.1245	0.1889	≈ 1.0	1594

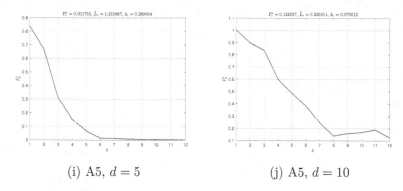

(i) A5, $d = 5$ (j) A5, $d = 10$

Figure 7.4: (*Continued*)

2. Algorithmic test (entropy-based estimation for PDFs of randomized model parameters [149])

Consider an array of input and output data for some dynamic nonlinear object. Denote by $\mathbf{y} = \{y[m], \ldots, y[m+s]\}$ the vector (packet) of output data and by $\mathbf{x}\{x[0], \ldots, x[m], \ldots, x[m+s]\}$ the vector (packet) of input data. Assume that the input measurements are perfect whereas the output measurements have errors that are simulated by a random vector $\bar{\xi} = \{\xi[m], \ldots, \xi[m+s]\}$ with independent elements. Let the elements of this vector be of the interval type, i.e.,

$$\xi[j + m] \in \Xi_j = [\xi_j^-, \xi_j^+], \text{ for all } j = \overline{0, s}. \tag{7.86}$$

In addition, there exist PDFs $q_j(\xi[j + m])$, $j = \overline{0, s}$, on these intervals.

The relation between the input and output packets is described by a discrete dynamic randomized model with second-order structured nonlinearities and an additive random noise of the form

$$v[j+1] = \sum_{h=1}^{R} \left(\sum_{n=0}^{m} w^{(h)}[n] x[j + m - n] \right)^h + \xi[j+m], \quad j = \overline{0, s}, \tag{7.87}$$

which has memory of capacity $m = 1$, $s = 1$ and $R = 2$.

The discrete random weight functions (random pulse characteristics) $w^{(1)}[n] = w^{(2)}[n] = 0$, $n > 1$, reflect the dynamic properties of the linear and nonlinear blocks of this model. Introduce the notations

$$a_0^{(1)} = w^{(1)}[0], \ a_1^{(1)} = w^{(1)}[1]; \quad \mathbf{a}^{(1)} = \{a_0^{(1)}, a_1^{(1)}\};$$
$$a_0^{(2)} = w^{(2)}[0], \ a_1^{(2)} = w^{(2)}[1]; \quad \mathbf{a}^{(2)} = \{a_0^{(2)}, a_1^{(2)}\};$$
$$\mathbf{x}[j] = \{x[j + 1], x[j]\}. \tag{7.88}$$

Table 7.2: Parameters of intervals.

Parameters	$\alpha_0^{(1)}$	$\alpha_1^{(1)}$	$\beta_0^{(1)}$	$\beta_1^{(1)}$
Values	0.5	1.0	1.0	2.0

Choose $\mathbf{x}[0] = \{0.39; 0.19\}$ and $\mathbf{x}[1] = \{0.93; 0.39\}$.

The random parameters $a_n^{(h)}$ take their values within the intervals

$$\mathcal{A}_n^{(h)} = [-A_n^{(h)}, A_n^{(h)}], \quad A_n^{(h)} = \beta_n^{(h)} \exp(-\alpha_n^{(h)} n). \tag{7.89}$$

The parameters of the above-mentioned intervals are given in Table 7.2.

The PDFs $P^{(1)}\left(\mathbf{a}^{(1)}\right)$ and $P^{(2)}\left(\mathbf{a}^{(2)}\right)$ are defined on the intervals (7.89).

According to the RML algorithm, the entropy-optimal PDFs of the model parameters and measurement noises have the form

$$\tilde{P}^{(h)}(\mathbf{a}) = \frac{1}{\mathcal{R}^{(h)}(\theta)} \exp\left(-\sum_{j=0}^{s} \theta_j V_j^{(h)}\left(\mathbf{a}^{(h)}, \mathbf{x}[j]\right)\right), \quad h = \overline{1, R};$$

$$\tilde{q}_j(\xi) = \frac{1}{\mathcal{Q}_j(\theta_j)} \exp\left(-\theta_j \xi\right), \ j = \overline{0, s}, \tag{7.90}$$

where

$$\mathcal{R}^{(h)}(\theta) = \int_{\mathcal{A}^{(h)}} \exp\left(-\sum_{j=0}^{s} \theta_j V_j^{(h)}\left(\mathbf{a}^{(h)}, \mathbf{x}[j]\right)\right) d\mathbf{a}^{(h)},$$

$$\mathcal{Q}_j(\theta) = \int_{\Xi_j} \exp\left(-\theta_j \xi_j\right) d\xi_j, \quad j = \overline{0, s}, \tag{7.91}$$

and

$$V_j^{(1)}(\mathbf{a}, \mathbf{x}) = a_0^{(1)} x[j+1] + a_1^{(1)} x[j],$$

$$V_j^{(2)}(\mathbf{a}, \mathbf{x}) = \left(a^{(2)} x[j+1]\right)^2 + \left(a^{(2)} x[j]\right)^2 +$$

$$+ 2a_0^{(2)} a_1^{(2)} x[j+1] x[j]. \tag{7.92}$$

The Lagrange multipliers θ_0 and θ_1 are calculated by solving the system of equations

$$\Phi_j(\theta) = \sum_{h-1}^{R} \frac{\tilde{\mathcal{R}}_j^{(h)}(\theta)}{\mathcal{R}^{(h)}(\theta)} + \frac{\tilde{\mathcal{B}}_j(\theta_j)}{\mathcal{B}_j(\theta_j)} - y[j+1] = 0, \quad j = \overline{0, s}, \tag{7.93}$$

Table 7.3: Results of computations.

Var	J_τ^*	θ_τ^*	ε_τ	$\hat{r}_\tau$	$\hat{P}_\tau$	t, s
1	0.0025	-2.4457	0.0025	0.1200	≈ 1.0	1576
		-0.0537				
2	0.0001	-2.4042	0.0001	0.0051	≈ 1.0	1573
		-0.0634				

where the integral components have the form

$$\tilde{\mathcal{R}}_j^{(h)}(\theta) = \int_{\mathcal{A}^{(h)}} V_j^{(h)}\left(\mathbf{a}^{(h)}, \mathbf{x}[j]\right) \exp\left(-\sum_{i=0}^{1} \theta_i V_i^{(h)}\left(\mathbf{a}^{(h)}, \mathbf{x}[j]\right)\right) d\mathbf{a}^{(h)},$$

$$\tilde{\mathcal{B}}_j(\theta) = \int_{\Xi_j} \xi_j \exp\left(-\theta_j \xi_j\right) d\xi_j. \tag{7.94}$$

The measured values of the object's output vector are $y[1] = 0.5$ and $y[2] = 2.2$.

For solving equations (7.93), minimize the residual function

$$J(\theta) = \|\Phi(\theta)\|_{L_2} \tag{7.95}$$

on the cube Π_θ.

The results of computations are presented in Table 7.3 and also illustrated in Fig. 7.5. Due to the existing integral components (7.94), the objective function was defined algorithmically. For this collection of tests, the algorithm was implemented in MATLAB using Parallel Computing Toolbox. The integral components (7.94) were calculated by `integral2` function whereas the minimum value by the GFS algorithm. All computations were performed with 32 working processors (often called workers) on the same computing system as in the case of analytical tests. The solution of equations (7.95) was obtained for the two search domains Π_θ: $\theta \in [-10, 10] \times [-10, 10]$ for Variant 1, and $\theta \in [-3, -2] \times [-0.5, 0.5]$ for Variant 2; see Table 7.3.

7.3 ON CALCULATION OF MULTIDIMENSIONAL INTEGRALS USING MONTE CARLO METHOD

The equations of the Lagrange multipliers that are induced by the empirical balance constraints contain the integral components (7.13, 7.14) on the domains with unit volumes.

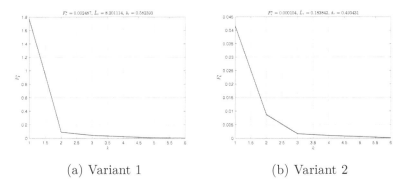

(a) Variant 1   (b) Variant 2

Figure 7.5

These integrals will be calculated using the Monte Carlo method with a random generator as follows. For each element of a vector $\mathbf{w} \in I^n \subset R^n$, it produces sequences of random numbers of length M to form $N = M^n$ random vectors (points) $\mathbf{w}^{(i)}$, $i = \overline{1, N}$. The details can be found in [147].

A major difficulty of the Monte Carlo method is to estimate the length M of the sequence of random numbers (the number N of random points) that is sufficient for integrals calculation with a given error δ. Obviously, such an estimate has a random character and should be equipped with a corresponding probability.

For obtaining these estimates, two implementation strategies of the Monte Carlo method will be compared with one another. The first strategy involves a *uniform orthogonal lattice* $\mathbb{S}$ constructed in I^n from the uniform lattices $\mathbb{S}_k$ for each element w_k with the step

$$h = \frac{1}{M}. \tag{7.96}$$

The elementary cubes in I^n have the same volumes

$$v = h^n = \left(\frac{1}{M}\right)^n. \tag{7.97}$$

Then equalities (7.13, 7.14) can be written as

$$\tilde{\mathbb{B}}(\mathbf{z}) \approx v \sum_{i=1}^{N} \phi(\mathbf{w}^i, \mathbf{z}),$$

$$\tilde{\mathbb{G}}(\mathbf{z}) \approx v \sum_{i=1}^{N} \varphi(\mathbf{w}^i, \mathbf{z}), \tag{7.98}$$

where

$$\phi(\mathbf{w}^i, \mathbf{z}) = \exp\left(-E(\mathbf{z}, \mathbf{w}^{(i)})\right),$$
$$\varphi(\mathbf{w}^i, \mathbf{z}) = E(\mathbf{z}, \mathbf{w}^{(i)}) \exp\left(-E(\mathbf{z}, \mathbf{w}^{(i)})\right). \tag{7.99}$$

The second strategy is based on a *nonuniform (random) orthogonal lattice* $\hat{\mathbb{S}}$ constructed from the product of nonuniform (random) lattices $\hat{\mathbb{S}}_k$ for each kth element of the vector $\mathbf{w}$. In contrast to the uniform lattice, the step of the nonuniform one makes up

$$h_k^{(i)} = w_k^{(i+1)} - w_k^{(i)}, \tag{7.100}$$

thereby representing a *random variable*. The elementary cubes in the lattice $\hat{\mathbb{S}}$ have *random* volumes $v^{(1)}, \ldots, v^{(N)}$, i.e.,

$$v^{(i)} = \prod_{k=1}^{n} h_k^{(i)}, \qquad i = \overline{1, N}. \tag{7.101}$$

According to the mean value theorem, the *exact* values of the integrals on the nonuniform lattice (7.100, 7.101) are given by

$$\mathbb{B}(\mathbf{z}) = \sum_{i=1}^{N} v^{(i)} \phi(\zeta^{(i)}, \mathbf{z}),$$
$$\mathbb{G}(\mathbf{z}) = \sum_{i=1}^{N} v^{(i)} \varphi(\xi^{(i)}, \mathbf{z}), \tag{7.102}$$

where $\zeta^{(i)}$ and $\xi^{(i)}$ denote the mean points of the elementary cubes of the lattice $\hat{\mathbb{S}}$ for the integrals (7.13) and (7.14), respectively.

Define the error between the exact (7.102) and approximate (7.98) values of these integrals as

$$|\mathbb{B}(\mathbf{z}) - \tilde{\mathbb{B}}(\mathbf{z})| = |\sum_{i=1}^{N} v^{(i)} \phi(\zeta^{(i)}, \mathbf{z}) - v \sum_{i=1}^{N} \phi(\mathbf{a}^{(i)}, \mathbf{z})| \le$$
$$\le \sum_{i=1}^{N} |v^{(i)} - v| |\phi(\zeta^{(i)}, \mathbf{z}| +$$
$$v \sum_{i=1}^{N} |\phi(\zeta^{(i)}, \mathbf{z}) - \phi(\mathbf{a}^{(i)}, \mathbf{z})| = \epsilon(N), \tag{7.103}$$

and

$$|\mathbb{G}(\mathbf{z}) - \tilde{\mathbb{G}}(\mathbf{z})| = |\sum_{i=1}^{N} v^{(i)} \varphi(\zeta^{(i)}, \mathbf{z}) - v \sum_{i=1}^{N} \varphi(\mathbf{a}^{(i)}, \mathbf{z})| \le$$

$$\le \sum_{i=1}^{N} |v^{(i)} - v||\varphi(\zeta^{(i)}, \mathbf{z})| +$$

$$v \sum_{i=1}^{N} |\varphi(\zeta^{(i)}, \mathbf{z}) - \varphi(\mathbf{a}^{(i)}, \mathbf{z})| = \kappa(N). \tag{7.104}$$

First, let us derive upper bounds $\tilde{\epsilon}(N, n)$ and $\tilde{\kappa}(N, n)$ for $\epsilon(N)$ and $\kappa(N)$, and then solve the inequalities

$$\epsilon(N) \le \tilde{\epsilon}(N) \le \delta, \qquad \kappa(N) \le \tilde{\kappa}(N) \le \delta. \tag{7.105}$$

Here δ is a given error. We should bear in mind that $\epsilon(N)$ and $\kappa(N)$ are random variables. Therefore, inequalities (7.105) may hold with some probability only.

1. Consider inequality (7.3) and estimate the first term:

$$\sum_{i=1}^{N} |v^{(i)} - v||\phi(\zeta^{(i)}, \mathbf{z})| \le \max_{\zeta \in I^n} |\phi(\zeta, \mathbf{z})| \sum_{i=1}^{N} |v^{(i)} - v|. \tag{7.106}$$

By the definition of the function $\phi(\mathbf{w}, \mathbf{z})$, under fixed $\mathbf{z}$ (7.98) write

$$|\phi(\mathbf{w}, \mathbf{z})| \le \max_{\mathbf{w} \in I^n} \exp(|E(\mathbf{w}, \mathbf{z})|) \le \exp(G), \tag{7.107}$$

where

$$G = (\sum_{t=1}^{T} A^{(t)})^2, \qquad A^{(t)} = \max_{(n_1, \dots, n_t)=1} \sum_{n_t}^{n} |A_{n_1, \dots, n_t}|. \tag{7.108}$$

Next, estimate the difference between the random elementary volume $v^{(i)}$ (7.101) and the elementary volume v of the uniform lattice.[2] Then

$$|v^i - v| = |\prod_{k=1}^{n} h_k^i - h^n| \le h^{(n-1)} \max_i |h_k^i - h|. \tag{7.109}$$

---

[2]Our analysis is based on the following estimation for the differential of the function $f(y) = \prod_{k=1}^{m} y_k$:

$$|f(y + \delta y) - f(y)| \le \delta \max_k |f_k'(y)| \max_k |y_k|.$$

Note that the variable h_k^i is random. Consider the event

$$\mathcal{V}^k = \{\max_i |h_k^i - h| < \gamma\}, \tag{7.110}$$

and also the union of the events $\mathcal{V}_1, ..., \mathcal{V}_m$,

$$\mathcal{V} = \bigcup_{k=1}^{n} \mathcal{V}^k. \tag{7.111}$$

Here γ specifies a small positive value. A lower bound for the probability of the event $\mathcal{V}$ will be established below.

Choosing $\gamma = h^2$, write

$$\max_i |v^i - v| \leq h^{(m-1)}\gamma = h^{(m+1)}. \tag{7.112}$$

Thus, by (7.106–7.108) the inequality

$$|\sum_{i=1}^{N}(v_i - v)\phi(\zeta^i, \mathbf{z})| \leq N\left(\frac{l}{N}\right)^{n+1} \exp(G). \tag{7.113}$$

holds with the probability of the event (7.111).

2. Find a lower bound for the probability of the event (7.111). Consider the difference $|h_k^{(i)} - h|$ and take a small positive value γ such that $|h_k^{(i)} - h| > \gamma$ for all steps $h_k^{(i)}$ of the random lattice $\hat{\mathbb{S}}$. Calculate the probability of this event. If the number of random points is large (small h) and they have a uniform spread, then the distance between these points must be close to h.

The event $\max_i |h_k^{(i)} - h| > \gamma$ means that it is possible to find a pair of random variables $(\xi, \eta) \in [0, 1]$ with the following property: if one of them (e.g., ξ) takes a value $y \in [0, 1]$, then the other $\eta \neq \mathcal{L} = [y + h - \gamma, y + h + \gamma]$. Hence, the interval $\mathcal{L}$ of length 2γ that has center at the point $y + h$ is empty. Actually, assume the random variable η gets into the interval $\mathcal{L}$, i.e., $\eta = \xi - h \pm \delta$, where $0 < \delta < \gamma$; in this case, it follows that $|h_k^{(i)} - h| < \gamma$.

Due to the independence of the uniformly distributed random variables, the probability of the event $\max_i |h_k^{(i)} - h| > \gamma$ is the same for all elements of the vector $\mathbf{a}$ and does not exceed $[1 - 2\gamma + \psi(N)]^N$, where $\psi(N)$ gives an empirical probability correction for a specific random generator, which is found experimentally. An analysis of the standard random generators indicates that the function $\psi(N)$ almost vanishes starting from $N = 1000$. Therefore,

$$\tilde{P}(\max_i |h_k^{(i)} - h| > \gamma) = \tilde{P}(\mathcal{V}^k) = (1 - 2\gamma)^N. \qquad (7.114)$$

The probability of the event $\mathcal{V}$ (7.111) has the form

$$\tilde{P}(\mathcal{V}) = \tilde{P}(N, n) = (1 - 2\gamma)^{Nn}. \qquad (7.115)$$

The probability of the opposite event $\bigcup_{k=1}^{m}(\max_i |h_k^{(i)} - h| < \gamma)$ makes up

$$P(\mathcal{V}) = 1 - (1 - 2\gamma)^{Nn}. \qquad (7.116)$$

Choosing $\gamma = h^2$, from (7.116) we establish that the probability of the event (7.111) is

$$P(\mathcal{V}) > 1 - ([-2\left(\frac{l}{N}\right)^2]^{Nn}) = P(N, n). \qquad (7.117)$$

Consequently, inequality (7.113) holds with a probability not smaller than $P(N, m)$.

3. Consider the second term in (7.3). The following estimate is valid:

$$v \sum_{i=1}^{N} |\phi(\zeta^i, \mathbf{z}) - \phi(\mathbf{w}^i, \mathbf{z})| \le \left(\frac{1}{N}\right)^u \max_{(\zeta, \mathbf{w}) \in I^n} |\phi(\zeta, \mathbf{z}) - \phi(\mathbf{w}, \mathbf{z})|. \quad (7.118)$$

The expression under the summation sign defines the modulus of continuity of the function $\phi(\mathbf{w}, \mathbf{z})$; see [61]. Since this function is differentiable,

$$\max_{(\zeta, \mathbf{w}) \in I^n} |\phi(\zeta, \mathbf{z}) - \phi(\mathbf{w}, \mathbf{z})| \le \max_{\mathbf{w} \in I^n; k \in [1,n]} |\frac{\partial \phi(\mathbf{w}, \mathbf{z})}{\partial w_k}| \max_{(\zeta, \mathbf{w}) \in I^n} \|\zeta - \mathbf{w}\|. $$
$$(7.119)$$

The distance between any random point and its nearest neighbors does not exceed the diagonal of the elementary cube, i.e.,

$$\max_{(\zeta^i, \mathbf{w}^i) \in I^n; i \in [1,N]} \|\zeta^i - \mathbf{w}^i\| \le \sqrt{\sum_{k=1}^{n}(\zeta_k^i - w_k^i)^2} \le \sqrt{n} \max_k \delta_k^i, \qquad (7.120)$$
$$\delta_k^i = \zeta_k^i - w_k^i.$$

Take a sufficiently small positive value γ. According to (7.100) and the definition of ζ^i as the mean point of the elementary cube of the nonuniform random lattice, consider the chain of inequalities

$$\max_k \delta_k^i < \max_k h_k^i < \gamma. \tag{7.121}$$

Let $\gamma = h^2$, where h is the step of the uniform lattice. The latter inequality holds with the probability $P(N)$ (7.117).

Estimate the value

$$\frac{\partial \phi}{\partial w_k} = -E(\mathbf{w}, \mathbf{z}) \exp(-E(\mathbf{w}, \mathbf{z})) \frac{\partial E}{\partial w_k}, \tag{7.122}$$

where

$$\frac{\partial E}{\partial w_k} = \sum_{t=1}^{T} \sum_{(n_1,\ldots,n_t)=1}^{n} A_{(n_1,\ldots,n_t)}(\mathbf{z}) \prod_{r=1, n_r \neq k}^{t} x_{n_r}. \tag{7.123}$$

Using (7.106–7.108), write

$$\max_{\mathbf{w} \in I^n} \max_k |\frac{\partial \phi}{\partial w_k}| \leq G^2 \exp(G). \tag{7.124}$$

Then the estimate (7.119) takes the form

$$\max_{(\zeta, \mathbf{w}) \in I^n} |\phi(\zeta, \mathbf{z}) - \phi(\mathbf{w}, \mathbf{z})| \leq \sqrt{m} \left(\frac{l}{N}\right)^{2+n} G^2 \exp(G) = \frac{n}{N} \sqrt{n} G^2 \exp(G). \tag{7.125}$$

Substitute (7.107) and (7.119) into (7.3) to establish that the inequality

$$|\mathbb{B}(\mathbf{z}) - \tilde{\mathbb{B}}(\mathbf{z})| \leq N \exp(G) \left(\frac{l}{N}\right)^m \left[1 + \sqrt{m} \left(\frac{l}{N}\right)^2 G^2\right] = \tilde{\epsilon}(N, m) \tag{7.126}$$

holds with a probability not smaller than $P(N, m)$ (7.117).

4. Now switch to inequalities (7.3). They are structurally similar to inequalities (7.3), and the only difference consists in the function $\varphi(\mathbf{w}, \mathbf{z})$ (7.98). Hence,

$$\sum_{i=1}^{N} |v^i - v| |\varphi(\zeta, \mathbf{z})| \leq \max_{\zeta \in I^n} |\varphi(\zeta, \mathbf{z})| \sum_{i=1}^{N} |v^i - v| \leq N \left(\frac{l}{N}\right)^{n+1} \max_{\zeta \in I^n} |\varphi(\zeta, \mathbf{z})|. \tag{7.127}$$

By the definition of the function $\varphi(\mathbf{a}, \mathbf{z})$ (7.98),

$$\max_{\zeta \in I^n} |\varphi(\zeta, \mathbf{z})| \leq \max_{\mathbf{w} \in I^n} |E(\mathbf{w}, \mathbf{z})| \exp(E(\mathbf{w}, \mathbf{z})) = G \exp(G), \qquad (7.128)$$

where G is given by equalities (7.108). For the second term in (7.3), the estimate takes the form

$$\max_{(\zeta, \mathbf{w})} |\varphi(\zeta, \mathbf{z}) - \varphi(\mathbf{w}, \mathbf{z})| \leq \max_{\mathbf{w} \in I^n, k \in [1,n]} |\frac{\partial \varphi}{\partial w_k}| \max_{(\zeta, \mathbf{w}) \in I^n} \|\zeta - \mathbf{w}\|. \qquad (7.129)$$

According to the definition of the function $\varphi(\mathbf{w}, \mathbf{z})$ (7.98), also write

$$\frac{\partial \varphi}{\partial w_k} = \frac{\partial E(\mathbf{w}, \mathbf{z})}{\partial w_k} \exp(-E(\mathbf{w}, \mathbf{z}))(1 - E^2(\mathbf{w}, \mathbf{z})), \qquad (7.130)$$

which yields the estimate

$$\max_{\mathbf{w} \in I^n, k \in [1,n]} |\frac{\partial \varphi}{\partial w_k}| \leq G(1 + G^2) \exp(G). \qquad (7.131)$$

Thus, the inequality

$$|\mathbb{G}(\mathbf{z}) - \tilde{\mathbb{G}}(\mathbf{z})| \leq N \left(\frac{l}{N}\right)^{n+1} G \exp(G) \left[1 + \frac{l}{N}(1 + G^2)\right] = \tilde{\kappa}(N, n) \qquad (7.132)$$

takes place with a probability not smaller than $P(N, m)$ (7.117).

An upper bound for the value N (or M) can be calculated from the condition

$$N^* = \max(\tilde{\epsilon}(N, n) = \delta; \ \tilde{\kappa}(N, n) = \delta). \qquad (7.133)$$

7.4 MULTIPLICATIVE ALGORITHMS WITH P-ACTIVE VARIABLES

Many problems arising in RML and ML procedures are reduced to systems of nonlinear equations, e.g., the balance equations (7.9–7.12). Let us illustrate a method for solving such equations that are associated with the entropy maximization problems on polyhedrons [140].

Consider a system of equalities that are interpreted as the balance constraints between data $\mathbf{q} \in R^r$ and an output $\mathbf{y} \in R^n$ of some linear parametric model:

$$T\mathbf{y} = \mathbf{q}, \quad (T, \mathbf{y}) \geq 0, \ \mathbf{q} \geq \varepsilon > 0, \qquad r < n. \qquad (7.134)$$

Here T denotes a matrix of full rank r. For the system (7.134), there exists a set of solutions and a unique solution among them that maximizes the entropy

$$H(\mathbf{y}) = -\sum_{i=1}^{n} y_i \ln \frac{y_i}{ea_i}, \qquad e = 2.71, \tag{7.135}$$

with $a_1, \ldots, a_n$ acting as parameters.

Consider the following constrained optimization problem for the entropy (7.135):

$$H(\mathbf{y}) \Rightarrow \max, \tag{7.136}$$
$$T\mathbf{y} = \mathbf{q}.$$

The corresponding Lagrange function has the form

$$L(\mathbf{y}, \bar{\lambda}) = H(\mathbf{y}) + \sum_{k=1}^{r} \lambda_k \left(q_k - \sum_{i=1}^{n} t_{ki} y_i \right). \tag{7.137}$$

The first-order optimality conditions are given by

$$\nabla_{y_i} L(\mathbf{y}, \bar{\lambda}) = \frac{\partial H(\mathbf{y})}{\partial y_i} - \sum_{k=1}^{r} \lambda_k t_{ki} = 0, \quad i = \overline{1, n};$$

$$\nabla_{\lambda_k} L(\mathbf{y}, \bar{\lambda}) = q_k - \sum_{i=1}^{n} t_{ki} y_i = 0, \quad k = \overline{1, r}. \tag{7.138}$$

The first group of equations yields a relation between the variables $\mathbf{y}$ and the Lagrange multipliers $\bar{\lambda}$ in the form

$$y_i(\bar{\lambda}) = \frac{a_i}{\sum_{k=1}^{r} \exp(t_{ki} \lambda_k)}, \qquad i = \overline{1, n}. \tag{7.139}$$

This relation allows us to eliminate the primal variables $\mathbf{y}$ from the definition of the Lagrange function (7.137), leading to the following expression for the dual function (which depends on the dual variables—the Lagrange multipliers):

$$\tilde{L}(\bar{\lambda}) = L[\mathbf{y}(\bar{\lambda}), \bar{\lambda}]. \tag{7.140}$$

The elements of its gradient are

$$\nabla_{\lambda_k} \tilde{L}(\bar{\lambda}) = q_k - \sum_{i=1}^{n} y_i(\bar{\lambda}), \qquad k = \overline{1, r}. \tag{7.141}$$

In terms of the dual function, the second group of the optimality conditions (7.138) can be written as

$$\nabla_{\lambda_k} \tilde{L}(\bar{\lambda}) = 0, \qquad k = \overline{1, r}. \tag{7.142}$$

Hence, the balance constraints take the form

$$\tilde{B}_k(\bar{\lambda}) = \frac{1}{q_k} \sum_{i=1}^{n} t_{ki} \frac{a_i}{\sum_{j=1}^{r} \exp\left(\lambda_j t_{ji}\right)} = 1, \quad \bar{\lambda} \in R^r, \qquad k = \overline{1, r}. \tag{7.143}$$

Often it seems convenient to transform these equations using the nonnegative variables $z_k = \exp(\lambda_k)$, $k = \overline{1, r}$, known as the exponential Lagrange multipliers. Then equations (7.143) are reduced to

$$B_k(\mathbf{z}) = \frac{1}{q_k} \sum_{i=1}^{n} t_{ki} a_i \left(\prod_{j=1}^{r} z_k^{-t_{ji}} \right) = 1, \quad \mathbf{z} \in R_+^r, \qquad k = \overline{1, r}. \tag{7.144}$$

Now, for the dual function (7.140), the Hessian matrix has the form

$$\Gamma(\bar{\lambda}) = [\nabla_{\lambda_j, \lambda_l} \tilde{L}(\bar{\lambda}) \,|\, (j, l) = \overline{1, r}], \tag{7.145}$$

where

$$\nabla_{\lambda_j, \lambda_l} \tilde{L}(\bar{\lambda}) = -\sum_{i=1}^{n} t_{kj} \frac{\partial y_j}{\partial \lambda_l}. \tag{7.146}$$

The primal variables expression (7.139) implies the inequality

$$\frac{\partial y_j}{\partial \lambda_l} \leq 0, \tag{7.147}$$

holding for all $\lambda \in R^r$. Therefore, the Hessian matrix (7.143) consists of nonnegative elements. Taking into account (7.145), the quadratic form $K = \langle \Gamma \bar{\lambda}, \bar{\lambda} \rangle$ is strictly positive definite for any $\bar{\lambda} \in R^r$. This means that the dual function $\tilde{L}$ is strictly convex. The system of equations 7.143 guarantees boundedness of its solution, i.e., there exists a constant $c > 0$ such that $|\bar{\lambda}| \leq c < \infty$. Consequently, $\tilde{L} \geq L_{\min}$.

For solving equations (7.138), utilize the multiplicative algorithm with p-active variables as follows.

Initial step:

$$\mathbf{z}^0 > \mathbf{0}.$$

Iterative step:

$$z_{k_1(s)}^{(s+1)} = z_{k_1(s)}^s B_{k_1(s)}^\gamma(\mathbf{z}^s),$$
$$\cdots = \cdots\cdots\cdots,$$
$$z_{k_p(s)}^{(s+1)} = z_{k_p(s)}^s B_{k_p(s)}^\gamma(\mathbf{z}^s) \tag{7.148}$$
$$z_k^{(s+1)} = z_k^s, \quad k \neq k_1(s), \dots, k_p(s),$$
$$(k, k_1(s), \dots, k_p(s)) \in \overline{1, n}.$$

Further analysis will need an additive form of the algorithm (7.148) in the variables $\lambda_k = \ln z_k$. This form is written in the following way.

Initial step:

$$\bar{\lambda}^0 \in R^n.$$

Iterative step:

$$\lambda_{k_1(s)}^{(s+1)} = \lambda_{k_1(s)}^s + \gamma \ln \tilde{B}_{k_1(s)}(\bar{\lambda}^s),$$
$$\cdots = \cdots\cdots\cdots,$$
$$\lambda_{k_p(s)}^{(s+1)} = \lambda_{k_p(s)}^s + \gamma \ln \tilde{B}_{k_p(s)}(\bar{\lambda}^s) \tag{7.149}$$
$$\lambda_k^{(s+1)} = \lambda_k^s, \quad k \neq k_1(s), \dots, k_p(s),$$
$$(k, k_1(s), \dots, k_p(s)) \in \overline{1, n},$$

where $\tilde{B}_k(\bar{\lambda}) = B_k(\exp(\bar{\lambda}))$, $k = \overline{1, r}$.

A major difficulty for such algorithms is to select active variables. This selection can be organized by a cyclical rule, e.g., for $p = 1$,

$$k_1(s) = s(\bmod n + 1). \tag{7.150}$$

A more efficient approach involves residual feedback, i.e.,

$$\vartheta_k(\mathbf{z}^s) = |1 - B_k(\mathbf{z}^s)|, \qquad k = \overline{1, n}. \tag{7.151}$$

Introduce the following notations:

$$\frac{r}{p} = I + \delta, \qquad \rho = s[(mod)\,(I + 1)],$$
$$I = \left[\frac{r}{p}\right], \qquad 0 \leq \delta \leq p - 1. \tag{7.152}$$

Here $[\cdot]$ indicates the integer part of value $\cdot$. Consider the index sets

$$K = \{1, \dots, r\}, \qquad K_m(s) = \{k_1(s), \dots, k_m(s)\}, \tag{7.153}$$

where

$$m = \begin{cases} 1, \ldots, p \text{ if } \rho < I; \\ 1, \ldots, \delta \text{ if } \rho = I. \end{cases} \tag{7.154}$$

Let

$$\begin{aligned} P_{m-1}(s) &= \left[\bigcup_{v=1}^{\rho} K_p(s-v) \right] \bigcup K_{m-1}(s), \\ G_{m-1}(s) &= K \setminus P_{m-1}(s), \end{aligned} \tag{7.155}$$

where $K_0(s) = P_0(s) = G_0(s) = \emptyset$.

Now we can state the selection rule of active variables: *the packet of active variables is formed by the p variables whose indexes correspond to the maximum residuals*, i.e.,

$$k_m(s) = \arg \max_{i \in G_{m-1}(s)} \vartheta_i(\mathbf{z}^s). \tag{7.156}$$

This rule yields the chain of inequalities

$$\vartheta_{k_p(s)}(\mathbf{z}^s) < \vartheta_{k_{p-1}(s)}(\mathbf{z}^s) < \cdots \vartheta_{k_1(s)}(\mathbf{z}^s). \tag{7.157}$$

Application of the rule (7.156, 7.157) makes all variables active in $(I+1)$ iterations, and this situation occurs again and again with period $(I+1)$.

Our convergence analysis of the algorithm (7.149) will begin with an important notion as follows.

Definition *The algorithm (7.149) is called G_λ-convergent if there exist a subset $G_\lambda \in R^n$ and positive scalars $a(G_\lambda), \gamma$ such that, for all $\gamma \in (0, a(G_\lambda))$ and $\bar{\lambda}^0 \in G_\lambda$, the algorithm converges to the solution $\bar{\lambda}^*$ of equation (7.143) with the linear rate.*

A similar definition can be introduced for the algorithm (7.148), in which the variables are the exponential Lagrange multipliers $\mathbf{z} \in R^r_+$.

Consider a subsequence $\bar{\lambda}^{(I+1)s}$ of the elements that are generated by the algorithm (7.149) and also an auxiliary system of nonlinear differential equations that is obtained from (7.149) as $\gamma \to 0$, i.e.,

$$\frac{d\lambda_k}{dt} = \ln \tilde{B}_k(\bar{\lambda}), \qquad \bar{\lambda}(0) = \bar{\lambda}^0, \qquad k = \overline{1, r}. \tag{7.158}$$

Lemma 1 *The singular point $\bar{\lambda}^*$ of the system (7.158) is asymptotically stable for any initial deviations $\bar{\lambda}^0 \in R^r$.*

Proof. Consider the dual function (7.140) and define its derivative along the trajectory of the system (7.158):

$$\frac{d\tilde{L}}{dt} = \sum_{k=1}^{r} \frac{d\tilde{L}}{d\lambda_k} \frac{d\lambda_k}{dt} = \sum_{k=1}^{r} [1 - \tilde{B}_k(\bar{\lambda})] \ln \tilde{B}_k(\bar{\lambda}). \tag{7.159}$$

Then $\frac{d\tilde{L}}{dt} < 0$ for all $\tilde{B}_k(\bar{\lambda}) \neq 1$ and $\frac{d\tilde{L}}{dt} = 0$ for $\tilde{B}_k(\bar{\lambda}) = 1$. Since there exists a unique minimum point of the dual function, its gradient vanishes only at this point. Hence, $\frac{d\tilde{L}}{dt} < 0$ for all $\bar{\lambda} \in R^r$, and $\frac{d\tilde{L}}{dt} = 0$ for $\bar{\lambda} = \bar{\lambda}^*$.

Therefore, $\tilde{L}$ (7.140) is a Lyapunov function for equation (7.158), and this property holds for any $\bar{\lambda}^0 \in R^r$. ■

The convergence conditions of the iterative process (7.149) are defined by **Theorem 1**. *Assume that:*

(a) the functions $\Phi_k(\bar{\lambda}) = \ln \tilde{B}_k(\bar{\lambda})$, $(k = \overline{1, n})$, *are continuously differentiable;*

(b) the Jacobian matrix $J(\bar{\lambda}^*) = \left[\frac{\partial \Phi_k(\bar{\lambda})}{\partial \lambda_i}\right]_{\bar{\lambda}^*}$ *is a Hurwitz matrix;*

(c) the set G_λ *is compact;*

(d) there exists $\varepsilon > 0$ *such that the inequality* $\|\bar{\lambda}(t) - \bar{\lambda}^*\| \leq \varepsilon$, *where* $\bar{\lambda}(t)$ *is the solution of the system (7.156) with the initial condition* $\bar{\lambda}(0) = \bar{\lambda}^0$, *holds for all* $\bar{\lambda}^0 \in G_\lambda$ *starting from some* $t > 0$.

Then the process generated by the algorithm (7.149) is G_λ-*convergent.*

Under the hypotheses of this theorem, the algorithm (7.149) gives a Euler polygon approximation for the solution of the system (7.158).

Proof. Choose a countable everywhere dense set of points $\{\zeta_s\}$ in G_λ. Let t_1 be the first time when the solution of equation (7.158) with the initial condition ζ_1 gets into the $\varepsilon/2$-neighborhood of the point $\bar{\lambda}^*$; this time t_1 exists by condition (d). Choose a neighborhood $\mathcal{G}_1$ of the point ζ_1 so that the points from it will reach the $3\varepsilon/4$-neighborhood of the point $\bar{\lambda}^*$ along the trajectory of the system (7.158) after the time t_1. Note that condition (a) guarantees that the solution is continuously dependent on the initial conditions.

Choose a point ζ_2 among the ones not belonging to $\mathcal{G}_1$. Like for the point ζ_1, define t_2 and a neighborhood $\mathcal{G}_2$ for the point ζ_2. Continue this process to obtain a cover of the set G_λ by open sets $\mathcal{G}_1, \mathcal{G}_2, \ldots$ Due to condition (c), from these sets it is possible to choose a finite subcover $\mathcal{G}_{i_1}, \ldots, \mathcal{G}_{i_m}$. Let $T = \min\{t_{i_1}, \ldots, t_{i_m}\}$. Finally, using condition (a), take a sufficiently small value $\gamma > 0$ so that on the interval $0, T$ the solution of the system (7.158) is approximated by the Euler polygons with the accuracy $\varepsilon/4$.

CHAPTER 8

GENERATION METHODS FOR RANDOM VECTORS WITH GIVEN PROBABILITY DENSITY FUNCTIONS OVER COMPACT SETS

The final stage of the RML procedure is to generate an ensemble of random vectors or trajectories using the entropy-optimal randomized model. Here of crucial importance are the methods that allow to generate these random objects with given PDFs of arbitrary form. In the general statement, the generation problem of random sequences and random vectors is classical for the Monte Carlo method. There exist many methods for generating "standard" sequences of random numbers with given probabilistic characteristics (uniform, Gaussian, Poisson distributions, etc. [159, 168]). However, when the matter concerns the sequences obeying arbitrary-form PDFs and also the sequences of high-dimensional random vectors, just a few methods can be suggested. At the beginning

DOI: 10.1201/9781003306566-8

of this chapter, we will give a brief survey of generation methods, which does not claim to be exhaustive and provides an insight into the main concepts underlying these methods. Our attention will be focused on a method for generating random vectors with PDFs defined on compact sets. It approximates a given PDF by n-dimensional piecewise constant functions, in which the step depends on the Lipschitz constant of this PDF. A major aspect of random object generation is to perform a statistical check whether the probabilistic properties of the generated objects match the given PDF.

8.1 A SURVEY OF GENERATION METHODS FOR RANDOM OBJECTS

The problem to generate random variables, vectors and trajectories has a growing significance because the Monte Carlo method is a real tool for solving different problems of numerical integration, optimization, control and machine learning, including randomized machine learning. Therefore, it would be useful for the reader to get acquainted with some generation methods that are frequently mentioned in the literature. This survey consists of two parts, the first being devoted to the generation of random variables and the second to the generation of random vectors.

8.1.1 Random variables

1. Continuous random variables
    Consider a random variable ξ that takes values within an interval $[a, b]$ and has a continuous PDF $w(x)$. Its probability distribution function has the form

$$F(x) = \int_a^x w(y)dy. \tag{8.1}$$

The range of the function F is the interval $[0, 1]$. Define a random variable η obeying the uniform distribution on the interval $[0, 1]$ and study the equation

$$F(\xi) = \eta. \tag{8.2}$$

The solution of this equation—a random variable ξ—has the PDF $w(x)$, see [168].
    2. The method of inverse functions
    Consider a random variable ξ distributed on an interval $[a, b]$ with a continuous PDF $w(x)$. Its probability distribution function $F(x)$ has the form (8.1). Assume that it possesses the inverse function $G(y) = F^{-1}(y)$. As established in [159, 168], the random variable $\xi = G(\eta)$, where η is given by (8.2), has the PDF $w(x)$. This result applies to the PDFs with discontinuities of the first kind.
    3. The composition method of inverse functions
    Let a random variable ξ be distributed on an interval $[a, b]$ with a PDF $w(x)$ (possibly, with discontinuities of the first kind). In addition, assume that this PDF

represents the sum of PDFs $w_i(x)$ with the same properties, i.e.,

$$w(x) = \sum_{i=1}^{n} p_i w_i(x), \qquad p_i \geq 0, \qquad \sum_{i=n}^{n} p_i = 1. \tag{8.3}$$

The random variables ξ_i with the PDFs $w_i(x)$ are generated using one of the above-mentioned methods. The non-random variables $p_1, \ldots, p_n$ take values within the unit interval. They will be interpreted as probabilities; to this end, introduce an integer random variable Y such that $Pb(Y = i) = p_i$. Then the random variable ξ can be written as

$$\xi = \sum_{i=1}^{n} \xi_i \Delta(Y = i), \qquad \Lambda(Y = i) = 1 \text{ if } Y = i; \; \Lambda(Y = i) = 0 \text{ if } Y \neq i. \tag{8.4}$$

4. The acceptance-rejection method (AR method)

This method was developed by S. Ulam and J. von Neumann. In contrast to the previous methods, it is based on simulations of generated random variables, including their checking subject to a given PDF. Let us study the main idea of the AR method. Consider a random variable ξ distributed on an interval $[a, b]$ with a PDF $w(x)$. Introduce the number

$$c = \max\{w(x), x \in [a, b]\}. \tag{8.5}$$

Generate a pair of independent random variables η, γ obeying the uniform distribution on the intervals $[a, b]$ and $[0, c]$, respectively. If $\gamma < w(\eta)$, then $\xi = \eta$. Otherwise, repeat the generation procedure.

A pair (γ, η) corresponds to a point on the (y, x) plane, which contains the set $\mathcal{W} = \{(y, x) : y \leq w(x), x \in [a, b]\}$. The points generated by the AR method are uniformly filling the rectangle $[0, c] \times [a, b]$. However, many of them do not belong to the requisite set $\mathcal{W}$. This leads to a natural desire to put the set $\mathcal{W}$ into a closer set. A general approach to solve this problem was proposed in [159].

5. Generation of common PDFs: several examples

The exponential PDF. Consider a random variable ξ distributed with the PDF $w(x) = \lambda \exp(-\lambda x), x \geq 0$. The corresponding probability distribution function has the form

$$F(x) = 1 - \exp(-\lambda x).$$

The inverse function is

$$G(y) = \frac{1}{\lambda} \ln(1 - y).$$

Generating the random variable Y with the uniform distribution on the interval $[0, 1]$, obtain the random variable

$$\xi = \frac{1}{\lambda} \ln(1 - Y).$$

The Gaussian PDF. A Gaussian random variable is $\xi \sim N(\mu, \sigma^2)$, where μ denotes its mean and σ^2 its variance. Assume that a generator of a standard Gaussian random variable $\Xi \sim N(0, 1)$ is available. Then

$$\xi = \mu + \Xi \sigma.$$

The generation methods of random variables with the Gaussian, gamma and beta PDFs distributions (with continuous and discrete "sharpness" parameters) were described, e.g., in the book [159]. These methods are based on generators of uniformly distributed random variables on the interval $[0, 1]$.

8.1.2 Random vectors

1. Independent elements

If a random vector $\bar{\xi}$ consists of independent elements ξ_i, then its PDF and probability distribution function have a multiplicative structure, i.e.,

$$w(\mathbf{x}) = \prod_{i=1}^{n} w_i(x_i), \qquad F(\mathbf{x}) = \prod_{i=1}^{n} F_i(x_i). \tag{8.6}$$

Therefore, each element ξ_i can be generated independently by solving the equation

$$F_i(\xi_i) = \eta_i, \qquad i = \overline{1, n}, \tag{8.7}$$

where $\eta_1, \ldots, \eta_n$ are the independent random variables that obey the uniform distribution on the interval $[0, 1]$.

2. Dependent elements

If a random vector $\bar{\xi}$ has dependent elements ξ_i, then its PDF can be written as the product of the conditional PDFs:

$$w(\mathbf{x}) = w_1(x_1)w(x_2 \,|\, x_1) \cdots w_n(x_n \,|\, x_1, \ldots, x_{n-1}). \tag{8.8}$$

Let us introduce the conditional probability distribution functions

$$F_i(x_i \,|\, x_1, \ldots, x_i) = \int_{-\infty}^{x_i} w_i(x \,|\, x_1, \ldots, x_i)dx, \qquad i = \overline{1, n}. \tag{8.9}$$

Consider the system of equations

$$F_1(\xi_1) = \eta_1,$$
$$F_2(\xi_2 \,|\, \xi_1) = \eta_2,$$
$$\cdots\cdots\cdots\cdots$$
$$F_n(\xi_n \,|\, \xi_1, \ldots, \xi_{n-1}) = \eta_{n-1}. \tag{8.10}$$

Note that $(\eta_1, \ldots, \eta, n)$ are the independent random variables with the uniform distribution on the interval $[0, 1]$. Solving these equations sequentially, find a collection of the desired random variables, i.e., a vector $\bar{\xi} = \{\xi_1, \ldots, \xi_n\}$ with the PDF $w(\mathbf{x})$ [168].

3. The multidimensional AR method

Consider a given PDF $w(\mathbf{x})$, where the vector $\mathbf{x} \in G \subset R^n$. Denote

$$c = \max(w(\mathbf{x}), \mathbf{x} \in Q). \tag{8.11}$$

Consider a set Q in space $R^{(n+1)}$ that contains the points $\{y \leq w(\mathbf{x}, \mathbf{x} \in Q)\}$, and also a cylindrical set with height c and base Q, i.e.,

$$W = \{c, Q\}. \tag{8.12}$$

Following the general idea of the AR method (see above), generate a random variable γ with the uniform distribution on the interval $[0, c]$ and also a random vector $\bar{\eta}$ with the uniform distribution on the set Q. This yields a random point $(\gamma, \bar{\eta})$ in the set W (8.12). If it gets into the set Q, then $\mathbf{x} = \bar{\eta}$. Otherwise, the generation procedure is repeated.

Clearly, the efficiency of this method essentially depends on the configuration of the set that contains the set $\mathcal{G}$. It seems that the cylindrical sets are not best. Therefore, the generation method under study is sensitive to the quality of approximation for the set $\mathcal{G}$.

4. The Metropolis-Hastings (MH) method

Unlike the previously considered ones, this method involves a specially designed Markov chain for generating each element of a sample with a given PDF. Assume that it is necessary to generate an element of a sample with a PDF $w(\mathbf{x})$. Since the generation procedure is based on a Markov chain, each new value $\mathbf{x}^{(s+1)}$ of this element depends on its preceding value $\mathbf{x}^s$ only. Consider an auxiliary conditional PDF $q(\mathbf{y}\,|\,\mathbf{x}^s)$. A reasonable choice is using a PDF with a rather simple generation of random objects. As a rule, the Gaussian PDF works well.

At step s of the MH method, one should generate a random vector $\mathbf{y}^s$ and calculate the parameter

$$u = \frac{w(\mathbf{y}^s)}{w(\mathbf{x}^s)}\,\frac{q(\mathbf{x}^s)\,|\,\mathbf{y}^s}{q(\mathbf{y}^s)\,|\,\mathbf{x}^s}. \tag{8.13}$$

The transition rule has the form

$$\begin{aligned} u \le 1, \qquad \mathbf{x}^{(s+1)} &= \mathbf{y}^s, \\ u > 1, \qquad \mathbf{x}^{(s+1)} &= \mathbf{x}^s. \end{aligned} \tag{8.14}$$

Similarly to the AR method, the efficiency of the MH method depends on the form of the auxiliary function. In particular, if the Gaussian distribution is used as the auxiliary one, then its parameters are adjusted for a better approximation of a given PDF. For details, the reader is referred to [31, 32, 64, 120, 121].

8.2 DIRECT GENERATION METHOD FOR RANDOM VECTORS WITH GIVEN PROBABILITY DENSITY FUNCTION

This method uses an approximation of a given PDF by a step function, with a subsequent generation of the uniformly distributed independent random vectors inside a randomly chosen elementary cube. Consider a PDF $w(\mathbf{x})$, $\mathbf{x} \in Q \subset R^n$.

8.2.1 Unit cube Q

Introduce a uniform lattice with step h for each coordinate of the unit cube Q. Then this cube is partitioned into the elementary cubes of volumes $V_i = h^n$. Let $(1/h)$ be an integer value. Then the number of elementary cubes is $N = (1/h)^n$ and $i = \overline{1, N}$. For each elementary cube, introduce another lattice with step η. As a result, each elementary cube consists of inner subcubes of volumes $v_i^j = \eta^n$. Denote their number by $M = (1/\eta)^n$, $j = \overline{1, M}$. Assume that the total volume of all elementary cubes coincides with the total volume of all their subcubes. Choose an arbitrary point $\mathbf{x}_i^j$ in each inner subcube and define the value

$$C_i = \frac{1}{M} \sum_{j=1}^{M} w(\mathbf{x}_i^j). \tag{8.15}$$

For a sufficiently large number M,

$$C_i \sim \frac{1}{V_i} \int_{V_i} w(\mathbf{x}) d\mathbf{x}. \tag{8.16}$$

Therefore,

$$C_i \sim w(\mathbf{x}) \quad \text{if } \mathbf{x} \in V_i. \tag{8.17}$$

Denote by $\bar{\xi}^i$, $i = \overline{1, N}$, the independent random vectors obeying the uniform distribution on the elementary cubes. The density of such a vector is $1/V_i$ inside an elementary cube and 0 outside it. These vectors can be generated using the method (8.6, 8.7).

Let β be an integer random variable that does not depend on $\{\bar{\xi}^i\}$ and takes values $1, \ldots, N$ with the probabilities $p_i = C_i V_i$. Then the generated vectors have the form

$$\mathbf{x} = \sum_{i=1}^{N} \bar{\xi}^i \Lambda(\beta = i), \tag{8.18}$$

where $\Lambda(A)$ means the indicator function of a set A. The PDF of the vector $\mathbf{x}$ can be written as

$$P(\mathbf{x} = d\mathbf{x}) = \sum_{i-1}^{N} P(\mathbf{x} = d\mathbf{x} \,|\, \beta = i) P(\beta = i) =$$

$$\sum_{i=1}^{N} C_i d\mathbf{x} \Lambda(V_i) \sim \sum_{i=1}^{N} w(\mathbf{x}) d\mathbf{x} \Lambda(V_i). \tag{8.19}$$

8.2.2 Compact set Q

1. Q—parallelepiped Π

Consider the set $Pi = \{\mathbf{x} : \mathbf{a} \leq \mathbf{x} \leq \mathbf{b}\}$, where $\mathbf{a}$ and $\mathbf{b}$ are given vectors. Denote the lengths of their edges by $l_i = b_i - a_i$, $i = \overline{1, n}$. Introduce a nonuniform lattice on this set, which can be designed either with a fixed number K of elements for each coordinate or with a fixed step h. In the former case, there are different steps for all coordinates, i.e.,

$$h_i = l_i / K; \tag{8.20}$$

in the latter case, different numbers of elements for all lattices, i.e.,

$$K_i = l_i/h. \tag{8.21}$$

In fact, the fixed-step statement has been considered above. Hence, we will study the former case with a variable step. Because the lattice has the same number of elements for each coordinate, the number of all elementary parallelepipeds Π_i is $N = K^n$. The elementary parallelepipeds have the same volume

$$V = \prod_{i=1}^{n} h_i. \tag{8.22}$$

Inside each parallelepiped, introduce another lattice with M elements for each coordinate. Then for each coordinate the lattice step is

$$\eta_i = h_i/M, \tag{8.23}$$

and the volume of the jth inner subparallelepiped π_i^j makes up

$$v = \prod_{i=1}^{n} \eta_i. \tag{8.24}$$

(here j denotes the index of an inner subparallelepiped inside the ith elementary parallelepiped). The number of all inner subparallelepipeds π_i^j inside the elementary parallelepiped Π_i is

$$L = M^n. \tag{8.25}$$

Choose an arbitrary point $\mathbf{x}_i^j$ in π_i^j and calculate the value

$$C_i = \frac{1}{M} \sum_{j=1}^{M} w(\mathbf{x}_i^j). \tag{8.26}$$

Under a sufficiently large number M,

$$C_i \sim \frac{1}{V_i} \int_{V_i} w(\mathbf{x}) d\mathbf{x}. \tag{8.27}$$

The remainder of this generation procedure is organized in the same way as for the unit cube; see above.

2. Arbitrary compact set Q

In this case, put the set Q into the parallelepiped $\Pi = \{\mathbf{x} : \mathbf{a} \leq \mathbf{x} \leq \mathbf{b}\}$. The edges can be defined arbitrarily, but in such a way that

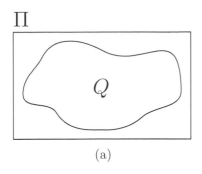

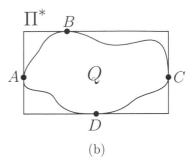

Figure 8.1

$Q \subset \Pi$. A more efficient approach to generate random vectors is putting the set Q into *a hull in form of the parallelepiped* Π^*. These possibilities are illustrated in Figs. 8.1a and 8.1b (*a*—arbitrary parallelepiped Π, *b*—hull). Clearly, hull construction requires a preliminary calculation of the extreme points (A, B, C, D).

Recall that the PDF $w(\mathbf{x})$ is defined on the set Q. Introduce an auxiliary PDF of the form

$$W(\mathbf{x}) = \begin{cases} w(\mathbf{x}) & \text{if } \mathbf{x} \in Q, \\ 0 & \text{if } \mathbf{x} \in \Pi \setminus Q. \end{cases} \tag{8.28}$$

It is immediately obvious that the parallelepiped Π forms the domain of definition of the PDF $W(\mathbf{x})$. We will use the generation procedure of item 2 for the PDF $W(\mathbf{x})$. Note that $W(\mathbf{x}) = 0$ on the elementary parallelepipeds Π_i and also on the inner subparallelepipeds π_i^j inside the set $\Pi^* \setminus Q$. Hence, $P(\mathbf{x} = d\mathbf{x}) = 0$ by (8.2.1).

8.3 APPROXIMATION OF GIVEN PROBABILITY DENSITY FUNCTION

The suggested generation procedure for random vectors with a given probability density function has such important aspects as approximation *accuracy* and lattice *step*.

1. Estimation of approximation accuracy

Consider a PDF $w(\mathbf{x})$, where $\mathbf{x} \in Q \subset R^n$. The set Q is compact, i.e., has a nonzero volume $V_Q \neq 0$. Assume that the set Q is put into an n-dimensional parallelepiped Π. The function $w(\mathbf{x})$ is square integrable.

To generate random vectors with the PDF $w(\mathbf{x})$, it will be approximated by a piecewise constant function $g(\mathbf{x})$, $\mathbf{x} \in \Pi$. The quality of this approximation will be measured by the functional

$$J(A) = |\int_A (w(\mathbf{x}) - g(\mathbf{x}))\, d\mathbf{x}|, \tag{8.29}$$

where A indicates an arbitrary set in Π.

Define a uniform lattice with a step h_k, $k = \overline{1, n}$, for each coordinate on the set Π of the volume V_Π. Each element of this lattice is an elementary parallelepiped Π_i, $i = \overline{1, N}$, with the sides h_k. Without loss of generality, assume that

$$N = \frac{V_\Pi}{V_{\Pi_i}} = \frac{V_\Pi}{\prod_{k=1}^n h_k}. \tag{8.30}$$

Moreover, let this value be integer. Note that all the elementary parallelepipeds Π_i have the same volumes. In each set Π_i, replace the function $w(\mathbf{x})$ with a constant B_i. Performing this operation for all the elementary parallelepipeds, obtain a piecewise constant linear approximating function $g(\mathbf{x})$.

Consider a certain set $A \in \Pi$. It is covered by the union of the elementary parallelepipeds, i.e.,

$$A = \bigcup_{i=1}^N \Pi_i. \tag{8.31}$$

Then, using the Cauchy–Bunyakovsky–Schwarz inequality, write the upper bound

$$J(A) \le \sqrt{\int_A (w(\mathbf{x}) - g(\mathbf{x}))^2\, d\mathbf{x} \int_A \Lambda(A) d\mathbf{x}}. \tag{8.32}$$

Here $\Lambda(A)$ denotes the indicator function of the set A. It follows from (8.32) that

$$\int_A (w(\mathbf{x}) - g(\mathbf{x}))^2\, d\mathbf{x} \le \sum_{i=1}^N \int_{\Pi_i} (w(\mathbf{x}) - B_i)^2\, d\mathbf{x}. \tag{8.33}$$

Hence,

$$\sup_A J(A) \le \sum_{i=1}^N \int_{\Pi_i} (w(\mathbf{x}) - B_i)^2\, d\mathbf{x}. \tag{8.34}$$

As a result, the best estimates of the constants $B_1, \dots, B_N$ can be derived by minimizing the functional

$$\sum_{i=1}^N \int_{\Pi_i} (w(\mathbf{x}) - B_i)^2\, d\mathbf{x} \Rightarrow \min_{B_1,\dots,B_N} \tag{8.35}$$

subject to the condition

$$V_{\Pi_i} \sum_{i=1}^{N} B_i = 1. \tag{8.36}$$

This equality expresses the normalization condition of the PDF. The solution of the minimization problem is given by

$$B_i^* = \frac{1}{V_{\Pi_i}} \int_{\Pi_i} w(\mathbf{x}) d\mathbf{x}. \tag{8.37}$$

Thus, the described approximation procedure yields the optimal piecewise constant PDF $g^*(\mathbf{x})$ with the approximation accuracy

$$\sup_A J(A) \leq \sqrt{V_{\Pi}} \sqrt{\sum_{i=1}^{N} \int_{\Pi_i} (w(\mathbf{x}) - B_i^*)^2 \, d\mathbf{x}} \leq \delta. \tag{8.38}$$

2. Optimization of lattice step (estimation of Lipschitz constant)

The procedure suggested below employs an estimate of the Lipschitz constant for the function $w(\mathbf{x})$, which is assumed to be continuous. We have $B_i^* = w(\mathbf{x}_i^*)$, where $\mathbf{x}_i^* \in \Pi_i$ but the point $\mathbf{x}_i^*$ is unknown. Let the function $w(\mathbf{x})$ satisfy the Lipschitz condition with a constant L, i.e.,

$$\max_{\mathbf{x} \in \Pi_i} (w(\mathbf{x}) - B_i^*)^2 \leq L^2 \|\mathbf{x} - \mathbf{x}_i^*\|. \tag{8.39}$$

The integral form of this inequality on the intervals $[0, h_1] \times [0, h_2] \times \cdots \times [0, h_n]$ can be written as

$$\int_{|pi_i} (w(\mathbf{x}) - B_i^*)^2 \, d\mathbf{x} \leq L^2 \int_0^{h_1} \cdots \int_0^{h_n} (x_1^2 + \cdots + x_n^2) \, dx_1 \cdots dx_n =$$
$$L^2 \left(\frac{h_1^3}{3} h_2 h_3 \cdots h_n + \frac{h_2^3}{3} h_1 h_3 \cdots h_n + \cdots + \frac{h_n^3}{3} h_1 h_2 \cdots h_{n-1} \right) = L^2 \varrho. \tag{8.40}$$

The variable ϱ is independent of the index i of an elementary parallelepiped because all of them have the same sizes. Taking into account (8.40), transform inequality (8.34) into

$$\sup_A J(A) \leq L \sqrt{V_{\Pi} \sum_{i=1}^{N} \varrho} = L \sqrt{V_{\Pi} N \varrho}. \tag{8.41}$$

Then

$$V_{\Pi} = N V_{\Pi_i} = N \prod_{j=1}^{n} h_j. \tag{8.42}$$

Hence, using (8.40), the right-hand side of inequality (8.41) can be reduced to

$$L\sqrt{V_\Pi N\varrho} = L\sqrt{V_\Pi}N\varrho = L\sqrt{V_\Pi^2}\frac{\varrho}{\prod_{j=1}^n h_j} = \frac{L}{\sqrt{3}}V_\Pi\sqrt{\sum_{j=1}^n h_j^2}. \quad (8.43)$$

Denote $h = \max_{1 \le j \le n} h_j$. Then it follows from (8.41, 8.43) that

$$\sup J(A) \le L\sqrt{n/3}hV_\Pi. \quad (8.44)$$

For a given approximation accuracy δ of the PDF $w(\mathbf{x})$, the corresponding maximum lattice step is

$$h = \frac{\delta}{L\sqrt{\frac{n}{3}}V_\Pi}. \quad (8.45)$$

The maximum step for the longest edge of the parallelepiped allows us to find the number N of points and, using it, the steps h_j for all other coordinates.

3. Rough estimate of step

The estimation method of the Lipschitz constant L and the maximum step h for elementary parallelepipeds design rests on the assumption that the integral in (8.37) can be calculated analytically (without errors). Unfortunately, this is not always the case. A given PDF often has a complex structure, which requires numerical calculation of the integral in formula (8.37). Such calculations are performed using an inner lattice with a step η_i that is considerably smaller than the step h_i of the main lattice.

The step η_i can be chosen using a simple technique as follows. First, specify a certain step η_i^j, and calculate the number $M_i^j = l_i/\eta_i^j$ of points in the integral sum S_i^j. Then repeat these operations, and check the variability of the resulting integral sums. If $|S_i^j - S_i^{(j-1)}| \le \delta$, where δ is a given accuracy, then stop further partition of the inner lattice. The integral can be estimated by

$$\int_{\Pi_i} \left(w(\mathbf{x}) - \hat{B}_i^*\right)^2 d\mathbf{x} \le \hat{J}_i = S_i^{j^*} + \delta/2. \quad (8.46)$$

Here j^* denotes the index of the inner lattice for which the calculations are stopped. Then, according to (8.38),

$$\sup_A J(A) \le \sqrt{V_\Pi}\sqrt{\sum_{i=1}^N \hat{J}_i + N\delta/2}. \quad (8.47)$$

The constants $\hat{B}_1^*, \ldots, \hat{B}_N^*$ and the estimates $\hat{J}_1, \ldots, \hat{J}_N$ are obtained numerically.

Because all the calculations have a rough character, it makes sense to consider an accuracy estimate based on another metric. Let $A \subseteq \bigcup_{i=1}^N Q_i$. Since V_{Q_i} is independent of i,

$$\sup_A J(A) \leq V_Q \sum_{i=1}^N \max_{\mathbf{x} \in Q_i} |w(\mathbf{x} - C_i)|. \tag{8.48}$$

The optimal constants $C_i, i = \overline{1, N}$, are yielded by solving the optimization problem

$$\sum_{i=1}^N \max_{\mathbf{x} \in Q_i} |w(\mathbf{x} - C_i)| \Rightarrow \max_{C_1, \ldots, C_N}. \tag{8.49}$$

The solution has the form

$$\tilde{C}_i = \frac{\max_{\mathbf{x} \in Q_i} w(\mathbf{x}) + \min_{\mathbf{x} \in Q_i} w(\mathbf{x})}{2}. \tag{8.50}$$

Recall that these constants are used to construct an approximating function for the PDF. Hence, they must be normalized in the following way:

$$C_i^* = \frac{\tilde{C}_i}{V_Q \sum_{i=1}^N \tilde{C}_i}. \tag{8.51}$$

In the final analysis, the error can be estimated as

$$\sup_A J(A) \leq V_Q \sum_{i=1}^N \max_{\mathbf{x} \in Q_i} |w(\mathbf{x} - C_i^*)|. \tag{8.52}$$

Consequently, we have derived two estimates for the approximation error, (8.47) and (8.52). In applications, it is necessary to calculate both estimates and compare them with a given accuracy. If at least one of the estimates fails, calculations should be repeated with a smaller lattice step.

CHAPTER 9

INFORMATION TECHNOLOGIES OF RANDOMIZED MACHINE LEARNING

By now the computer systems of various performance levels have evolved towards a heterogeneous structure that is characterized by an availability of computing devices of different types. A standard system includes a universal multi-core processor in combination with a massive-parallel architecture accelerator (also called a computational processor or a co-processor). An overwhelming majority of modern high-performance systems is based on the concept of horizontal scaling, integrating thousands of heterogeneous computation modules within a uniform organization.

The heterogeneous architecture of computation modules considerably complicates an efficient utilization of computer systems, which is often comprehended as an almost 100% load of all computing devices. Here a main difficulty consists in architectural features and, in many cases, in significant distinctions of the application programming interfaces (API) available for computing devices of different types. For massive-parallel accelerators, the former factor causes the need for parallel data processing and use of corresponding algorithms. The latter factor predetermines a high complexity of the software tools implementing different models of programming, in terms of their development and integration.

DOI: 10.1201/9781003306566-9

Therefore, a topical problem of contemporary computer science is to suggest new approaches to the development and implementation of the algorithms and technologies intended for a wide range of heterogeneous computer systems.

In the recent times, machine learning has achieved a definite level of maturity, being now applicable to many practical problems that had been considered unsolvable or far too computationally intensive. Much of the current advancements in this science lie in the field of software engineering, due to a high computational complexity of most real problems. These are high-dimensional optimization problems that demand efficient computational algorithms.

In some aspects, randomized machine learning imposes even tougher requirements to computational efficiency, which is associated with the difficult mathematical problems in the foundation of this theory. Thus, an efficient implementation of RML methods comes to parallel calculations using heterogeneous structure computing systems.

9.1 ARCHITECTURE OF MODERN COMPUTER SYSTEMS

The present-day computer systems based on von Neumann's architecture may have different configuration depending on the number and type of processors, the capacity and structure of memory, and the presence of accelerators (coprocessors). These systems can be classified using a logical structure of memory, which defines memory access rules for the program codes executed by a processor. Such an approach discriminates between two classes of the systems, namely, the ones with shared and distributed memory. Of course, this classification suffers from certain drawbacks but yields a convenient and efficient representation of computer systems for the development and analysis of system-oriented algorithms.

Within this classification, the address space of memory accessed by processors is either uniform or separate. This property predetermines an appropriate *programming model* for utilization of computer system resources.

Note that the classification under study is not unique. There exist technologies allowing a user to treat a computer system with physically distributed memory as a system with a uniform address space. Conversely, a system with distributed memory can be emulated on a system with shared memory.

Two standard programming models have become widespread in these classes to date, namely, the model based on *shared memory* and the model based on *message passing*. The former model has the advantages of a uniform address space owing to a direct usage of shared memory for data transmission between processors. The latter model incorporates a data blocking mechanism, and data blocks are sent to other processors in the form of messages. For the time being, a detailed organization of these mechanisms is not so important: note that message transfer is a slower operation than direct memory access. This property must be taken into account in the course of algorithm development.

First and foremost, further advancement of computer systems aims at reducing the time required to solve the existing problems or at formulating and solving new (or current) of larger scale. Historically, computer performance has been improved using two approaches that finally formed the concepts of *vertical* and *horizontal scaling*. Vertical scaling is implemented by increasing the performance of one computation module whereas horizontal scaling by increasing the number of computation modules.

As a rule, a computation module represents a single- or multi-processor system with shared memory, i.e., a computer system with a uniform address space that can be accessed by any processor. Such a general description covers common single-processor workstations and also rather large multi-processor computer systems.

At the initial phase, the computation modules were developed within the concept of vertical scaling, since a rapid progress of microelectronics yielded faster and faster processors and memory modules. As soon as the physical limits were achieved, making further speed-up of a single processor impossible, the evolution moved towards horizontal scaling, i.e., multi-core processors. Note that horizontal scaling was also used for increasing the number of processors with identical access to shared memory.

An accumulated experience in the design of large multi-processor systems with shared memory demonstrated that many processors cannot be utilized with good efficiency within the framework of a common memory model. Therefore, the only way for a theoretically unrestricted growth of computer performance is horizontal scaling applied to the computation modules operating in the shared memory model. Obviously, at the upper level a system with this architecture represents a system with distributed memory. Most modern high-performance computer systems are designed using this principle and architecture.

As a result, a parallel execution of programs at different levels of computer systems became a basic approach to improve their performance. A real feasibility of program code parallelization led to the uprise of a new

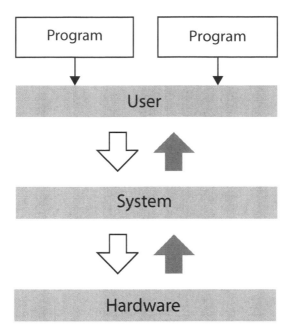

Figure 9.1

science called parallel computing. This science studies hardware tools for a parallel execution of programs and also a wide circle of issues related to the efficient exploitation of such systems.

The modern computer systems have a support of program parallelism, at the level of hardware and also at the level of system software (operating system). Users of such a computer system are developing programs that are implemented and executed using a functionality provided by its operating system. Thus, we can identify three main layers for a program that implements a given algorithm; see Fig. 9.1.

The lower layer answers for hardware support (*hardware layer*); in accordance with von Neumann's architecture, this layer can be described by a command handling node (processor) and storage memory for commands and data. Of course, the real systems have a more complicated organization, but this simple diagram suffices for our level of detail. The second layer, called the *system layer*, forms system software required for an abstract operation of all hardware tools by a user regardless of their functionality. Finally, user (or application software) occupies the upper level of the architecture, which is known as the *user layer*.

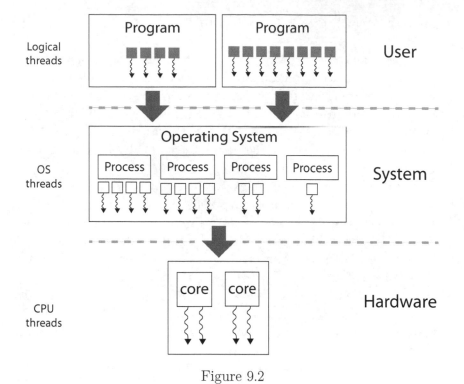

Figure 9.2

Now let us discuss the elements of this architecture in more depth, as illustrated in Fig. 9.2. The commands executed by a processor have a certain organization and are coming sequentially, thereby forming *a thread of execution*. This is a major concept for parallelization at the hardware level of a computer system [45].

Parallelism is implemented through a simultaneous execution of several commands or threads. Here a widespread approach is the so-called *simultaneous multi-threading* (SMT), in particular, the Hyper-Threading Technology developed by Intel. This approach allows to execute several threads on a single processor (processor core), thereby implementing not physical but logical parallelization. Nevertheless, a considerable increase in performance can be gained because in most programs the execution process also includes latent operations (working with memory and storage devices). Waiting for their completion, a processor may execute several commands from another thread. If there exist many available processors or cores (*processing elements* (PEs) in the terminology of large-scale computer systems), then it is possible to organize a real parallelism for threads.

A user manages hardware tools (in our case, processor threads) through an application programming interface (API) provided by corresponding software tools. Therefore, for a user processor threads as physical objects are represented in the form of logical objects independent of hardware support. These objects implement a program code created by a user of a computer system. The main objects of this class are threads and processes of an operating system.

A *process* is a basic element of execution for an operating system that is implemented by a special program object. All contemporary operating systems can run several processes via a simultaneous execution. A program code is executed in a processor, and a real parallelism is impossible for a single processor. However, in this case, a logical parallelism can be organized.

For safe execution, the processes of an operating system are isolated from each other in computer memory. In particular, this feature imposes some restrictions on their interaction, which are implemented using special mechanisms. A concrete form of these mechanisms bears computational cost, as it is necessary to guarantee a safe functioning of all processes. Thus, a process is a "heavy" object consuming considerable computational resources for their creation, functioning and elimination. At the same time, a *thread* is a "light" program object that operates within the framework of the resource sharing model of a process (memory, file descriptors, etc.). Hence, threads do not need much resources for their functioning. A process contains at least one thread.

Being fundamental objects at the system level, the threads and processes also define an abstraction level for a user-friendly development of algorithms and programs. Therefore, the system-level architecture of a computer system that is implemented by an operating system has a hierarchical organization. In accordance with this architecture, a major task for the system level is *to map* system objects into hardware ones, i.e., to support an execution of the system (logical) threads within the hardware threads directly on a processor.

The contemporary operating systems are multitasking, which means that they can support many simultaneously running processors by sharing the address space of memory among them to guarantee their safe execution. This leads to a parallel execution of *tasks* and hence a logical parallelism. A physical parallelism can be achieved if a computing device (processor) features a real execution of a process or thread.

Consequently, for a successful parallelization of several processes a computer system must have several computing devices (processors or cores).

Note that multitasking can be implemented with a single computing device, e.g., based on time sharing: an operating system executes each process during a fixed period of time, then suspends it and switches to another process. Clearly, this approach gives no real increase in performance, and the advantages of parallelism are not gained.

9.2 UNIVERSAL MULTITHREADED ARCHITECTURE

An important aspect of software development consists in the design and analysis of corresponding algorithms that are implemented by software products. The general architecture of a computer system described above (particularly, the system level) predetermines possible methods to analyze and develop application software at the user level. The development of efficient algorithms becomes even a more topical problem for the computer systems that support parallelism [45, 69, 183].

Historically, the algorithms intended for parallel computer systems were analyzed by considering the lowest level of implementation, i.e., the decomposition of "calculations" into elementary operations with direct analogs in a programming language. After such a detailed specification, further analysis was focused on hardware tools (the hardware level) with a detailed specification down to processors and their internal structure in order to find a most efficient utilization of a concrete computer system.

The detailed approach to the development and analysis of algorithms led to the appearance of different theories, with a main goal to suggest a universal analysis procedure for parallel algorithms. In this context, note an original method with a *model of computations* or a *model of programs*, which employs graph theory techniques for modeling and analysis [62, 183].

The progress of the computer systems in a wide sense, i.e., the whole range of hardware and software technologies, has achieved a stage in which many algorithms and approaches become standard. In other words, there exist efficient implementations for them in form of software tools or technologies supported by all basic architectures of modern computer systems. Also note that a large number of such technologies are commercial, which guarantees a high level of quality, reliability and efficiency.

As indicated by the history of many fields of science, owing to the development of tools it is possible to consider new problems on a new layer of abstraction defined by these tools. Therefore, "higher-level" approaches are needed to develop and analyze the algorithms intended for modern computer systems. In view of the general architecture of a computer system (see Section 8.1), we believe that the desired approaches should involve the user layer: in this case, the algorithms can be efficiently formulated and further analyzed under an available level of abstraction.

The computations within an algorithm are the result of operation of a corresponding program. In literature there is almost no separation between the concepts of program and algorithm. Nevertheless, we will discriminate between them, treating an algorithm as an abstract thing that includes abstract operations and operands, whereas a program implementing this algorithm deals with real operations provided by its execution environment.

Consider a virtual thread as a basic abstract object that implements computations. This object lies at the user level of the computer system architecture. Here all computations have an abstract sense. It is also possible to further specify and study these concepts and associated objects, depending on the level of detail of a given algorithm.

By assumption, the user layer of a computer system provides a certain set of such virtual threads and also the feasibility of a parallel execution for them within the shared memory model, thereby defining a virtual *multithread execution environment*. Depending on the specifics of a task (algorithm), such a general parallelism-oriented representation of a computational environment facilitates a more efficient analysis of different characteristics of the algorithms on its basic elements (on the one hand) and also an efficient mapping of the user layer into the system layer (on the other hand).

Interestingly, nowadays similar ideas are prevailing in different technologies, and hence the abstractions are "bound" to the corresponding implementation objects. For example, note well-known program platforms such as Java, MPI, OpenMP, and others. Here the abstractions, which can be treated as the virtual threads under consideration, have a unique correspondence to the program objects implementing them. Particularly, this applies to the Java and OpenMP threads as well as to the MPI processes. The suggested approach yields necessary and sufficient tools with various implementation capabilities owing to its abstraction.

This approach can be easily scaled, vertically (with an increase in the volume of computations in a thread, causing a higher granularity of computations) and also horizontally (with an increase in the

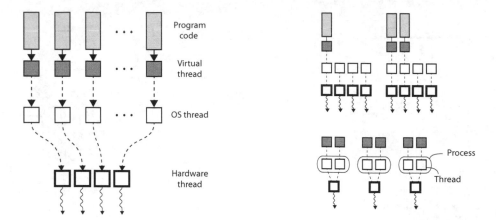

Figure 9.3

number of threads). In addition, it is possible to model a computer system with different memory structure, e.g., distributed memory, by considering several virtual execution environments.

The described architecture of a virtual computer system, further referred to as *the universal multithread architecture (UMA)*, is high-level and can be efficiently implemented on all computer systems using the existing technologies. Due to this universalism, the architecture may serve for the development and analysis of parallel algorithms (see Fig. 9.3).

Modern computer systems often have a heterogeneous structure, i.e., their computation modules contain computing devices of different types. The main types of such devices are universal central processors and computational coprocessors (accelerators) with a massive-parallel architecture. However, all these devices involve threads as a basic element of command execution; for a user, threads represent virtual (logical) objects that are mapped into hardware threads during execution.

It is possible to work with an accelerator in several modes, which have logical and physical aspects dictated by implementations on concrete devices or families of devices. For example, the well-known massive-parallel architectures, MIC [7] or CUDA [6], utilize a processor in the offload mode as follows: a current computation job is "offloaded" to a processor and performed on it, and then the results are sent back to a calling program. Such an organization allows for an asynchronous operation of a device from different threads of a calling program.

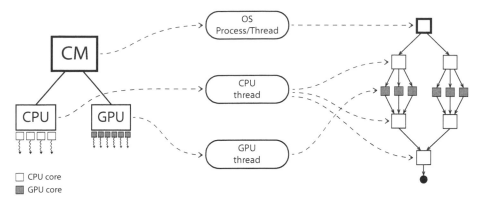

Figure 9.4

For the systems designed as an aggregate of computation modules with a heterogeneous architecture, there are three levels to implement parallelism (see Fig. 9.4), namely,

- the level of computation modules, which is implemented on the system layer by the processes of an operating system;

- the level of central processor, which is implemented by the processor threads;

- the level of accelerator, which is implemented by the accelerator threads.

Consequently, parallelism has a hierarchical three-level structure, which is typical for modern computer systems. The universal multithread architecture lies at a higher level of abstraction, providing a universal abstract interface for the development and analysis of parallel algorithms. In addition, the main objects of this interface can be implemented (described) by the objects occupying different levels in the three-level architecture.

9.3 INFORMATION TECHNOLOGIES OF RANDOMIZED MACHINE LEARNING

Information technologies include computer methods and tools that are intended for data storage, handling, extraction and transmission [103]. The matter concerns hardware tools (computers, storage devices, telecommunication equipment) and also software tools if they form a complete technology with a requisite user functionality. Hardware tools are often divided into computer hardware and communication hardware; software tools, into system and application software. According to this structural classification, the information technologies of RML occupy the application software level.

Any technology aims at solving some practical tasks or problems. We will comprehend solution in a general sense because in many cases initial problems are associated with an abstract subject area. This state of the things arises in fundamental research, as subject areas are mostly defined by an abstract framework (e.g., different mathematical constructions). The expected results—the solution of a problem—must also belong to the same subject area.

The implementation results of an information technology are obviously related to the subject area of computer systems, representing artifacts of this subject area. Thus, the information technologies intended for solving definite problems can be treated as tools for a corresponding subject area, supplementing the basic tools (on the one part) and forming an intermediate layer for a higher-level technology (on the other part).

A machine learning procedure (see Chapter 1 and Fig. 1.3) can be described as an abstract technology with the following stages:

1. development of parameterized model (PM);

2. parameter estimation (PE);

3. model testing (MT);

4. model implementation (MI).

The second and third stages together form *model learning (ML)*.

This technology rests on the abstract mathematical methods introduced above, which are the basic tools. Hence, there exists a correspondence between mathematical tools and abstract technological stages, as illustrated in Fig. 9.5. At the level of mathematical tools, the PM stage is associated with the mathematical modeling methods discussed earlier. The model learning stage is implemented through the statement of corresponding optimization problems. According to the diagram, the model implementation stage has no counterpart in mathematical tools. Also, note that for real problems, the resulting optimization problems cannot be solved analytically, and the only alternative consists in a numerical solution.

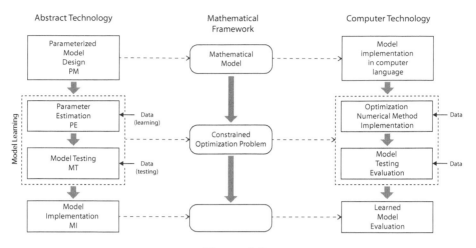

Figure 9.5

The solution of these problems lies in the subject area of computer (information) technologies, which supplement the necessary tools for solving an initial problem. In particular, they implement the stages of model learning and implementation.

Consider these tools in detail, as demonstrated in Fig. 9.6. The elements of the technology under study are implemented through the program objects of an execution environment, depending on the architecture of a current or target computer system. There exist a rich variety of concrete technologies available for modern computer systems. Hence, the approach with a maximum possible abstraction from their specifics seems most productive here.

1. Model Development (MD)

This block constructs a parameterized model in accordance with the existing ideas of a modeled phenomenon. As mentioned earlier, it is possible to outline a narrow circle of model types for a successful solution of real problems. This block outputs a developed model (M).

2. Model Learning (ML)

The model learning block receives the model M as its input data. The main tasks of this block are to estimate the model parameters (block PE) and to test the model itself (block MT). Both blocks operate using data (D), which are represented in computer form at the implementation stage of the technology and can be divided into learning data (LD) and test

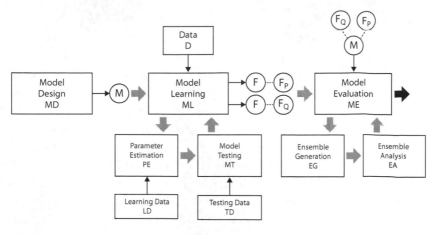

Figure 9.6

data (TD). This block outputs optimal PDFs (F), namely, the PDFs F_P of the model parameters and the PDFs F_Q of the measurement noises.

The ML block is most "loaded" due to its tasks, in the first place, solution of the optimization problem. We have discussed the inherent difficulties of this problem in the previous chapters of the book. Hence, the ML block must incorporate efficient global optimization algorithms or have access to them.

3. Model Implementation (MI)

The model implementation block yields a learned model, i.e., the model M equipped with the functions F_P and F_Q calculated in the preceding stage. Model implementation consists in generating an output ensemble (EG block), with further calculation of its characteristics in the ensemble analysis (EA) block. For model implementation and testing (see the preceding stage), the ensemble is generated by producing a large number of realizations of the random variables (sampling) with given PDFs. Therefore, these blocks must have access to the corresponding generation algorithms of random variables with arbitrary PDFs.

The described technology of RML can be implemented in the form of separate program modules communicating with each other in a manual mode or through a larger-scale technology. Such technologies are implemented within *information systems* intended for solving specific practical tasks. A comprehensive treatment of all issues related to the architecture of information systems goes beyond the scope of this book.

We will dwell on some theoretical aspects only, considering them in a most general statement.

An information system (IS) is often defined as a system that integrates different resources (human, technical, financial) and serves for data storage, retrieval and processing [3, 39]. A major component of an information system consists in software resources, which can be classified by analogy with the architecture of computer systems. Since the system-level software is standardized and widespread, the most significant part of an IS subject to its tasks is implemented on the user level as application software. Note that users will be understood in an abstract sense, as an information system can be utilized by humans or other systems. Such an approach is conventional in program system design and development [59, 78].

A user interacts with an IS through a user interface (see Fig. 9.7), which will be accordingly considered in an abstract sense. This interface determines a collection of admissible operations for system users, i.e., represents an abstract "control panel" of a given system. Thus, system functionality is separated from control functionality, at the logical level and, as a rule, at the physical level.

A user interface controls a program system and, obviously, demonstrates the results of operation in a requisite user-friendly form. Once again, we will adhere to an abstract consideration as follows. If a user is a human, then a user-friendly interface consists of images, graphs, numbers displayed in an appropriate way; this also applies to the "control panels," which are now organized, e.g., through a window interface with control elements such as push buttons, checkboxes, entry fields, etc. For external systems, window interfaces are not required but we still have the same principles as for human users: data are represented in a form dictated by the input specifications of an external system, and control is implemented by linking control elements on both sides.

Under the described approach, it is necessary to achieve the logical (and often physical) separation of a system functionality into loosely coupled modules for a higher reliability and smaller complexity of program system design. There exist several efficient methods for solving this problem, e.g., the Model-View-Controller (MVC) design. This software architectural pattern identifies three abstract subsystems (components) communicating through corresponding interfaces, namely, Model, View, and Controller [99]; see the diagram in Fig. 9.7. Here the main idea is that a user controls a program system with a functionality provided by Controller and reads information about the system operation

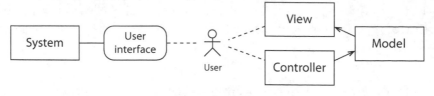

Figure 9.7

with View. This architectural pattern implies a possible multiplicity of information representations (views) and control devices, depending on problem specifics. The conceptual functionality of a program system is mostly concentrated in Model.

The described general scheme is prevailing in almost all fields associated with program systems. Due to abstractiveness, it can be implemented on any computer systems using software technologies.

A significant feature of the approach is its universalism with respect to system components. In other words, the approach works well for the whole system and also for its components whenever they are designed and implemented in an intelligent way. Particularly, the RML technology implemented as a program component can be a component for a larger system, being integrated into the latter through the interfaces of its control and view subsystems.

Speaking most generally, the RML component can be represented as an object that receives data and "knowledge" at its input and then implements a randomized model (RM). In this context, all the things mentioned are considered as the program realizations of the corresponding abstract objects: the knowledge is encapsulated into the model object (see Fig. 9.6) whereas the randomized model into the learned model object, with the optimal PDFs defined by the RML program component.

For the structural diagram of the RML technology in Fig. 9.6, the most resource-intensive operations are

- optimization;

- generation of the random variables with given PDFs;

- analysis of the output ensemble of the randomized model.

Therefore, these operations should be implemented using high-performance computer systems with the potential of parallel computing.

Still they may have structural differences due to the architectural properties of computer systems, which predetermine a universal approach to their operation for users. We will accordingly separate out the basic configurations of computer systems as follows:

- a workstation, which is equipped with a single multi-core central processor and a single massive-parallel accelerator (in modern conditions, a graphics adaptor);

- a multi-processor computing module or several such modules, each containing several central processors and possibly several accelerators;

- a computer system of the supercomputer class, which contains thousands or tens of thousands of computing modules.

This approach leads to the implementation problem of program modules for realizing the potential of parallel computing for different configurations of computer systems (see Fig. 9.8). The universal multithread architecture serves as a base for the efficient algorithms for solving these issues; moreover, it would reduce the program complexity of such algorithms and related technologies under an intelligent implementation using the system-level software technologies.

9.4 IMPLEMENTATION OF PACKET ITERATIONS

As mentioned earlier, a major difficulty to implement the RML technology is to solve the global optimization problem for the entropy-optimal PDFs of a randomized model. Consider some general implementation aspects of the global optimization method described in Chapter 6.

The method of packet iterations calls for a large amount of computations. However, they are similar and involve very many *data elements*, for which it is necessary to calculate the values of definite functions. Therefore, *parallelism by data* seems natural here. Our idea is to implement the calculations for each data element using a virtual thread in accordance with the UMA.

The data elements in each stage of the algorithm are implemented through different objects. In the generation stage, they have the serial number (index) of a random point; in the stages of filtration and selection, the data elements are these points themselves (see Fig. 9.9).

While implementing the algorithm on specific equipment, it is possible to map virtual threads into hardware threads in different ways;

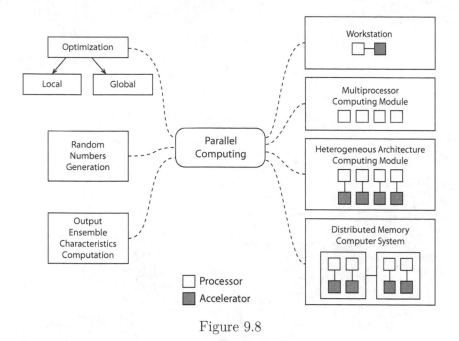

Figure 9.8

some of them are illustrated in Fig. 9.9. The mapping process consists in implementing virtual threads in the real system-level objects of a computer system, i.e., threads and processes. There exists different methods for such an implementation, e.g., a "manual" control of this process involving some characteristics of a desired mapping of virtual threads into real ones. Another approach employs an automatic mapping, which is performed either with system-level software or intermediate user-level software that implements the algorithm.

Universal processors support a small number of threads; in this case, an execution of virtual threads in real ones (through their implementation in the form of processes or threads) can be organized in groups so that the virtual threads of each group are executed sequentially. Massive-parallel accelerators support a simultaneous execution of hundreds of threads but virtual threads are mapped into the real threads of an accelerator in a similar way, despite that their number is extremely higher. In practice the mapping mechanism is completely automated and thread scheduling is performed by the system-level software of a computer system.

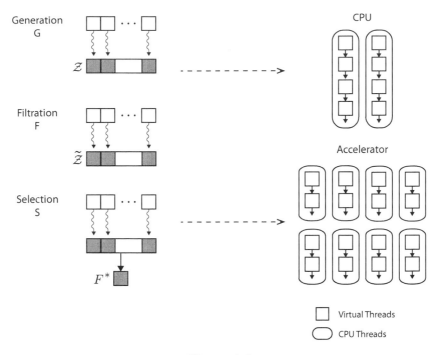

Figure 9.9

Under the multithread approach, the global optimization algorithm involves very many virtual threads related to a corresponding number of data elements in each stage of the algorithm. The algorithm was successfully implemented on the OpenMP and CUDA platforms, which demonstrated an efficiency of this approach. Note that its major advantage lies in good scalability.

CHAPTER 10

ENTROPY CLASSIFICATION

Classification forms a central problem in machine learning and data analysis. An object is classified by associating it with one of two or several categories depending on its characteristics (attributes). The binary (two-class) classification problem acts as a basic statement for further generalization to the case of several classes in real applications.

This chapter is devoted to a solution of the binary classification problem within the randomized machine learning framework. We will briefly survey the existing classification methods (as well as some of their drawbacks) and then describe in detail the entropy classification method, in particular, the application of this method, for a series of numerical examples.

10.1   STANDARD CLASSIFICATION METHODS

Object classification is a learning problem with a "learner." In other words, there exists a learning sample with marked answers that is used by an algorithm for identifying most significant attributes and associating their values with an appropriate class.

Classification algorithms may differ by their treatment of a learning sample, representation of classification errors, convergence rate, and so on. In this section, we will outline most widespread classification algorithms, namely,

- Decision Tree (DT);

- k-Nearest Neighbor (kNN);

DOI: 10.1201/9781003306566-10

- Naive Bayes;

- linear classifiers (Logistic Regression, Support Vector Machines, etc.).

The interested reader may find more information about these and other classification methods in the classical books on machine learning, e.g., [10, 118, 180, 194] in Russian and [21, 57] in English.

10.1.1 Decision Tree

Decision Tree is a branching hierarchical structure of attribute values for an object. Branching stops as soon as further division of attribute values into groups does not refine information, i.e., does not yield different classes. Therefore, in the final analysis each leaf on the last layer of a decision tree contains the members of the same class.

The concept of Decision Tree was pioneered by Breiman and his colleagues [23]. With the course of time, the original algorithm has been continuously developed and modified. Nowadays, Decision Tree is a most popular classification method in the world, and its modification C4.5 ranks first among the top-10 data analysis algorithms [193].

For decision tree design, a major difficulty is to choose an proper branching criterion, i.e., a sequence in which the attributes should be considered. Since optimal decision tree design represents an NP-complete problem [75], branching criteria are often based on "greedy" heuristics, e.g., information gain [117] and Gini impurity [23, 57].

Decision trees have an clear meaning and a simple interpretation of results. Moreover, they do not require special data preparation and are applicable to large amounts of data without auxiliary tools. However, decision tree design suffers from considerable disadvantages as follows.

In the first place, decision trees can be relearned due to an excessively detailed partition. This effect may take place even for relatively simple problems [110]. Second, decision tree design is very sensitive to input data: even small changes in a learning sample may dramatically modify the resulting structure of a decision tree, especially in the case of data spreads.

Thus, decision trees are easy to implement and often serve as a rough approximation of solution but other classification algorithms yield better results.

10.1.2 k-Nearest Neighbor

As a matter of fact, k-Nearest Neighbor (kNN) belongs to the group of metric classifiers that are based on a distance between the objects of a learning sample. This method employs the following ideology: each new object is associated with a most widespread class of its neighbors. Note that one or several neighbors can be considered (e.g., k neighbors), with equal weights or different weights depending on the distance to a classified object [37, 194].

Like other metric classifiers, k-Nearest Neighbor proceeds from the hypothesis of compactness, which claims that similar objects tend to the same class rather than to different classes under an appropriately selected measure of similarity. That is, classes form compactly localized subsets in the space of objects [14].

In applications the distance between objects is often defined as the Euclidean distance, which has a high efficiency for numerical attributes. On the other hand, for the problems with categorial (particularly, binary) attributes, k-Nearest Neighbor may give worse results than other classification algorithms, e.g., Decision Tree [33].

Another significant factor restricting the applicability of this classifier is related to the curse of dimensionality [19]. If a problem has a high dimension and the distance between objects is calculated as the sum of deviations over all attributes, then by the law of large numbers these sums take close values with a high probability. In other words, in a multidimensional space all objects are almost at the same distance from each other, which complicates the choice of an optimal number k of neighbors.

In terms of convergence rate, k-Nearest Neighbor demonstrates good results only for relatively small samples. Because the solution rests on known classes for nearest neighbors, for classification of each new object it is necessary to store and calculate distances from this object to all other objects in a learning sample, an extremely difficult procedure for very large amounts of data ($N >> 10^3$) [198].

Consequently, despite an intuitive interpretation and a high efficiency in the space of numerical attributes, k-Nearest Neighbor (as well as the other metric classifiers) can be used for a narrow range of classification problems.

10.1.3 Naive Bayes

The naive Bayesian classifier (in short, Naive Bayes or Simple Bayes) is a special case from a general group of the Bayesian classifiers based on well-known Bayes' theorem and the posterior probability maximization principle [10,180]. For each object under classification, first it is necessary to calculate likelihood functions over each possible class. Using these functions and the probabilities of occurrence for the representatives of different classes, the next step is to calculate the posterior probabilities of the classes themselves. In the final analysis, an object is associated with the class that has the highest posterior probability [44,57].

In contrast to the algorithms considered earlier, the Bayesian classifiers allow one to construct a probabilistic model $p(c|x_1, x_2 \ldots x_n)$ for a variable of class c that depends on the variable attributes $x_1, x_2 \ldots x_n$. Bayes' theorem can be formally stated in the following way:

$$p(c|x_1, x_2 \ldots x_n) = \frac{p(c)p(x_1, x_2 \ldots x_n|c)}{p(x_1, x_2 \ldots x_n)}, \tag{10.1}$$

where the numerator $p(c)p(x_1, x_2 \ldots x_n|c)$ gives the joint probability of the model $p(c, x_1, x_2 \ldots x_n)$; the denominator $Z = p(x_1, x_2 \ldots x_n)$, a normalization constant independent of the variable c.

Naive Bayes derived its name because it involves the hypothesis of independent object's attributes: in this case, their joint probability can be written as a simple product of all components. Under this hypothesis, formula (10.1) gets simplified to

$$p(c|x_1, x_2 \ldots x_n) = \frac{1}{Z} p(c) \prod_{i=1}^{n} p(x_i). \tag{10.2}$$

Naive Bayes is a classification rule that chooses the class with the maximum posterior probability, i.e.,

$$C_{naive} = \arg \max_c p(c|x_1, x_2 \ldots x_n). \tag{10.3}$$

Therefore, each new object is classified using information about the whole learning sample, more specifically, the likelihood function and the prior probabilities of classes. However, in practice an analytical form of these functions turns out to be unknown, and the likelihood function and the probabilities of classes are empirically restored by a learning sample (note that generally the classes may have different probabilities).

Despite a deviation from the optimal Bayesian classifier [44] and the need for an empirical restoration of the functions, Naive Bayes still remains a widespread and efficient classification algorithm, although it tends to be relearned (especially for small learning samples) [155].

10.1.4 Linear classifiers

A linear classifier is a classification algorithm that constructs a linear separating surface in the space of attributes. For a binary classification of two-dimensional objects (two attributes), the separating surface represents a line; for three-dimensional ones, a plane; in the general case, a hypersurface [57, 180].

Let the objects from a set X be described by numerical attributes $f_j : X \to R, j = 1, \ldots, n$, in space R^n. Denote by $Y = \{-1, 1\}$ the set of class labels in the case of binary classification. Then a linear classifier is a classification algorithm $a : X \to Y$ of the form

$$a(x, w) = sign\left(\sum_{j=1}^{n} w_j f_j(x) - w_0\right) = sign \langle x, w \rangle, \qquad (10.4)$$

where $\langle x, w \rangle$ gives the scalar product of an object's attribute description and a weight vector $w = (w_0, w_1, \ldots, w_n)$.

Accordingly, a major goal of machine learning is to "adjust" the weight vector: given a learning sample $X^m = \{(x_1, y_1), \ldots, (x_m, y_m)\}$, it is necessary to construct an algorithm of the above form that minimizes the empirical risk functional

$$Q(w) = \sum_{i=1}^{m} [a(x_i, w) \neq y_i] \to min_w. \qquad (10.5)$$

For solving the optimization problem (10.5), introduce the margin

$$M(x_i) = y_i \langle x_i, w \rangle. \qquad (10.6)$$

This is a numerical value that characterizes the level of object's immersion into a corresponding class. The smaller is the value $M(x_i)$, the closer is the object x_i to the separating surface of the classes and hence the higher is the probability of classification errors. Since the margin $M(x_i)$ takes negative values only in case of errors, the empirical risk functional can be written as

$$Q(w) = \sum_{i=1}^{m} [M(x_i) < 0]. \qquad (10.7)$$

Therefore, for solving the problem (10.5), we have to find a maximum feasible subsystem in a given system of inequalities. This problem is NP-complete and may possess infinitely many solutions because even a minimum number of errors can be realized on different subsets of objects. However, in most applications there is no practical sense to obtain an exact solution of this problem (to say nothing of all solutions). For proper learning of a classifier, it suffices to get an approximate solution near enough to exact value [187].

For a rough solution the threshold loss function is replaced with different continuous approximations

$$[M < 0] \leq L(M), \tag{10.8}$$

with a subsequent minimization of an upper bound for the empirical risk functional, i.e.,

$$Q(w) \leq \tilde{Q}(w) = \sum_{i=1}^{m} L\left(M(x_i) < 0\right). \tag{10.9}$$

Introducing a continuous approximation for the loss function, we can employ well-known numerical optimization methods, such as gradient descent and convex programming techniques, which considerably speeds up the learning process. Moreover, in a series of cases, with a continuous approximation it is possible to improve the generalizing capability of a linear classifier [105].

Let us mention the common approximation functions $L(M)$ and the associated methods.

- **Fisher's Linear Discriminant** is a quadratic approximation of the form
$$L(M) = (1 - M)^2. \tag{10.10}$$

  Used under the condition that the center vector of a corresponding class is subtracted from the attribute vector of each object [52].

- **Single-Layer Perceptron** is an approximation by the sigmoidal function
$$L(M) = \frac{2}{1 + e^{\alpha M}}. \tag{10.11}$$

  Often used in neural networks but requires additional heuristics in order to avoid relearning [122, 187].

- **Logistic Regression** is a logarithmic approximation of the form

$$L(M) = \log(1 + e^{-M}).\qquad(10.12)$$

Allows to calculate the probabilities of belonging to different classes. Under rather strong hypotheses on the likelihood functions of the classes, empirical risk minimization with this loss function is equivalent to likelihood maximization [73, 187].

- **Support Vector Machines** is a piecewise linear approximation of the form

$$L(M) = (1 - M)_+.\qquad(10.13)$$

Used together with quadratic regularization. The corresponding optimization problem is solved by special quadratic programming methods [57, 180].

- **Exponential Approximation** is given by

$$L(M) = \exp(-M).\qquad(10.14)$$

Used in boosting algorithms, e.g., AdaBoost [56]. Suffers from accentuating on large negative margins, i.e., has a strong sensitivity to data spreads.

10.2   COMPOSITION ALGORITHMS

As evidenced by the development of machine learning subject to applications, still there is no unique classification algorithm that would yield best results for any size of learning samples, any dimensionality of the attribute space and any form of the separating surface. This conclusion has motivated research in the field of ensemble machine learning methods, also called composition algorithms.

The key idea of composition algorithms is to unite the forecasts of several basic classifiers, each constructed using an available learning sample [163]. Perhaps, a first example of this approach was the Condorcet jury theorem (1784), which states the following: the probability of a correct majority voting-based decision by a jury in which each individual juror has an independent opinion and chooses a correct outcome with a probability greater than 0.5 in increasing with the number of jurors and tends to 1.

Another example of the ensemble approach is the wisdom of crowds (1906), a principle discovered by F. Galton as the result of the following contest. The crowd at a county (nearly 800 people) fair accurately guessed the weight of an ox when their individual guesses were averaged. The average was closer to the ox's true butchered weight than the estimates of most crowd members.

Consider the most widespread methods to construct composition algorithms for classification problems, namely, bagging [22] stacking [192] and boosting [56].

10.2.1 Bagging

The idea of bagging (bootstrap aggregating) is to reproduce a small learning sample by generating a collection of new random data arrays for further learning. Each sample is then used for the learning of several independent classifiers, and the results are aggregated for obtaining a final high-accuracy forecast.

Bagging algorithms may differ in the methods used to generate subsamples and in the methods used to aggregate or weight the results of basic classifiers. Traditional aggregation methods include simple majority voting, mean and median values, weighted averaging for extracting high-accuracy classifiers, and others [79].

A necessary prerequisite for bagging is the choose different basic classifiers with uncorrelated errors. In this case, the errors of the basic classifiers compensate each other, which yields a much better forecast [22].

In this context, also note the random subspace method (RSM) [163], which differs from standard bagging in the following feature: random samples are generated also using random subsets from an attribute description of objects. This method has a high efficiency under redundant spurious attributes and a relatively small size of a learning sample.

10.2.2 Stacking

The concept of stacking further develops the idea of an aggregated generalization for the results yielded by basic classifiers, but with a separate auxiliary model that is learned on their forecasts.

As a rule, for stacking a learning sample is divided into two subsamples, the first being intended for the learning of independent basic classifiers whereas the second for their testing. After that, the forecasts of the basic classifiers and the true answers are used as input data for the learning of a meta-classification model [192].

Among the benefits of stacking, we mention a possible design of nonlinear combinations of basic classifiers for high-accuracy forecasting.

10.2.3 Boosting

Boosting is an iterative sequential improvement procedure for a composition of basic classifiers by constructing a weighted learning subsample using an initial sample. At each new iteration, higher weights are assigned to the replicates yielding incorrect results for a previous composition. The algorithm is therefore accentuated on most difficult replicates from a learning sample.

The concept of boosting also employs basic classifiers but, unlike bagging, at each successive iteration a learning sample is defined by the classification results at the previous iterations (not randomly). Such an approach turns out to be much more efficient even with weak basic classifiers [25].

Due to simple implementation, universalism and, above all, good capability for generalization, boosting algorithms still remain most popular machine learning methods, together with Support Vector Machines (SVM) and neural networks. A drawback of boosting is traditionally tough requirements applied to the size of a learning sample (10^4–10^6 replicates) [187].

10.3 ENTROPY CLASSIFICATION

Now let us describe the entropy-robust approach to the learning of classifiers. This approach differs from the traditional machine learning problems in the optimization statement and also fruitfully combines the advantages of the classical individual classifiers and ensemble methods.

10.3.1 Problem formulation

Consider a given collection of objects to be distributed between two classes. Divide all available data into two samples, namely, a learning sample $\mathcal{E} = \{e_1, \ldots, e_h\}$ and a testing sample $\mathcal{T} = \{t_1, \ldots t_r\}$. The percentage of objects in these samples (subcollections) will be discussed below.

The objects in both subcollections have a vector representation; for each object, the elements of a corresponding vector are variables-attributes characterizing it, i.e.,

$$\mathcal{E} = \{\mathbf{e}^{(1)}, \dots, \mathbf{e}^{(h)}\}, \qquad \mathcal{T} = \{\mathbf{t}^{(1)}, \dots, \mathbf{t}^{(r)}\}. \tag{10.15}$$

Our analysis will be confined to the quantitative attributes only but the suggested method can be extended to the case of categorial attributes.

The vectors of both subcollections have the same dimension: $(\mathbf{e}^{(i)}, \mathbf{t}^{(j)}) \in R^n$. The subcollection $\mathcal{E}$ is used for learning whereas the subcollection $\mathcal{T}$ for testing.

The objects in both subcollections are labeled in accordance with their belonging to an appropriate class: if an object e_l or t_k is a member of the first (second) class, then it has label 1 (label 0, respectively). Hence, the learning subcollection is characterized by a vector of answers $\mathbf{y} = \{y_1, \dots, y_h\}$ composed of ones and zeros; the testing subcollection, by an answer vector $\mathbf{z} = \{z_1, \dots, z_r\}$. The indexes of elements in these vectors correspond to the indexes of objects in the learning and testing samples.

The classifier learning problem is to design an algorithm $a : \mathcal{E} \to y$ that restores the class of an object using its attribute description based on a learning sample.

10.3.2 Learning stage

Due to the availability of a learning subcollection, we can hypothesize the existence of a function (some decision rule) $F : \mathcal{E} \to \mathbf{y}$. The learning problem consists in choosing an appropriate form and parameters of an approximate function $\hat{F}(\mathbf{a})$. The function $\hat{F}(\mathbf{a})$ represents a decision rule model.

Following the entropy classification concept, a decision rule is described by a randomized parametric model (RPM), i.e., a model with randomized parameters $\mathbf{a}$. It has the input vectors $\{\mathbf{e}^{(1)}, \dots, \mathbf{e}^{(h)}\}$ and the output $\hat{\mathbf{y}}(\mathbf{a})$ that depends on the randomized parameters $\mathbf{a}$.

As such a model choose a single-layer perceptron [63], i.e.,

$$\hat{y}^{(i)}(\mathbf{a}) = \text{sigm}\left(\langle \mathbf{e}^{(i)}, \mathbf{a} \rangle\right), \qquad i = \overline{1, n}, \tag{10.16}$$

where

$$\begin{aligned}
\text{sigm}(x_i) &= \frac{1}{1 + \exp[-\alpha(x_i - \Delta)]}, \\
x_i &= \left(\langle \mathbf{e}^{(i)}, \mathbf{b} \rangle\right), \\
\mathbf{a} &= \{\mathbf{b}, \alpha, \Delta\}.
\end{aligned} \tag{10.17}$$

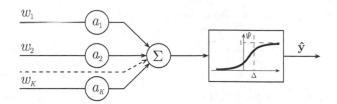

Figure 10.1: Structure of single-layer neural network.

A graph of the sigmoidal function with the slope α and threshold Δ parameters is shown in Fig. 10.1. The values of the function $\operatorname{sigm}(x)$ within the interval $[1/2, 1]$ correspond to the first class; within the interval $[0, 1/2)$, to the second class.

In the randomized model (10.16, 10.17), the parameters $\mathbf{a} = \{a_1, \ldots, a_{(n+2)}\}$ are of the interval type and their probabilistic properties are characterized by a joint probability density function $P(\mathbf{a})$ on a set $\mathcal{A}$, i.e.,

$$a_k \in \mathcal{A}_k = [a_k^-, a_k^+], \qquad k = \overline{1, n+2},$$

$$\mathcal{A} = \bigotimes_{k=1}^{n+2} \mathcal{A}_k. \tag{10.18}$$

Since the parameters $\mathbf{a}$ have a random character, for each object e_i there is an ensemble $\hat{\mathcal{Y}}^{(i)}$ of random numbers $\hat{y}^{(i)}(\mathbf{a})$ from the interval $[0, 1]$. Define its mean value as

$$\mathcal{M}\{\hat{y}^i(\mathbf{a})\} = \int_{\mathcal{A}} P(\mathbf{a})\operatorname{sigm}\left(\langle \mathbf{e}^{(i)}, \mathbf{a}\rangle\right) d\mathbf{a}, \qquad i = \overline{1, h}. \tag{10.19}$$

Consequently, according to the general RML procedure [144, 150], the classification problem can be written as follows:

$$\mathcal{H}[P(\mathbf{a})] = -\int_{\mathcal{A}} P(\mathbf{a}) \ln P(\mathbf{a}) d\mathbf{a} \Rightarrow \max \tag{10.20}$$

subject to

• the PDF normalization conditions

$$\int_{\mathcal{A}} P(\mathbf{a}) d\mathbf{a} = 1 \tag{10.21}$$

and

- the output balance constraints

$$\int_{\mathcal{A}} P(\mathbf{a})\operatorname{sigm}\left(\langle \mathbf{e}^{(i)}, \mathbf{a}\rangle\right) d\mathbf{a} = y_i, \qquad i = \overline{1, h}, \qquad (10.22)$$

where $P(\mathbf{a})$ belongs to the class $\mathbb{C}^1$ of all continuously differentiable functions.

Equations (10.20–10.22) make up a functional entropy-linear programming problem, which possesses an analytical solution—the entropy-optimal PDF $P^*(\mathbf{a}\,|\,\theta)$ parameterized by the Lagrange multipliers θ, i.e.,

$$P^*(\mathbf{a}\,|\,\theta) = \frac{\exp\left[-\sum_{i=1}^{h}\theta_i \hat{y}^{(i)}(\mathbf{a})\right]}{\mathcal{P}(\theta)}, \qquad (10.23)$$

where $\hat{y}^{(i)}(\mathbf{a})$ is given by (10.2) and

$$\mathcal{P}(\theta) = \int_{\mathcal{A}} \exp\left[-\sum_{i=1}^{h}\theta_i \hat{y}^{(i)}(\mathbf{a})\right]. \qquad (10.24)$$

The Lagrange multipliers are calculated from the balance constraints (10.22). Here a possible approach lies in the L^2-norm minimization of a general residual function for equations (10.22).

10.3.3 Testing stage

This stage utilizes the testing subcollection $\mathcal{T}$ of objects that are characterized by the answer vector $\mathbf{z} = \{z_1, \ldots, z_r\}$ with a known classification (their belonging to the first or second class). The vector $\mathbf{z}$ will be used for a quality assessment of the testing procedure. The latter itself involves the subcollection $\mathcal{T}$ and also the decision rule model yielded by the learning stage.

It is necessary to test the randomized decision rule (10.16, 10.17) with the entropy-optimal PDF of the model parameters according to formulas (10.23, 10.24). The procedure performs a sequence of Monte Carlo simulations of size N, generating in each simulation a random vector $\mathbf{a}$ that obeys the entropy-optimal PDF $P^*(\mathbf{a})$.

Assume that, as the result of such simulations, the first object from the testing subcollection has been assigned N_1 times to the first class and $N - N_1$ times to the second class, ..., the kth object has been assigned N_k times to the first class and $N - N_k$ times to the second class, and so on. Under a sufficiently large number of simulations, define the empirical probabilities

$$
\begin{aligned}
p_1^{(1)} &= \frac{N_1}{N}, \ldots, p_1^{(k)} = \frac{N_k}{N}, \ldots, \ldots \\
p_2^{(1)} &= \frac{N - N_1}{N}, \ldots, p_2^{(k)} = \frac{N - N_k}{N}, \ldots, \ldots
\end{aligned}
$$

$$(10.25)$$

Therefore, for each object i from the testing sample, the testing algorithm can be written in the following form.

Step 1-i. In accordance with the optimal PDF, generate a set of output values for the entropy-optimal decision rule model (10.23, 10.24), which contains N random numbers from the interval $\{0, 1\}$.

Step 2-i. For each realization of the model (i.e., for each parameter vector from the ensemble of size N), calculate the decision rule values using formulas (10.16, 10.17). If the resulting value exceeds $1/2$, then the object $t^{(i)}$ is assigned to the first class; otherwise, to the second class.

Step 3-i. Find the empirical probabilities of belonging to these classes by formula (10.25).

With this procedure each object can be assigned to one of the two classes with a definite probability, which reflects an inherent uncertainty of the available data and also of the approximate decision rule model.

Note that, as a rule, the objects of the learning and testing subcollections are considered to be independent replicants. Under this assumption, Step 1-i can be organized as a common preliminary procedure to generate a set of random variables, which will be used in the same way for all testing objects.

Transition to hard classification can be performed by introducing a threshold probability above which an object is assigned to a corresponding class. The number of objects for a correct hard classification depends on this threshold. As easily established, some objects are not classified for the thresholds exceeding 0.5 but the results of classification have a higher reliability. On the other hand, for the thresholds less than 0.5, all objects are classified (as the sum of both probabilities is 1) but the results may resemble guessing if the probabilities of belonging are close to each other.

Table 10.1

i	$e_1^{(i)}$	$e_2^{(i)}$	$e_3^{(i)}$	$e_4^{(i)}$
1	0.11	0.75	0.08	0.21
2	0.91	0.65	0.11	0.81
3	0.57	0.17	0.31	0.91

10.3.4 Numerical examples

1. Classification of four-dimensional objects

Consider the objects characterized by four attributes, i.e., the elements of the vectors $\mathbf{e}$ and $\mathbf{t}$.

Let the learning sample consist of three objects, each being described by four attributes (see Table 10.1).

The randomized decision rule model (10.16, 10.17) has the parameters $\alpha = 1.0$ and $\Delta = 0$. The learning answer vector is $\mathbf{y} = \{0.18; 0.81; 0.43\}$ (as before, $y_i < 0.5$ corresponds to the second class and $y_i \geq 0.5$ to the first class). The Lagrange multipliers for the entropy-optimal PDF (10.23) take the values $\bar{\theta}^* = \{0.2524; 1.7678; 1.6563\}$. The parameters belong to the intervals $a_i \in [-10, 10]$, $i = \overline{1, 4}$.

For this learning subcollection, the entropy-optimal PDF has the form

$$P^*(\mathbf{a}, \bar{\theta}) = \frac{\exp\left(-\sum_{i=1}^{3} \theta_i y_i(\mathbf{a})\right)}{\mathcal{P}(\bar{\theta})},$$

$$y_i(\mathbf{a}) = \left(1 + \exp(-\sum_{k=1}^{4} e_k^{(i)}, a_k)\right)^{(-1)}. \tag{10.26}$$

The two-dimensional cutset of the PDF $P^*(\mathbf{a}, \bar{\theta}^*)$ with fixed values of the third and fourth dimensions is illustrated in Fig. 10.2.

Let us demonstrate the mechanism to obtain forecasts for new objects using the learned model. At this stage, utilize a subcollection $\mathbb{T} = \{t_1, \ldots, t_r\}$ of objects in which each element is characterized by a vector $\mathbf{t}^{(j)} \in R^{(4)}$. First, generate an array of dimensions (500×4) consisting of four-dimensional random vectors $\mathbf{t}^{(i)}$, $i = \overline{1, 500}$, with independent uniformly distributed elements on the interval $[0, 1]$. Then apply the classification algorithm described in the testing stage subsection.

The empirical probabilities $p_1^{(i)}, p_2^{(i)}$ of belonging of an object t_i to the first and second classes are presented in Fig. 10.3.

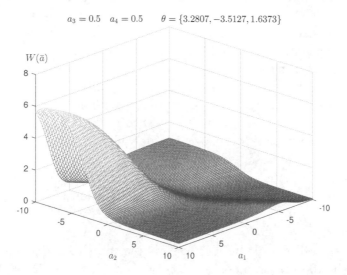

Figure 10.2: Two-dimensional cutset of entropy-optimal PDF $P^*(\mathbf{a}, \bar{\theta}^*)$.

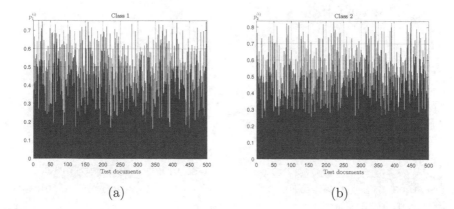

Figure 10.3: Empirical probabilities of belonging to different classes for Example 1.

For each replicant, the soft classification procedure has assigned the probabilities of object's belonging to the first and second classes. Note that the sum of these probabilities is 1 for each object.

Further decision-making (hard classification, i.e., assignment of each object to an appropriate class) is possible by introducing some threshold of "reliable classification." For example, choose a threshold of 0.75; then all objects belonging to the first class with a probability higher than 0.75 (or the ones belonging to the second class with a probability less than 0.25) are assigned to the former class with some level of reliability.

The level of reliability is specified for each new classification problem individually. It represents one of the hyperparameters of the algorithm and is often adjusted using cross validation or sliding control [118, 187].

2. Classification of two-dimensional objects

Consider a more visual example of classification of two-dimensional random objects.

Let the objects of the learning sample from Example 1 be characterized by the first two components of their attribute description. The learning sample consists of three objects, each being described by two attributes (see Table 10.1). The values of the parameters α and Δ, as well as the intervals of the random parameters $\mathbf{a}$, are the same as in Example 1.

Solving the system of equations (10.22) with the entropy-optimal PDF (10.23) yields the Lagrange multipliers $\bar{\theta}^* = \{9.6316, -18.5996, 16.7502\}$. For this learning subcollection, the entropy-optimal PDF $P^*(\mathbf{a}\,|\,\bar{\theta})$ has the form

$$P^*(\mathbf{a}) = \frac{\exp\left(-\sum_{i=1}^{3}\theta_i y_i(\mathbf{a})\right)}{\mathcal{P}(\bar{\theta})},$$

$$y_i(\mathbf{a}) = \left(1 + \exp(-\sum_{k=1}^{2} e(i)_k, a_k)\right)^{(-1)}. \tag{10.27}$$

The entropy-optimal PDF $P^*(\mathbf{a}, \bar{\theta}^*)$ is shown in Fig. 10.4.

By analogy with Example 1, use a sample of 500 random objects described by a two-dimensional vector with the independent uniform distribution on the interval $[0, 1]$. The other parameters are the same as in Example 1. The empirical probabilities $p_1^{(i)}, p_2^{(i)}$ of belonging of an object t_i to the first and second classes are presented in Fig. 10.5.

Like for Example 1, the graphs in Fig. 10.5 have a considerable spread of the empirical probabilities of belonging. This is caused by an intentionally small size of the learning sample. However, these results are much better than random guessing.

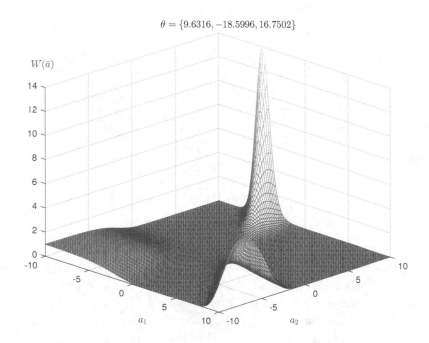

Figure 10.4: Two-dimensional entropy-optimal PDF $P^*(\mathbf{a}, \bar{\theta}^*)$.

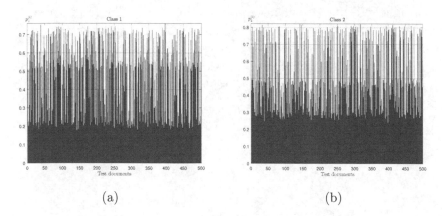

Figure 10.5: Empirical probabilities of belonging to different classes for Example 2.

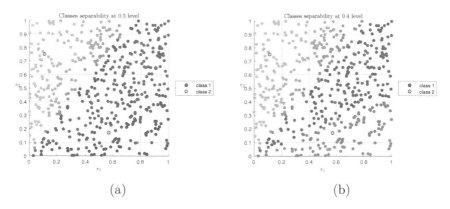

(a)   (b)

Figure 10.6: Classification diagram for Example 2.

To show this, place all random objects under classification on the same diagram and choose a level of reliability for their hard classification. Red color will be used for all objects related to the first class and green color for the ones related to the second class. In addition, color intensity will indicate how large has been the empirical probability for a given object in the case of soft classification. The hard classification results of the objects are presented in Fig. 10.6 for thresholds 0.5 (top) and 0.4 (bottom).

The classification results on this diagram testify that the limits between the groups of objects from the same class are refined depending on the level of reliability. This allows us to separate out the domains of the bright red- and green-color points (the first and second classes). In both cases, the objects of a corresponding class are located around the learning objects inside the black circles.

Consequently, a proper variation of the level of reliability gives a possible mechanism for the statistical treatment of entropy soft classification results.

CHAPTER 11

RANDOMIZED MACHINE LEARNING IN PROBLEMS OF DYNAMIC REGRESSION AND PREDICTION

The restoration of relations between two groups of variables is a basic class of problems arising in econometric research. Generally, these problems are treated as *static regression problems*, which involve retrospective data. In many cases one needs to restore a relation and also to forecast its future dynamics. As "future" data are unavailable for some period of time, such forecasts assist in making more or less reasonable decisions. An obvious recommendation here is to use *dynamic regressions*, which have a remarkable property of looking into the future. In other words, given some initial conditions at a time t_0, a dynamic regression generates trajectories on a time interval $[t_0, t]$, where $t > t_0$. A dynamic regression rests on a parameterized dynamic model with a two-stage procedure as follows. First, like for a static regression, it is necessary to estimate regression parameters based on retrospective data on a time interval $[t_-, t_0]$, where $t_- < t_0$. In fact, this stage implements machine

DOI: 10.1201/9781003306566-11

learning of the model. Then, the resulting model acts as a generator of forecasted trajectories on a second time interval $[t_0, t_+]$.

In this chapter, we will apply RML procedures for estimating the probabilistic characteristics of parameterized dynamic models. Recall that the general concept of randomized machine learning employs randomized parameters with entropy-optimal PDFs. Due to this circumstance, a randomized dynamic model generates an ensemble of entropy-optimal[1] forecasted random trajectories.

11.1 RESTORATION OF DYNAMIC RELATIONSHIPS IN APPLICATIONS

Many real processes are dynamic in the sense that at each time their state depends on the past states on some time interval, often considered as memory. Researchers often discriminate between processes with finite and infinite memory. The latter case is convenient for mathematical analysis whereas the hypothesis about finite memory is efficient for a numerical study of a process. In the sequel, the term "processes" will indicate dynamic processes.

A process $Y(t)$ is generated by some object in accordance with an internal transformation mechanism of observable input actions $X(\tau)$, $\tau \in \mathcal{T}$ and unobservable factors $F(t)$ with hypothetical properties. Recall that a rather general descriptive formalism for a dynamic process is a functional $\mathcal{B}[X(\tau), F(\tau); \tau \in \mathcal{T}]$ that characterizes a relation between the object's output at a time t and its input $x(\tau)$ and factors $F(\tau)$ on a time interval $\mathcal{T} = [t_-, t]$, i.e.,

$$Y(t) = \mathcal{B}[X(\tau), F(\tau); \tau \in \mathcal{T}], \qquad t \geq t_-. \tag{11.1}$$

The need to model the relation (11.1) occurs in many applications. Consider some of them in detail.

11.2 RANDOMIZED MODEL OF WORLD POPULATION DYNAMICS

The state of World population is characterized by its size $E(t)$ measured in *billions of people* on a calendar time. The population size $v(t) = \frac{dE(t)}{dt}$ varies due to the impact of fertility and mortality processes, since World

---

[1]In this context, we mean the trajectories generated by a dynamic model with entropy-optimal random parameters.

population is an isolated system. Fertility and mortality are measured by the numbers of newborns B and decedents M per unit time (the flows of newborns and decedents, respectively):

$$B(E) = bE, \qquad M(E) = mE. \tag{11.2}$$

The birth b and mortality m coefficients change during observation, i.e., are time-varying functions

$$b = b(t), \qquad m = m(t). \tag{11.3}$$

As a first approximation, consider the linear functions

$$b(t) = b_0 + b_1 t, \qquad m(t) = m_0 + m_1 t. \tag{11.4}$$

Consequently, the flows of fertility and mortality depend on time:

$$B(t) = (b_0 + b_1 t)E(t), \qquad M(t) = (m_0 + m_1 t)E(t). \tag{11.5}$$

The difference between these flows, the so-called reproduction flow, determines the dynamics of the population size, which are described by the following differential equation with variable coefficients:

$$\frac{dE(t)}{dt} = (r + ut)E(t), \qquad r = b_0 - m_0, \quad u = b_1 - m_1; \quad E(t_0) = E_0. \tag{11.6}$$

Here t_0 and E_0 denote the initial time and initial population size, respectively; the parameter r characterizes the constant component of the population reproduction rate and parameter u its rate of change over time.

Equation (11.6) with the initial condition E_0 has the solution

$$E(t) = E_0 \exp\left((r + ut)t\right). \tag{11.7}$$

The measurements of population size are performed on a calendar time with a step h. According to the UN standard, $h = 5$ years. Then $t = t_0 + nh$, and

$$E[t_0 + nh] = E_0 \exp\left((r + uhn)nh\right) = E[t_0 + nh \,|\, r, u, E_0], \quad n = 0, 1, \dots. \tag{11.8}$$

The fertility and mortality coefficients are estimated through indirect measurements that contain errors. Thus, the resulting flow rate and its change over time are supposed to be random variables defined on the intervals

$$\mathcal{J}_r = [r^-, r^+], \qquad \mathcal{J}_u = [u^-, u^+]. \tag{11.9}$$

The values inside these intervals are characterized by a joint probability density function (PDF) $P(r, u)$ defined on the rectangle

$$\mathcal{J} = \mathcal{J}_r \bigcup \mathcal{J}_u. \tag{11.10}$$

The parameters of the model (11.8) are estimated by measuring the real population size $E_{real}[nh]$ on some time interval. These measurements have errors simulated by random noises $\xi[nh] = \xi_n$, which are specified on the corresponding intervals $\Xi_n = [\xi_n^-, \xi_n^+]$. The noises are supposed to be independent with a PDF $q_n(\xi_n)$, i.e., their joint PDF on the observation interval $[0, N]$ is given by

$$Q(\bar{\xi}) = \prod_{n=0}^{N} q_n(\xi[nh]). \tag{11.11}$$

Therefore, the *randomized* model of World population dynamics can be described by

$$v[t_0 + nh] = E[t_0 + nh \mid r, u, E_0] + \xi[nh], \qquad n = 0, 1, \dots \tag{11.12}$$

This RM is a basis of RML procedures for estimating the PDFs $P(r, u)$ (parameters) and $Q(\bar{\xi})$ (noises), testing and randomized forecasting of World population. In all stages, retrospective data are used:

- on the estimation interval $\mathcal{T}_{est} = [1960-1995]$ (see Table 11.1);

- on the testing interval $\mathcal{T}_{tst} = [1995-2015]$ (see Table 11.2). These are the UN forecast data announced in 1985 (E_{1985}^{prn}).

- on the forecasting interval $\mathcal{T}_{frc} = [2015-2050]$ (see Table 11.3).[2]

In all stages, the simulated data were generated with the corresponding initial values:

- on the estimation interval,

$$v[1960 + nh] = E[1960 + 5n \mid r, u, E_{real}^{est}[0]] + \xi[5n], \qquad n = [0, 7]; \tag{11.13}$$

---

[2]Data is available at `http://www.irbis.vegu.ru/repos/1002/Html/27.htm` and `http://data.un.org`.

Table 11.1: Estimation interval $\mathcal{T}_{est}$.

n	0	1	2	3	4	5	6
year	1960	1965	1970	1975	1980	1985	1990
E_{real}^{est}	3.026	3.358	3.691	4.070	4.449	4.884	5.320

Table 11.2: Testing interval $\mathcal{T}_{tst}$.

n	0	1	2	3	4
year	1995	2000	2005	2010	2014
E_{real}^{tst}	5.724	6.128	6.514	6.916	7.260
E_{1985}^{prn}	5,400	5,968	6,596	7,284	8,056

Table 11.3: Forecasting interval $\mathcal{T}_{frc}$.

n	0	1	2	3	4	5	6
year	2015	2020	2025	2030	2040	2050	2060
E_{UN}^{frc}	7.324	7.644	7.964	8.284	8.924	9.564	9.884

- on the testing interval,

$$v[1995 + nh] = E[1995 + 5n \,|\, r, u, E_{real}^{tst}[0]] + \xi[5n], \qquad n = [0, 4]; \tag{11.14}$$

- on the forecasting interval,

$$v[2015 + nh] = E[2015 + 5n \,|\, r, u, E_{real}^{prn}[0]] + \xi[5n], \qquad n = [0, 5]. \tag{11.15}$$

11.2.1 RML procedure for learning of World Population Dynamics Model

According to the concept of Randomized Machine Learning entropy-optimal PDFs of parameters and noises will have the following form:

- PDF of the reproduction coefficient r and its speed u

$$P^*(r, u) = \frac{1}{\mathcal{R}(\bar{\theta})} \prod_{n=0}^{7} p_n^*(r, u|\theta_n),$$

$$p_n^*(r, u|\theta_n) = \exp\left(-\theta_n E[1960 + nh \,|\, r, u, E_{real}^{est}[0]]\right); \tag{11.16}$$

- PDF of the noise

$$Q^*(\bar{\xi}) = \frac{1}{\mathcal{Q}(\bar{\theta})} \prod_{n=0}^{7} q_n^*(\xi[nh]|\theta_n), \quad q_n^*(\xi[nh]|\theta_n) = \exp\left(-\theta_n\xi[nh]\right),$$

(11.17)

where

$$\mathcal{R}(\bar{\theta}) = \int_{\mathcal{J}} \prod_{n=0}^{7} \exp(-\theta_n E[1960+$$

$$+nh \mid r, u, E_{real}^{est}[0]])drdu,$$

$$\mathcal{Q}(\bar{\theta}) = \prod_{n=0}^{7} \int^{\Xi_n} \exp(-\theta_n\xi[nh])d\xi[nh] =$$

$$= \prod_{n=0}^{7} \frac{1}{\theta_n} \left(\exp(-\theta_n\xi_n^-) - \exp(-\theta_n\xi_n^+)\right).$$

(11.18)

Lagrange multipliers are determined by the system of balance equations (1.9, 1.10):

$$\frac{1}{\mathcal{R}(\bar{\theta})} \int_{\mathcal{J}} E[1960 + nh \mid r, u, E_{real}^{est}[0]]\times$$

$$\times \prod_{n=0}^{7} \exp\left(-\theta_n E[1960 + nh \mid r, u, E_{real}^{est}[0]]\right) \, dr \, du+$$

$$+\frac{1}{\mathcal{Q}(\bar{\theta})} \int_{\Xi_n} \xi[nh]\times$$

$$\times \prod_{n=0}^{7} \exp\left(-\theta_n\xi[nh]\right) d\xi[nh] - E_{real}^{est}[nh] = G_n(\bar{\theta}) = 0,$$

(11.19)

$$n = \overline{0,7}.$$

To solve this system, the method of minimizing the residual is used.

$$J(\bar{\theta}) = \|\mathbf{G}(\bar{\theta})\| \Rightarrow \min.$$

(11.20)

11.2.2 Numerical results

1. General conditions

All stages of estimation, testing and forecasting are implemented on intervals for coefficients of reproduction and trend, which simulate both the growth and the fall of the values of these parameters

$$\mathcal{J}_r = [-0.025, -0.075], \qquad \mathcal{J}_u = [0.002, 0.001].$$

(11.21)

Table 11.4: Lagrange multipliers.

	0	1	2	3	4	5	6
$\bar{\theta}$	0.0002	−0.8127	−1.1151	−1.3610	−0.8435	−0.1806	1.4985

It is assumed that the values of the measurement noise for each observation are the same and belong to the interval

$$\Xi_n = [-0.5, 0.5], \qquad n = \overline{0,7}. \tag{11.22}$$

After the entropy-optimal PDFs of parameters and noise have been found, trajectories $E(t)$ with random parameters and noises with that PDFs are generated.

2. Estimation

To implement the RML procedure, the data of the world population dynamics are used (see Table 11.1). The residual function $J(\bar{\theta})$ (11.20) is a function of 8 variables and contains two-dimensional integral components, the values of which can be estimated only numerically. To calculate them, we used the [164] quadrature formulas, implemented in MATLAB *guad2d* function. The residual was minimized using the *lsqnonlin* function from MATLAB Optimization Toolbox. The obtained values of the Lagrange multipliers are given in the Table 11.4.

The entropy-optimal PDFs of the parameters $P^*(r, u)$ and noises $q_n^*(\xi[5n]), n = \overline{0,7}$ are shown in Fig. 11.1–11.2.

3. Testing

The model has been tested using the data from Table 11.2 on the interval 1995–2015. Population size has been evaluated by formula (11.14), where r and u are randomized parameters with the PDFs $P^*(r, u)$ and $q_n^*(\xi[5n]), n = \overline{0,4}$. To generate an ensemble of random variables, we used the 2-dimensional modification of the Ulam–von Neumann method (Acceptance-Rejection method) [159].

Each pair of the random values r and u defines a separate exponential growth curve; moreover, for each point $[nh]$ a random value of the noise is added according to its PDF. As a result, the constructed trajectory of world population dynamics is not an exponential function. In the testing stage,

$$E^{tst}[0] = E_{real}^{tst}[0]. \tag{11.23}$$

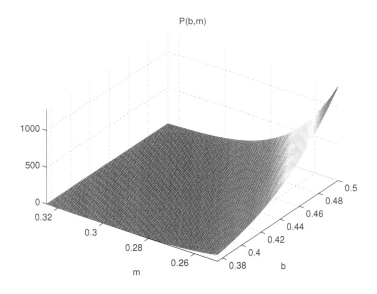

Figure 11.1: Entropy-optimal PDF $P^*(r, u)$.

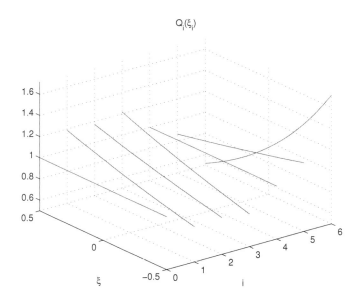

Figure 11.2: Entropy-optimal PDFs $q_n^*(\xi[5n]), n = \overline{0, 7}$.

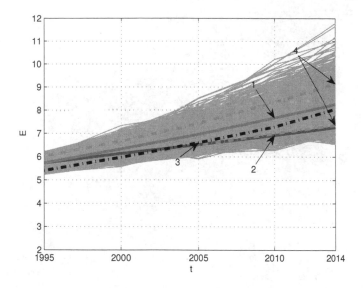

Figure 11.3: Ensemble of testing trajectories.

The ensemble is shown in Figure 11.3, where 1—the mean trajectories the mean trajectories of population dynamics; 2—real population dynamics on the testing interval 11.2, 3—population dynamics on the testing interval according to the UN forecast; 4—the boundaries of the variance.

Testing quality is assessed by the root-mean-square error (RMSE) between the mean and predicted trajectories

$$\delta = \sqrt{\sum_{i=0}^{4}(E^{tst}_{real}[ih] - E^{tst}[ih])^2}, \qquad (11.24)$$

and relative error

$$\varepsilon = \frac{\delta}{\|E^{tst}_{real}[ih]\| + \|E^{tst}[ih]\|}. \qquad (11.25)$$

For instance, for the UN forecast

$$\delta_{1985} = 0.9520, \qquad \varepsilon_{1985} = 0.0321. \qquad (11.26)$$

For the testing ensemble,

$$\delta_{RM} = 0.079, \qquad \varepsilon_{RM} = 0.003. \qquad (11.27)$$

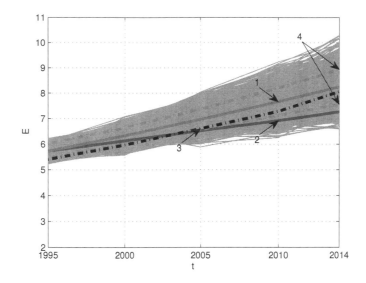

Figure 11.4: Ensemble of predicted trajectories.

Forecasting

The entropy-optimal randomized model was used to forecast the world population dynamics in the interval 2015–2050 years. Table 11.3 shows the UN-forecast data for the same period. Fig. 11.4 shows the ensemble of the predicted trajectories and (1)—the mean trajectory over the ensemble, (2)—the trajectory of the UN forecast, (3)—the variance boundaries.

11.3 RANDOMIZED FORECASTING OF DAILY ELECTRICAL LOAD IN POWER SYSTEM

The daily electrical load L of a power system depends on many various factors. The analysis below is restricted to one of the most significant external factors—the ambient temperature T. The daily temperature variations are fluctuating. These fluctuations affect electrical load, but with some time delay due the inertia of a power network supplying electrical energy from the generator to consumers.

11.3.1   Electrical Load Model

The electrical load model (the LT model) describes the dynamic relation between electrical load and ambient temperature while the ambient temperature model (the $T\xi$ model) describes the daily dynamics of ambient temperature. There exist quite a few versions of the LT model, albeit all being static, i.e., describing the relation between electrical load and ambient temperature at current time instants [49]. The daily temperature dynamics are fluctuating, and such fluctuations are described, in particular, by the periodic autoregressive model [13].

Note that the effect of ambient temperature on electrical load is dynamic, i.e., the change in load due to temperature at a given time instant depends on its value at a previous time instant. A similar property applies to ambient temperature fluctuations.

Therefore, following the general randomized approach, the LT model is designed as a first-order dynamic regression model with random parameters; the $T\xi$ model, as a second-order dynamic regression with a random parameter and a random input ξ. Then the $LT\xi$ model is the composition of the two models above.

In the class of linear models, the randomized dynamic regression load–temperature model (LT model) of the first order can be written in the form

$$L[n] = aL[n-1] + bT[n],$$
$$v[n] = L[n] + \mu[n], \qquad n = \overline{1,24}. \tag{11.28}$$

where random independent parameters a and b take values within intervals

$$a \in \mathcal{A} = [a^-, a^+], \; b \in \mathcal{B} = [b^-, b^+]. \tag{11.29}$$

The probabilistic properties are characterized by PDFs $P(a)$ and $F(b)$ defined on the sets $\mathcal{A}$ and $\mathcal{B}$, respectively. The random noise $\mu[n]$ that simulates electrical load measurement errors is of the interval type as well. In the general case, for each time instant the intervals may have different limits, i.e.,

$$\mu[n] \in \mathcal{M}_n = [\mu^-[n]], \mu^+[n]], \tag{11.30}$$

with PDFs $M_n(\mu[n]),], n = \overline{n^-, n^+}$.

Consider the $T\xi$ model. The fluctuating character of the daily temperature variations is described by the randomized dynamic regression model of the second order

$$
\begin{aligned}
\tau[n] &= c\left(2, 1\tau[n-1] - 1, 1\tau[n-2]\right), \quad c = \frac{1}{1+\omega^2}, \\
T[n] &= t + \tau[n] + \xi[n],
\end{aligned}
\tag{11.31}
$$

where ω is the temperature fluctuation frequency, t is the mean daily temperature. These parameters are random and take values within given intervals, i.e.

$$
\omega \in \Omega = [\omega^-, \omega^+], \qquad c \in \mathcal{C} = [c^-, c^+],
$$
$$
c^- = \frac{1}{1+(\omega^+)^2} > 0, \quad c^+ = \frac{1}{1+(\omega^-)^2} > c^- > 0.
\tag{11.32}
$$

The probabilistic properties of the parameters are characterized by PDFs $W(c)$ defined on corresponding intervals.

Equation (11.31) contains random noises described by independent random variables $\xi[n]$; in each measurement n, their values may lie in different intervals, i.e.,

$$
\xi[n] \in \Xi_n = [\xi_n^-, \xi_n^+].
\tag{11.33}
$$

The probabilistic properties of the random variable $\xi[n]$ are characterized by a PDF $Q_n(\xi[n])$, $n - \overline{n^-, n^+}$.

Thus, equations (11.28, 11.31) describing electrical load dynamics in a power system are characterized by the following PDFs:

- the LT model, by the PDFs $P(a)$ and $F(b)$ of the model parameters and the PDF $M_n(\mu[n])$, of the measurement noises, $n = \overline{n^-, n^+}$;

- the $T\xi$ model, by the PDFs $W(c)$ of the model parameters and the PDFs $Q_n(\xi[n])$ of the measurement noises , $n = \overline{n^-, n^+}$.

11.3.2 Learning dataset

For estimating the PDFs, the normalized real data from the GEF-Com2014 dataset (see [71]) on daily electrical load variations $0 \le L_r^{(i)}[n] \le 1$, mean daily temperature variations $0 \le t_r^{(i)} \le 1$ and temperature deviations $0 \le \tau_r^{(i)}[n] \le 1$ from the mean daily value can be used. (Here normalization means the reduction to the unit interval.)

The normalization procedure is performed in the following way:

$$L_r^{(i)}[n] = \frac{\hat{L}_r^{(i)}[n] - \hat{L}_{min}^{(i)}}{\hat{L}_{max}^{(i)} - \hat{L}_{min}^{(i)}},$$

$$\tau_r^{(i)}[n] = \frac{\hat{\tau}_r^{(i)}[n] - \hat{\tau}_{min}^{(i)}}{\hat{\tau}_{max}^{(i)} - \hat{\tau}_{min}^{(i)}},$$

$$t_r^{(i)} = \frac{1}{24} \sum_{n=1}^{24} \tau_r^{(i)}[n], \qquad (11.34)$$

where $\hat{L}_{min}^{(i)} = \min_n \hat{L}^{(i)}[n]$, $\hat{L}_{max}^{(i)} = \max_n \hat{L}^{(i)}[n]$, $\hat{\tau}_{min}^{(i)} = \min_n \hat{\tau}^{(i)}[n]$, $\hat{\tau}_{max}^{(i)} = \max_n \hat{\tau}^{(i)}[n]$.

Figure 11.5 shows graphs of the load $L_r^{(i)}[n]$, the temperature deviation $\tau_r^{(i)}[n]$ and data on daily-average temperature $t_r^{(i)}$ at 03.07.2016 ($i = 1$), 04.07.2016 ($i = 2$), 05.07.2016 ($i = 3$) as learning collection. Let us denote *learning interval* $\mathcal{T}_l = [n^-, n^+] = [1, 24]$.

According to (11.28), (11.31), the model variables and the corresponding real data on the learning interval $n \in \mathcal{T}_l$ are described by the vectors

$$\mathbf{L}^{(i)}(\mathcal{T}_l) = \{L^{(i)}[1], \ldots, L^{(i)}[24]\}, \mathbf{L}_r^{(i)}(\mathcal{T}_l) = \{L_r^{(i)}[1], \ldots, L_r^{(i)}[24]\},$$

$$\mathbf{L}^{(i)}(\mathcal{T}_l - 1) = \{L^{(i)}[0], \ldots, L^{(i)}[23]\},$$

$$\mathbf{L}_r^{(i)}(\mathcal{T}_l - 1) = \{L_r^{(i)}[0], \ldots, L_r^{(i)}[23]\},$$

$$\mathbf{V}^{(i)}(\mathcal{T}_l) = \{v^{(i)}[1], \ldots, v^{(i)}[24]\}, \mathbf{V}_r^{(i)}(\mathcal{T}_l) = \{v_r^{(i)}[1], \ldots, v_r^{(i)}[24]\},$$

$$\mathbf{T}^{(i)}(\mathcal{T}_l) = \{\tau^{(i)}[1], \ldots, \tau^{(i)}[24]\}, \mathbf{T}_r^{(i)}(\mathcal{T}_l) = \{\tau_r^{(i)}[1], \ldots, \tau_r^{(i)}[24]\},$$

$$\tilde{\mathbf{T}}^{(i)}(\mathcal{T}_l - 1, \mathcal{T}_l - 2) = \{2, 1\tau^{(i)}[0] - 1, 1\tau^{(i)}[-1], \ldots,$$

$$2, 1\tau^{(i)}[23] - 1, 1\tau^{(i)}[22]\},$$

$$\tilde{\mathbf{T}}_r^{(i)}(\mathcal{T}_l - 1, \mathcal{T}_l - 2) = \{2, 1\tau_r^{(i)}[0] - 1, 1\tau_r^{(i)}[-1], \ldots,$$

$$2, 1\tau_r^{(i)}[23] - 1, 1\tau_r^{(i)}[22]\},$$

$$\mu^{(i)}(\mathcal{T}_l) = \{\mu^{(i)}[1], \ldots, \mu^{(i)}[24]\}, \quad \xi^{(i)}(\mathcal{T}_l) = \{\xi^{(i)}[1], \ldots, \xi^{(i)}[24]\}.$$

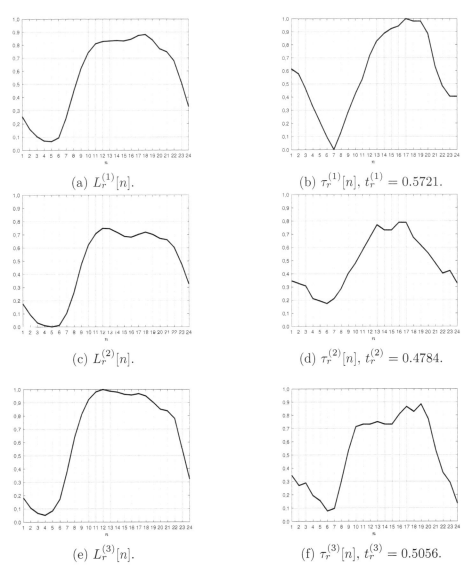

(a) $L_r^{(1)}[n]$.

(b) $\tau_r^{(1)}[n]$, $t_r^{(1)} = 0.5721$.

(c) $L_r^{(2)}[n]$.

(d) $\tau_r^{(2)}[n]$, $t_r^{(2)} = 0.4784$.

(e) $L_r^{(3)}[n]$.

(f) $\tau_r^{(3)}[n]$, $t_r^{(3)} = 0.5056$.

Figure 11.5: Load $L_r^{(i)}[n]$ and temperature deviation $\tau_r^{(i)}[n]$.

In terms of (11.3.2), the LT- and $T\xi$-models on the learning interval $\mathcal{T}_l$ have the form

$$
\begin{aligned}
\mathbf{L}^{(i)}(\mathcal{T}_l) &= a\mathbf{L}^{(i)}(\mathcal{T}_l - 1) + b\mathbf{T}^{(i)}(\mathcal{T}_l), \\
\mathbf{V}^{(i)}(\mathcal{T}_l) &= \mathbf{L}^{(i)}(\mathcal{T}_l) + \mu^{(i)}(\mathcal{T}_l), \\
\mathbf{T}^{(i)}(\mathcal{T}_l) &= c\,\tilde{\mathbf{T}}^{(i)}(\mathcal{T}_l - 1, \mathcal{T}_l - 2), \\
\mathbf{T}^{(i)}(\mathcal{T}_l) &= t + \tilde{\mathbf{T}}^{(i)}(\mathcal{T}_l) + \xi^{(i)}(\mathcal{T}_l).
\end{aligned}
\tag{11.35}
$$

The random parameters take values within the intervals

$$
\mathcal{A} = [0,05; 0,15], \quad \mathcal{B} = [0,5; 1,0], \quad \mathcal{C} = [0,75; 0,85]. \tag{11.36}
$$

The measurement noises take values within the intervals

$$
\mathcal{M}_n = [-0.1; 0,1], \quad \Xi_n = [-0,1; 0,1]. \tag{11.37}
$$

11.3.3 Entropy-optimal probability density functions of parameters and noises

According to the approach described in Section 5, for the LT model (11.28)–(11.30) the PDFs parameterized by the Lagrange multipliers $\theta^{(i)} = \{\theta_1^{(i)}, \ldots, \theta_{24}^{(i)}\}$ have the form

$$
\begin{aligned}
P_i^*(a, \theta^{(i)}) &= \frac{l_r^{(i)}(\theta)\exp\left(-a l_r^{(i)}(\theta)\right)}{\exp(-a^- l_r^{(i)}(\theta)) - \exp(-a^+ l_r^{(i)}(\theta))}, \\
F_i^*(b, \theta^{(i)}) &= \frac{h_r^{(i)}(\theta)\exp\left(-b h_r^{(i)}(\theta)\right)}{\exp(-b^- h_r^{(i)}(\theta)) - \exp(-b^+ h_r^{(i)}(\theta))}, \\
M_{i,n}^*(\mu[n]) &= \frac{\theta_n^{(i)}\exp\left(-\theta_n^{(i)}\mu[n]\right)}{\exp(-\mu^-[n]\theta_n^{(i)}) - \exp(-\mu^+[n]\theta_n^{(i)})}, \quad n = \overline{1,24}.
\end{aligned}
\tag{11.38}
$$

where

$$
l_r^{(i)}(\theta) = \sum_{n=1}^{24} \theta_n L_r^{(i)}[n-1], \quad h_r^{(i)}(\theta) = \sum_{n=1}^{24} \theta_n T_r^{(i)}[n]. \tag{11.39}
$$

The Lagrange multipliers $\theta^{(i)}$ are calculated by solving the system of balance equations

$$
L^{(i)}(\theta^{(i)}) + T^{(i)}(\theta^{(i)}) + M_n^{(i)}(\theta_n^{(i)}) = L_r^{(i)}[n], \quad n = \overline{1,24}, \tag{11.40}
$$

where

$$L^{(i)}(\theta^{(i)}) = \frac{1}{\exp(-a^- l^{(i)}(\theta^{(i)})) - \exp(-a^+ l^{(i)}(\theta^{(i)}))} \times$$
$$\times (\exp(-a^- l_r^{(i)}(\theta^{(i)}))(a^- l^{(i)}(\theta^{(i)}) + 1) -$$
$$- \exp(-a^+ l_r^{(i)}(\theta^{(i)}))(a^+ l^{(i)}(\theta^{(i)}) + 1)),$$

$$T^{(i)}(\theta^{(i)}) = \frac{1}{\exp(-b^- h^{(i)}(\theta^{(i)})) - \exp(-b^+ h^{(i)}(\theta^{(i)}))} \times$$
$$\times (\exp(-b^- h_r^{(i)}(\theta^{(i)}))(b^- h^{(i)}(\theta^{(i)}) + 1) -$$
$$- \exp(-b^+ h_r^{(i)}(\theta^{(i)}))(b^+ h^{(i)}(\theta^{(i)}) + 1)),$$

$$M_n^{(i)}(\theta_n^{(i)}) = \frac{1}{\theta_n^{(i)} \left(\exp(-\mu^- [n]\theta_n^{(i)}) - \exp(-\mu^+ [n]\theta_n^{(i)}) \right)} \times$$
$$\times (\exp(-\mu^- [n]\theta_n^{(i)})(\mu^- [n]\theta_n^{(i)} + 1) -$$
$$- \exp(-\mu^+ [n]\theta_n^{(i)})(\mu^+ [n]\theta_n^{(i)} + 1)). \tag{11.41}$$

Consider the $T\xi$ model. The corresponding entropy-optimal PDFs parameterized by the Lagrange multipliers have the form

$$W_i^*(c, \eta^{(i)}) = \frac{\tilde{h}_r^{(i)}(\eta) \exp\left(-c\tilde{h}_r^{(i)}(\eta)\right)}{\exp(-a^- \tilde{h}_r^{(i)}(\eta)) - \exp(-a^+ \tilde{h}_r^{(i)}(\eta))},$$

$$Q_{i,n}^*(\xi[n]) = \frac{\eta_n^{(i)} \exp\left(-\eta_n^{(i)}\xi[n]\right)}{\exp(-\xi^- [n]\eta_n^{(i)}) - \exp(-\xi^+ [n]\eta_n^{(i)})}, \quad n = \overline{1, 24}.$$

where

$$\tilde{h}_r^{(i)}(\eta) = \sum_{n=1}^{24} \eta_n (2T_r^{(i)}[n-1] - T_r^{(i)}[n-2]), \quad q^{(i)}(\eta^{(i)}) = \sum_{n=1}^{24} \eta_n^{(i)}. \tag{11.42}$$

The Lagrange multipliers $\eta^{(i)}$ are calculated by solving the system of balance equations

$$D^{(i)}(\eta^{(i)}) + N^{(i)}(\eta^{(i)}) + K_n^{(i)}(\eta_n^{(i)}) = T_r^{(i)}[n], \quad n = \overline{1, 24}, \tag{11.43}$$

where

$$D^{(i)}(\eta^{(i)}) = \frac{1}{\exp(-t^- q^{(i)}(\eta^{(i)})) - \exp(-t^+ q^{(i)}(\eta^{(i)}))} \times$$
$$\times (\exp(-t^- q^{(i)}(\eta^{(i)}))(t^- q^{(i)}(\eta^{(i)}) + 1) -$$
$$- \exp(-t^+ q^{(i)}(\eta^{(i)}))(t^+ q^{(i)}(\eta^{(i)}) + 1)),$$

$$N^{(i)}(\eta^{(i)}) = \frac{1}{\exp(-c^- \tilde{h}^{(i)}(\eta^{(i)})) - \exp(-c^+ \tilde{h}^{(i)}(\eta^{(i)}))} \times$$
$$\times (\exp(-c^- \tilde{h}_r^{(i)}(\eta^{(i)}))(c^- \tilde{h}^{(i)}(\eta^{(i)}) + 1) -$$
$$- \exp(-c^+ \tilde{h}_r^{(i)}(\eta^{(i)}))(c^+ \tilde{h}^{(i)}(\eta^{(i)}) + 1)),$$

$$K_n^{(i)}(\eta_n^{(i)}) = \frac{1}{\eta_n^{(i)} \left(\exp(-\xi^- [n]\eta_n^{(i)}) - \exp(-\xi^+ [n]\eta_n^{(i)}) \right)} \times$$
$$\times (\exp(-\xi^- [n]\eta_n^{(i)})(\xi^- [n]\eta_n^{(i)}) + 1) -$$
$$- \exp(-\xi^+ [n]\eta_n^{(i)})(\xi^+ [n]\eta_n^{(i)}) + 1)). \tag{11.44}$$

11.3.4 Results of model learning

Using the available data on daily variations of electrical load and ambient temperature for the three days indicated above, the balance equations (11.40), (11.3.3), (11.43)–(11.3.3) were formed. Their solution was determined by minimizing the quadratic residual between the left- and right-hand sides of the equations. Since the equations are significantly nonlinear, the resulting values of the Lagrange multipliers (see Table 11.5) correspond to a local minimum of the residual. All calculations were implemented in MATLAB; optimization was performed using the `fsolve` function.

Because the parameters of the LT model are independent, the joint PDFs $U_i^*(a, b) = P_i^*(a) F_i^*(b)$ of the parameters and noises have the form

$$\begin{aligned} U_1^*(a, b) &= 53,09 \exp(-9,72a) \exp(0,06b), \\ U_2^*(a, b) &= 55,49 \exp(-6,04a) \exp(0,58b), \\ U_3^*(a, b) &= 65,81 \exp(-6,09a) \exp(-0,81b), \end{aligned} \tag{11.45}$$

$$(a, b) \in [0,05; 0,15] \bigcup [0,5; 1,0], \quad \mu \in [-0,1,0,1], \quad i = \overline{1,24}.$$

Clearly, the PDFs are of exponential type. For $i = 1$, the graphs are shown in Fig. 11.6.

Table 11.5: Lagrange multipliers θ, η.

Time	$\theta^{(1)}$	$\theta^{(2)}$	$\theta^{(3)}$	$\eta^{(1)}$	$\eta^{(2)}$	$\eta^{(3)}$
1	−29.72	7009.28	1038.07	14.63	21.22	17.34
2	1.58	230.89	35.35	19.52	26.71	28.32
3	−4.09	369.96	26.23	35.91	31.60	26.33
4	−4.68	29.93	11.96	55.83	127.82	52.08
5	−7.21	24.25	1.03	96.85	642.35	110.94
6	−9.26	13.72	−15.76	592.99	7009.28	4729.52
7	−59.09	−5.96	−7009.28	7009.28	183.92	7009.28
8	−7009.28	−33.99	−767.99	48.21	39.94	23.16
9	−766.00	−1409.28	−22.91	66.58	12.28	−1.26
10	−50.90	−4229.90	−4.27	37.38	2.35	−19.78
11	−18.97	−45.22	3.72	22.51	−8.82	−22.73
12	−11.42	−15.07	9.17	7.16	−27.06	−23.06
13	−13.94	2.59	17.38	5.72	−172.25	−27.06
14	−17.62	5.82	14.94	2.83	−65.29	−23.02
15	−18.18	9.33	17.74	−0.30	−57.45	−23.15
16	−27.28	11.35	21.85	−1.24	−482.69	−47.78
17	−49.55	4.50	22.68	−5.49	−889.02	−130.49
18	−25.41	−7.09	29.39	−0.89	−28.12	−60.71
19	−8.20	−4.66	98.03	−4.23	−14.20	−270.17
20	0.95	−4.89	52.27	3.70	−6.41	−31.47
21	1.01	−16.37	8.15	21.16	2.48	−1.23
22	22.00	−8.24	7.45	17.85	12.15	14.86
23	2881.43	17.00	902.73	26.78	9.98	26.39
24	512.14	36.30	355.47	27.32	24.65	121.44
$l_r^{(i)}(\theta^*)$	9.71	6.04	6.09			
$h_r^{(i)}(\theta^*)$	0.06	0.58	0.81			
$\tilde{h}_r^{(i)}(\eta^*)$				0.41	0.05	0.19

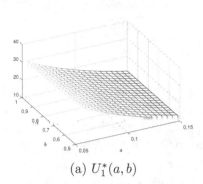

(a) $U_1^*(a, b)$

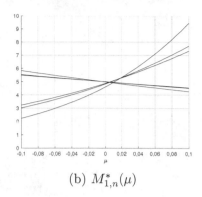

(b) $M_{1,n}^*(\mu)$

Figure 11.6

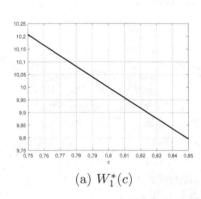

(a) $W_1^*(c)$

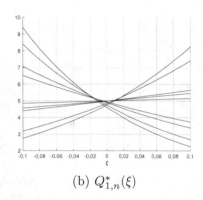

(b) $Q_{1,n}^*(\xi)$

Figure 11.7

For the $T\xi$ model, the PDFs of the parameters and noises have the form

$$
\begin{aligned}
W_1^*(c) &= 13,90 \exp(-0,41c), \\
W_2^*(c) &= 10,43 \exp(-0,05c), \\
W_3^*(c) &= 11,65 \exp(-0,19c),
\end{aligned}
$$

$$
c \in [0,75; 0,85], \quad \xi \in [-0,1; 0,1], \quad i = \overline{1,24}.
$$

For $i = 1$, the graphs can be seen in Fig. 11.7.

Thus, the randomized $LT\xi$ model generates random trajectories with the entropy-optimal PDFs of the model parameters and measurement noises:

$$\begin{aligned}
L[n] &= aL[n-1] + bT[n], \quad (P^*(a), F^*(b)), \\
v[n] &= L[n] + \mu[n], \quad M_n^*(\mu[n]); \\
\tau[n] &= c(2\tau[n-1] - \tau[n-2]), \quad W^*(c), \; i = \overline{1,3}, \quad (11.46) \\
T[n] &= t + \tau[n] + \xi[n], \quad Q_n^*(\xi[n]).
\end{aligned}$$

The corresponding ensembles are generated by the sampling procedure of the resulting PDFs of the parameters and noises using the acceptance-rejection (AR) method (also known as rejection sampling (RS); see [186]). During calculations, 100 samples for each parameter and 100 samples for each noise were used; in other words, the ensemble consisted of 10^4 trajectories.

11.3.5 Model testing

The adequacy of the model was analyzed by the self- and cross-testing of the LT and $T\xi$ models on the real load–temperature data for July 3–5, 2016 ($i = 1, 2, 3$). Self-testing means generating an ensemble of trajectories with the entropy-optimal parameters and noises for day i, calculating the mean (*mean*) and median (*med*) trajectories and also the variance pipe (*std*±) of the ensemble, and comparing the mean trajectory with the real counterparts by electrical load and ambient temperature *for the same day i*. The quality of approximation is characterized by relative errors,

$$\delta_L^{(i)} = \frac{\sum\limits_{n=1}^{24} \left(L_{mean}^{(i)}[n] - L_r^{(i)}[n] \right)^2}{\sum\limits_{n=1}^{24} (L_{mean}^{(i)}[n])^2 + \sum\limits_{n=1}^{24} (L_r^{(i)}[n])^2}, \quad i = \overline{1,3}, \quad (11.47)$$

in electrical load and

$$\delta_T^{(i)} = \frac{\sum\limits_{n=1}^{24} \left(T_{mean}^{(i)}[n] - T_r^{(i)}[n] \right)^2}{\sum\limits_{n=1}^{24} (T_{mean}^{(i)}[n])^2 + \sum\limits_{n=1}^{24} (T_r^{(i)}[n])^2}, \quad i = \overline{1,3}. \quad (11.48)$$

in ambient temperature.

Cross-testing represents a similar procedure in which the mean trajectories are compared with the real counterparts in terms of electrical load and ambient temperature for days $j \neq i$. The quality of approximation is characterized by relative errors,

$$\delta_L^{(i,j)} = \frac{\sum_{n=1}^{24} \left(L_{mean}^{(i)}[n] - L_r^{(j)}[n] \right)^2}{\sum_{n=1}^{24} (L_{mean}^{(i)}[n])^2 + \sum_{n=1}^{24} (L_r^{(j)}[n])^2}, \quad i = \overline{1,3}, \; i \neq j, \qquad (11.49)$$

in electrical load and

$$\delta_T^{(i,j)} = \frac{\sum_{n=1}^{24} \left(T_{mean}^{(i)}[n] - T_r^{(j)}[n] \right)^2}{\sum_{n=1}^{24} (T_{mean}^{(i)}[n])^2 + \sum_{n=1}^{24} (T_r^{(j)}[n])^2}, \quad i = \overline{1,3}, i \neq j. \qquad (11.50)$$

in ambient temperature.

Self-testing. For the LT model, the real ambient temperature data $T_r^{(i)}[n]$, as well as the entropy-optimal PDFs $P_i^*(a)$ and $F_i^*(b)$ of the parameters (a, b) and the PDFs $M_1^*(\mu[1]), \ldots, M_{24}^*(\mu[24])$ of the measurement noises $\mu[n]$, were used. The ensembles $\mathbb{L}^{(i)}$ were generated using the *sampling procedure* of the above PDFs. The mean trajectory $L_{mean}^{(i)}[n]$, the median trajectory $L_{med}^{(i)}[n]$ and also the trajectories $L_{std\pm}^{(i)}[n]$ corresponding to the limits of the variance graph were found. The errors $\delta_L^{(i)}$ were calculated. The resulting ensembles and relative errors $\delta_L^{(i)}$ for the three indicated days are demonstrated in Fig. 11.8.

The $T\xi$ model was tested by generating the ensemble $\mathbb{T}^{(i)}$ of random trajectories $T^{(i)}[n]$, $n = \overline{1,24}$ with the entropy-optimal PDFs $W^{(i)}(c)$ and $Q_1^*(\xi[1]), \ldots, Q_{24}^*(\xi[24])$ through sampling. The mean trajectory $T_{mean}^{(i)}[n]$, the median trajectory $T_{med}^{(i)}[n]$ and the trajectory $T_{std\pm}^{(i)}[n]$ corresponding to the limits of the variance pipe are calculated. The resulting ensembles and relative errors $\delta_T^{(i)}$ for the three days are shown in Fig. 11.9

Cross-testing. For cross-testing, the LT and $LT\xi$ models learned on the data for day i were used, and their mean trajectories were compared with the data for days $j \neq i$. The resulting errors are combined in Tables 11.6–11.8.

11.3.6 Randomized prediction of N-daily load

In the randomized prediction of the N-daily load, the $LT\xi$ model learned on the interval $\mathcal{T}_l$ was used. The quality of the forecast was characterized

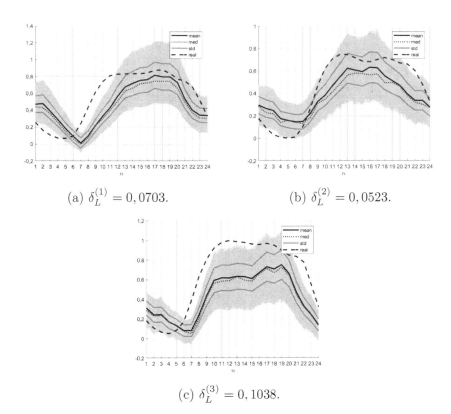

(a) $\delta_L^{(1)} = 0,0703.$

(b) $\delta_L^{(2)} = 0,0523.$

(c) $\delta_L^{(3)} = 0,1038.$

Figure 11.8: Ensembles of LT model.

Table 11.6: Values δ_L obtained by cross-testing of LT model. Mean value $\delta_L = 0.0530$.

i/j	1	2	3
1		0.0495	0.1052
2	0.0858		0.1428
3	0.0569	0.0364	

Table 11.7: Values δ_T obtained by cross-testing of $T\xi$ model. Mean value $\delta_L = 0.0757$.

i/j	1	2	3
1		0.1051	0.1079
2	0.1506		0.1185
3	0.1315	0.0676	

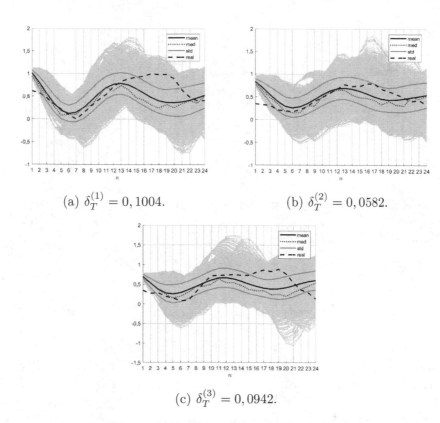

(a) $\delta_T^{(1)} = 0,1004.$ (b) $\delta_T^{(2)} = 0,0582.$

(c) $\delta_T^{(3)} = 0,0942.$

Figure 11.9: Ensembles of $T\xi$-model.

Table 11.8: Values δ_T obtained by cross-testing of $LT\xi$ model. Mean value $\delta_L = 0.1478$.

i/j	1	2	3
1		0.1437	0.2659
2	0.1756		0.2322
3	0.3475	0.1655	

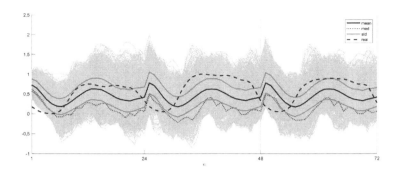

Figure 11.10: 24-h, 48-h and 72-h forecasts using $LT\xi$ model.

Table 11.9: Accuracy of 24-h, 48-h and 72-h forecasts using $LT\xi$ model.

$\delta_L^{(2)}$	$\delta_L^{(3)}$	$\delta_L^{(4)}$
0,1509	0,2515	0,2133

using the $LT\xi$ model with the entropy-optimal PDFs obtained on the real data for the *first* $(i = 1)$ day.

The 1-day $(n \in [1, 24])$, 2-day $(n \in [1, 48])$ and 3-day $(n \in [1, 72])$ ensembles were constructed by the sampling procedure of the above PDFs. For these ensembles, the mean trajectories $L_{mean}[n]$, the median trajectories $L_{med}[n]$ and also the limiting trajectories $L_{std\pm}[n]$ of the variance pipe were found. The forecast results were compared with the real data for July 3–7, 2006 $(i = \overline{1,4})$. The forecasting quality was characterized by the relative errors calculated similar to (11.49)–(11.50).

The resulting 24-h, 48-h and 72-h randomized forecasts of electrical load and their probabilistic characteristics (the mean and median trajectories, the limit trajectories of the variance pipes) are presented in Fig. 11.10. The errors, i.e., the deviations between the model forecasts and real data, can be seen in Table 11.9.

An important feature of the described method is the formation of predictive trajectories under "entropy-worst" measurement errors.

11.4 ENTROPY RANDOMIZED MODELLING AND FORECAST-ING OF THERMOKARST LAKE AREA EVOLUTION IN WESTERN SIBERIA

11.4.1 Thermokarst lakes and climate change

Permafrost zones, occupying a significant part of the earth's surface, are the localization of thermokarst lakes, which accumulate greenhouse gases (methane and carbon dioxide). These gases have a considerable contribution to global climate change. Climate warming may cause a positive inverse reaction: an increase in temperature will lead to melting of frozen rocks and additional release of methane and carbon dioxide produced by the vital activity of microorganisms that process thawed organic matter.

A thorough understanding of this global cycle and its catastrophic consequences motivates researchers to study the origin, evolution and transformation of thermokarst lakes in permafrost zones [92, 200]. The source data in such studies are acquired by the remote sensing of the earth surface and measurements of meteorological parameters [24, 89, 112, 113, 129, 154].

The available historical data allow tracing the past spatiotemporal evolution of thermokarst lakes. However, to forecast the spatiotemporal evolution, one needs *environmental models* (EnMd) adapted to historical data. For this purpose, EnMd are first trained through machine learning algorithms and then used to obtain forecasts.

The processes of formation and evolution of thermokarst lakes have been underinvestigated. Moreover, historical data about these processes, especially those acquired from satellites, contain significant errors. Due to all these factors, the predictive EnMd must operate under conditions of sufficiently high uncertainty. So, we consider *randomized forecasting* (RF) approach, an alternative approach compared to the traditional machine learning methods.

11.4.2 Thermokarst lakes of Western Siberia, tools and problems of their study

Remote studies of the spatiotemporal dynamics of thermokarst lakes (their number and size) in permafrost zones of Western Siberia are based on sensing of the earth surface using combinations of images of medium (30 m) and high resolution (from fractions of a meter to several meters). This combination allows detecting shallow lakes, which produce significantly more greenhouse gases than large ones.

The territory in the permafrost zone of Western Siberia studied in this paper has an area of about 1.05 million km². This territory includes three typical zones (Fig. 11.11): insular (*I*, southern), discontinuous (*D*, middle) and continuous (*C*, northern). Lakes in permafrost zones are characterized by an extremely wide range of their sizes, from several m² to hundreds of thousands of hectares. To explore this space, we use medium-resolution optical and radar images (Landsat, ERS-2, and ENVISAT), high-resolution images (Kanopus-V, BKA, and Alos), and ultra-high-resolution images (Resurs-P, QuickBird, and GeoEye-1). To minimize the seasonal fluctuations in the water level of the lakes, we select the images for a fairly short period of the summer season, from July to August. During this period, the ice cover on the lakes completely disappears, and the lakes can be identified when decoding images. We process the images using the standard means of geographic information systems, ArcGIS and ENVI.

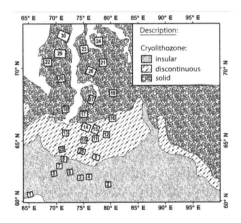

Figure 11.11: Diagram of test regions.

According to [24], for lakes with an area of 2 *ha* and more, the remote measurement error of lake areas based on satellite images does not exceed 5%. As reference images, we take the ultra-high resolution QuickBird images (0.7 *m²*). For illustration, a fragment of the Landsat-5 satellite image (July 13, 2007) is presented below. It shows the system of thermokarst lakes (Fig. 11.12a) and the result of its decoding (Fig. 11.12b).

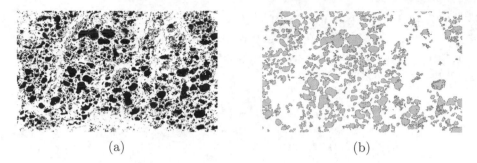

(a)   (b)

Figure 11.12: Satellite images.

An important problem for the evolution of thermokarst lakes is identifying the climatic parameters that affect their number and area. As is indicated by 1973–2007 data for 30 test domains in different permafrost zones, the average annual temperature and the annual precipitation have the greatest impact on the evolution of thermokarst lakes. For the Western Siberia territory under study, these climatic parameters were obtained using the ERA-40, ERA-INTERIM and APHRODITE JMA meteorological field reanalysis systems; for details, see [131, 133]. For the territory considered in this paper, the average annual temperature and the annual precipitation have a multidirectional relationship: an increase in the former is accompanied by a decrease in the latter precipitation [131]. A similar relationship was established in Alaska [154] and Eastern Siberia [130, 132, 195].

The total number of lakes in the permafrost zone of Western Siberia reaches 8.7 million, with a total area of about 6.4 million ha. Of these, nearly one half (44%) is occupied by ultra-small lakes (less than 500 m^2), which form merely 0.1% of the total area of all thermokarst lakes. Large lakes (more than 2×10^4 m^2) make up 86.6% of the total area, but their number does not exceed 3.6% of the total number of all thermokarst lakes.

11.4.3 Structures of randomized models of thermokarst lakes state

An important stage in the technology of randomized forecasting is to construct randomized models of thermokarst lakes area evolution as well as randomized models of the average annual temperature and the annual precipitation affecting this evolution. These variables will be modeled using linear dynamic randomized regressions [10] with random parameters (LDRR).

The temporal evolution of the lakes area $S[n]$ ($LDRR$-LA) is described by the following dynamic randomized regression equation with two influencing factors, the average annual temperature $T[n]$ and the annual sum of precipitation $R[n]$:

$$
\begin{aligned}
S[n] &= a_0 + \sum_{k=1}^{p} a_k S[n-k] + a_{(p+1)} T[n] + a_{(p+2)} R[n], \\
v[n] &= S[n] + \xi[n].
\end{aligned}
\tag{11.51}
$$

The parameters

$$
a_k \in \mathcal{A}_k = [a_k^-, a_k^+], \quad k = \overline{0, (p+2)},
$$

$$
\mathbf{a} = \{a_0, \ldots, a_p, a_{p+1}, a_{p+2}\} \in \mathcal{A} = \bigcup_{k=0}^{p+2} \mathcal{A}_k
\tag{11.52}
$$

are random with a PDF $P(\mathbf{a})$.

The variable $v[n]$ is the observed output of the model, and the values of the random measurement noise $\xi[n]$ at different time instants n may belong to different ranges:

$$
\xi[n] \in \Xi_n = [\xi^-[n], \xi^+[n]],
\tag{11.53}
$$

with a PDF $Q_n(\xi[n])$, $n = \overline{0, N}$.

To train this model (estimate the PDFs $P(\mathbf{a})$ and $Q_n(\xi[n])$), we utilize the existing collections of historical data on the lakes area, the average annual temperature, and the annual precipitation.

However, at the stage of lakes area forecasting, data on temperature and precipitation are not available. To generate necessary data at this stage, we construct auxiliary models of temperature ($LDRR$-T) and precipitation ($LDRR$-P) and supply random sequences $\mu[n]$ and $\zeta[n]$, respectively, to their inputs.

Consider the temperature model ($LDRR$-T):

$$
\begin{aligned}
T[n] &= b_0 + \sum_{k=1}^{p} b_k T[n-k] + \mu[n], \\
t[n] &= T[n] + \eta[n].
\end{aligned}
\tag{11.54}
$$

The random parameters of this model are of the interval type:

$$
b_k \in \mathcal{B}_k = [b_k^-, b_k^+], \quad k = \overline{0, p},
$$

$$
\mathbf{b} = \{b_0, \ldots, b_p\}, \quad \mathbf{b} \in \mathcal{B} = \bigcup_{k=0}^{p} \mathcal{B}_k.
\tag{11.55}
$$

The random input sequence is

$$\mu[n] \in \mathcal{M}_n = [\mu^-[n], \mu^+[n]], \quad n = \overline{0, N},$$

$$\mu = \{\mu[0], \ldots, \mu[N]\}, \quad \mu \in \mathcal{M} = \bigcup_{n=0}^{N} \mathcal{M}_n. \tag{11.56}$$

The random parameters and the input sequence are characterized by a joint PDF $W(\mathbf{b}, \mu)$.

The variable $t[n]$ is the observed output of the model, and the measurement noise $\eta[n] \in \mathcal{E}_n = [\eta^-[n], \eta^+[n]]$ is of the interval type with PDFs $E_n(\eta[n])$.

The precipitation model ($LDRR$-P) has the form:

$$
\begin{aligned}
R[n] &= c_0 + \sum_{k=1}^{p} c_k R[n-k] + \zeta[n], \\
r[n] &= R[n] + \chi[n].
\end{aligned}
\tag{11.57}
$$

The random parameters of this model are independent and of the interval type:

$$c_k \in \mathcal{C}_k = [c_k^-, c_k^+], \quad k = \overline{0, p},$$

$$\mathbf{c} = \{c_0, \ldots, c_p\}, \quad \mathbf{c} \in \mathcal{C} = \bigcup_{k=0}^{p} \mathcal{C}_k. \tag{11.58}$$

The random input sequence is

$$\zeta[n] \in \mathcal{Z}_n = [\zeta^-[n], \zeta^+[n]], \quad n = \overline{0, N},$$

$$\zeta = \{\zeta[0], \ldots, \zeta[N]\}, \quad \zeta \in \mathcal{Z} = \bigcup_{n=0}^{N} \mathcal{Z}_n. \tag{11.59}$$

The random parameters and the input sequence are characterized by a joint PDF $F(\mathbf{c}, \zeta)$.

The variable $r[n]$ is the observed output of the model, and the measurement noise $\chi[n] \in \mathcal{G}_n = [\chi^-[n], \chi^+[n]]$ is of the interval type with PDFs $G_n(\chi[n])$.

The variation in the average annual temperature and the annual precipitation are the input variables for $LDRR$-A. The general model of thermokarst lakes area evolution (LDRR) is shown in Fig. 11.13.

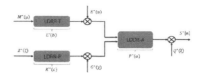

Figure 11.13: Structural diagram of LDRR.

11.4.4 Data on the state of thermokarst lakes

The state of regional thermokarst lakes is characterized by their total area $S^{(r)}[n]$ in region r, measured in (ha), and the factors influencing thermokarst formations—the average annual temperatures $T^{(r)}[n]$, measured in (C°), and the annual sums precipitation $R^{(r)}[n]$, measured in (mm), where n denotes calendar year.

Figure 11.11 shows a fragment of the Western Siberia map covering the range $65^{0}N$ to $-70^{0}N$ and $65^{0}E$ to $-95^{0}E$, including test domains in which temperature and precipitation were measured, and the areas of lakes were calculated based on the cosmic sensing of the earth surface.

The distribution of test regions by zones is as follows:

- zone C: 17, 18, 19, 20, 21, 22, 23, 24, 25, 26, 27, 30;

- zone D: 9, 10, 11, 12, 13, 14, 15, 16, 25, 26;

- zone I: 1, 2, 3, 4, 5, 6, 7, 8.

The data characterizing the state of these test domains were presented in [1]. To eliminate the scale effect, they are normalized, i.e., reduced to the interval $[0, 1]$, using the relations

$$S'_t[n] = \frac{S_t^{(r)}[n] - S_t^{min}}{S_t^{max} - S_t^{min}}, \quad T'_t[n] = \frac{T_t^{(r)}[n] - T_t^{min}}{T_t^{max} - T_t^{min}},$$

$$R'_t[n] = \frac{R_t^{(r)}[n] - R_t^{min}}{R_t^{max} - T_t^{min}}, \tag{11.60}$$

$$t = \overline{1, 30}, \quad n = \overline{0, 34}.$$

According to the tabulated data in [1], the data on lakes' surface area have many omissions. We restore them by the randomized reconstruction method of missing data described in [151].

Since the test domains are located in their zones rather compactly, the collections of normalized means for each data zone are formed, which will be used for training, testing, and forecasting:
—for zone C,

$$\bar{S}^{(r)}[n] = \frac{1}{12}\sum_{t \in C} S'_t[n], \ \bar{T}^{(r)}[n] = \frac{1}{12}\sum_{t \in C} T'_t[n],$$

$$\bar{R}^{(r)}[n] = \frac{1}{12}\sum_{t \in C} R'_t[n]; \tag{11.61}$$

—for zone D,

$$\bar{S}^{(r)}[n] = \frac{1}{10}\sum_{t \in D} S'_t[n], \ \bar{T}^{(r)}[n] = \frac{1}{10}\sum_{t \in D} T'_t[n], \bar{R}^{(r)}[n] = \frac{1}{10}\sum_{t \in D} R'_t[n];$$

—for zone I,

$$\bar{S}^{(r)}[n] = \frac{1}{8}\sum_{t \in I} S'_t[n], \ \bar{T}^{(r)}[n] = \frac{1}{8}\sum_{t \in I} T'_t[n], \bar{R}^{(r)}[n] = \frac{1}{8}\sum_{t \in I} R'_t[n].$$

We use these data to calculate the structural characteristics of the LDRR model, namely, the order p as well as the lower and upper limits of the ranges for the model parameters and measurement noises. The algorithm takes the form:

Algorithm 1 Parameters and measurement noises ranges.

1: **procedure** OLS_ESTIMATION(X, Y) ▷ X and Y as data matrices
2:     $B_{ols} = X \backslash Y$ ▷ Analog of $(X'X)^{-1} X'Y$ for Matlab
3:     $Y_{ols} = X B_{ols}$
4:     $S^2 = \sum (Y - Y_{ols})^2 / (length(Y_{ols}) - length(B_{ols}))$
5:     $SY^2 = \sum (Y - mean(Y))^2 / (length(Y) - 1)$
6:     $R^2 = 1 - S^2/SY^2$ ▷ Coefficient of determination
7:     $Sb = \sqrt{S^2 diag\left((X'X)^{-1}\right)}$
8:     $I_b = [B_{ols} - 3Sb, B_{ols} + 3Sb]$ ▷ Parameters ranges
9:     $I_e = [-3\sqrt{S^2}, 3\sqrt{S^2}]$ ▷ Noise ranges

Consequently, there are three time series for each zone, each containing 35 values. We divide them into two groups as follows:

- the training collections ($n = \overline{0, 24}$)

$$\mathfrak{C}_{tn}[n] = \left(S^{(r)}[n], T^{(r)}[n], R^{(r)}[n]\right),$$

Table 11.10: Structural characteristics of LDRR.

Model	Order p	Parameter	Parameter range lower limit	upper limit	Noise range lower limit	upper limit
			Zone C			
S	4	a_0	-0.5022	0.0747	-0.4280	0.4280
		a_1	-0.1366	0.5221		
		a_2	-0.4877	0.1982		
		a_3	-0.5277	0.1906		
		a_4	-0.4412	0.1905		
		a_5	0.4605	1.1440		
		a_6	0.1877	0.8777		
T	3	b_0	-0.0555	0.7512	-0.6920	0.6920
		b_0	-0.4863	0.6904		
		b_1	-0.2523	0.8608		
		b_2	-0.7525	0.4532		
R	6	c_0	0.2200	0.7412	-0.5489	0.5489
		c_1	-0.0420	1.1450		
		c_1	-0.5680	0.8050		
		c_2	-0.1511	1.1894		
		c_3	-0.9627	0.3813		
		c_4	-0.7177	0.7271		
		c_5	-1.0302	0.3402		

Table 11.11: Structural characteristics of LDRR.

Model	Order p	Parameter	Parameter range lower limit	upper limit	Noise range lower limit	upper limit
			Zone D			
S	4	a_0	-0.2180	0.4536	-0.3041	0.3041
		a_1	-0.2899	0.3146		
		a_2	-0.0934	0.5434		
		a_3	-0.3397	0.2687		
		a_4	-0.3727	0.2104		
		a_5	0.5365	1.1209		
		a_6	-0.3038	0.1809		
T	4	b_0	0.0064	0.9432	-0.6100	0.6100
		b_1	-0.6889	0.5007		
		b_1	-0.1742	1.0105		
		b_2	-0.8446	0.3379		
		b_3	-0.7874	0.4332		
R	4	c_0	-0.4305	0.6808	-0.5006	0.5006
		c_1	0.2060	1.3253		
		c_1	-0.7124	0.6210		
		c_2	-0.3067	1.0510		
		c_3	-0.9575	0.3755		

Table 11.12: Structural characteristics of LDRR.

Model	Order p	Parameter	Zone I			
			Parameter range		Noise range	
			lower limit	upper limit	lower limit	upper limit
S	5	a_0	−0.3191	0.1600	−0.2210	0.2210
		a_1	−0.2690	0.0804		
		a_2	−0.1893	0.1498		
		a_3	−0.1686	0.1679		
		a_4	−0.2009	0.1333		
		a_5	−0.2527	0.0863		
		a_6	0.9226	1.2945		
		a_7	−0.2073	0.2157		
T	4	b_0	0.0341	0.4649	−0.7110	0.7110
		b_1	−0.6822	0.4936		
		b_1	−0.1626	0.9894		
		b_2	−0.8443	0.3193		
		b_3	−0.8135	0.3945		
R	4	c_0	−0.3217	0.4938	−0.4661	0.4661
		c_1	0.1180	1.2049		
		c_1	−0.5728	0.6302		
		c_2	−0.3641	1.1894		
		c_3	−0.6541	0.4912		

$$\mathfrak{D}_{tn}[n] = \left(S^{(r)}[n], T^{(r)[n]}, R^{(r)}[n] \right),$$

and

$$\mathfrak{I}_{tn}[n] = \left(S^{(r)}[n], T^{(r)[n]}, R^{(r)}[n] \right)$$

for zones C, D and I, respectively;

- the testing collections ($n = \overline{25, 34}$)

$$\mathfrak{C}_{ts}[n] = \left(S^{(r)}[n], T^{(r)}[n], R^{(r)}[n] \right),$$

$$\mathfrak{D}_{ts}[n] = \left(S^{(r)}[n], T^{(r)[n]}, R^{(r)}[n] \right),$$

and

$$\mathfrak{I}_{ts}[n] = \left(S^{(r)}[n], T^{(r)}[n], R^{(r)}[n] \right)$$

for zones C, D, and I, respectively.

11.4.5 Entropy-Randomized machine learning of $LDRR$

11.4.5.1 Formation of data sets to train $LDRR$

The available dataset is divided in a proportion of 70% to 30% for model training and testing, respectively. The training procedure will be performed separately for $LDRR$-LA, $LDRR$-T and $LDRR$-P.

To train the model, we use the data from the training collection $(\mathfrak{C}_{tn}[n], \mathfrak{D}_{tn}[n], \mathfrak{I}_{tn}[n], \ n = \overline{0, 24})$.

From the historical data we construct the column vectors

$$\mathbf{S}^{(r)}_{(n-p)} = \{1, S^r[n-1], \dots, S^{(r)}[n-p]\},$$

$$\mathbf{S}^{(r)}_{(p)} = \{S^{(r)}[p], \dots, S^{(r)}[24]\},$$

$$\mathbf{T}^{(r)}_{(n-p)} = \{1, T^{(r)}[n-1], \dots, T^{(r)}[n-p]\},$$

$$\mathbf{T}^{(r)}_{(p)} = \{T^{(r)}[p], \dots, T^{(r)}[24]\}, \tag{11.62}$$

$$\mathbf{R}^{(r)}_{(n-p)} = \{1, R^{(r)}[n-1], \dots, R^{(r)}[n-p]\},$$

$$\mathbf{R}^{(r)}_{(p)} = \{R^{(r)}[p], \dots, R^{(r)}[24]\}, \tag{11.63}$$

$$n = \overline{p, 24},$$

and the matrices

$$S_p^{(r)} = \begin{pmatrix} \mathbf{S}_{(0)} \\ \cdots \\ \mathbf{S}_{(24-p)} \end{pmatrix}, \quad T_p^{(r)} = \begin{pmatrix} \mathbf{T}_{(0)} \\ \cdots \\ \mathbf{T}_{(24-p)} \end{pmatrix}, \quad R_p^{(r)} = \begin{pmatrix} \mathbf{R}_{(0)} \\ \cdots \\ \mathbf{R}_{(24-p)} \end{pmatrix}. \tag{11.64}$$

11.4.5.2   ERML algorithm

We apply the randomized machine learning method (see Section 3) to estimate the PDFs of the model parameters and measurement noises. The ERML algorithm for *LDRR-LA* has the form

$$\mathcal{H}[P(\mathbf{a}, Q(\xi)] = - \int_{\mathcal{A}} P(\mathbf{a}) \ln P(\mathbf{a}) \, d\mathbf{a} -$$

$$\sum_{n=p}^{24} \int_{\Xi_n} Q_n(\xi[n]) \ln Q_n(\xi[n]) \, d\xi[n] \Rightarrow \max \tag{11.65}$$

subject to the constraints

$$\int_{\mathcal{A}} P(\mathbf{a}) \, d\mathbf{a} = 1, \quad \int_{\Xi_n} Q_n(\xi[n]) \, d\xi[n] = 1,$$

$$\int_{\mathcal{A}} P(\mathbf{a}) D^{(r)}_{(n-p)} \, \mathbf{a} \, d\mathbf{a} + \int_{\Xi_n} Q(\xi[n]) \xi[n] \, d\xi[n] = S^{(r)}[n], \tag{11.66}$$

$$n = \overline{p, 24},$$

with the block matrix

$$D^{(r)}_{(n-p)} = \left[\mathbf{S}^{(r)}_{(n-p)}, T^{(r)}_{(p)}[n], R^{(r)}_{(p)}[n] \right]. \tag{11.67}$$

The solution of this problem, parametrized by the Lagrange multipliers $\theta - \{\theta_p, \ldots, \theta_{24}\}$, is given by

$$P^*(\mathbf{a}, \theta) = \frac{\exp\left(-\langle \theta, D^{(r)}_{(n-p)} \mathbf{a} \rangle\right)}{\mathcal{P}(\theta)},$$

$$\mathcal{P}(\theta) = \int_{\mathcal{A}} \exp\left(-\langle \theta, D^{(r)}_{(n-p)} \mathbf{a} \rangle\right) d\mathbf{a}. \qquad (11.68)$$

The measurement noise for *LDRR-LA* has the entropy-optimal PDFs

$$Q_n^*(\xi[n], \theta_n) = \frac{\exp(-\xi[n]\theta_n)}{\mathcal{Q}_n(\theta_n)},$$

$$\mathcal{Q}_n(\theta_n) = \int_{\Xi_n} \exp(-\xi[n]\,\theta_n) d\,\xi[n]. \qquad (11.69)$$

The Lagrange multipliers θ are calculated from the system of equations

$$\mathcal{P}^{-1}(\theta) \int_{\mathcal{A}} \exp\left(-\langle \theta, D^{(r)}_{(n-p)} \mathbf{a} \rangle\right) D^{(r)}_{(n-p)} \mathbf{a}\, d\,\mathbf{a} +$$

$$+ \mathcal{Q}_n^{-1}(\theta) \int_{\Xi} \exp(-\xi[n]\,\theta_n)\, d\xi[n] = S^{(r)}[n], \qquad (11.70)$$

$$n = \overline{p, 24}.$$

For *LDRR-T* and *LDRR-P*, the RML algorithm differs from (11.4.5.2, 11.4.5.2): it is necessary to estimate the joint PDFs $W(\mathbf{b}, \mu)$ and $F(\mathbf{c}, \zeta)$ of the model parameters and random input sequences as well as the PDFs $E(\eta)$ and $G(\chi)$ of the measurement noises.

Adapting the ERML algorithm for *LDRR-T*, we obtain the following optimization problem:

$$\mathcal{H}[W(\mathbf{b}, \mu), E(\eta)] = -\int_{\mathcal{B} \bigcap \mathcal{M}} W(\mathbf{b}, \mu) \ln W(\mathbf{b}, \mu)\, d\mathbf{b}\, d\mu -$$

$$- \sum_{n=p}^{24} \int_{\mathcal{E}_n} E_n(\eta[n]) \ln E_n(\eta[n])\, d\eta[n] \Rightarrow \max \qquad (11.71)$$

subject to the constraints

$$\int_{\mathcal{B} \bigcap \mathcal{M}} W(\mathbf{b}, \mu)\, d\mathbf{b}d\mu = 1, \quad \int_{\mathcal{E}_n} E_n(\eta[n])\, d\,\eta[n] = 1,$$

$$\int_{\mathcal{B} \bigcap \mathcal{M}} W(\mathbf{b}, \mu) \left[T^{(r)}_{(n-p)} \mathbf{b} + \mu[n] \right] d\,\mathbf{b}\, d\,\mu[n] +$$

$$+ \int_{\mathcal{E}_n} E_n(\eta[n])\eta[n]\, d\,\eta[n] = T^{(r)}[n], \qquad (11.72)$$

$$n = \overline{p, 24}.$$

The expressions (11.4.5.2, 11.4.5.2) incorporate the vector $\mu = \{\mu[p], \ldots, \mu[24]\}$.

Let $\vartheta = \{\vartheta_p, \ldots, \vartheta_{7+p}\}$ be the Lagrange multipliers for this problem. The linear model has random parameters and the auxiliary random sequence on the input. The solution of this problem, representing the optimal PDFs, can be therefore written as

$$W^*(\mathbf{b}, \mu, \vartheta) = L^*(\mathbf{b}, \vartheta)\, M^*(\mu, \vartheta), \quad M^*(\mu, \vartheta) = \prod_{n=p}^{24} M_n^*(\mu[n], \vartheta_n)$$

$$L^*(\mathbf{b}, \vartheta) = \frac{\exp\left(-\langle \vartheta, T_{(n-p)}^{(r)} \mathbf{b} \rangle\right)}{\mathcal{L}(\vartheta)},$$

$$\mathcal{L}(\vartheta) = \int_{\mathcal{B}} \exp\left(-\langle \vartheta, T_{(n-p)}^{(r)} \mathbf{b} \rangle\right) d\mathbf{b},$$

$$M_n^*(\mu[n], \vartheta_n) = \mathbb{M}_n^{-1}(\vartheta_n) \exp\left(-\vartheta_n \mu[n]\right),$$

$$\mathbb{M}_n(\vartheta_n) = \int_{\mathcal{M}_n} \exp\left(-\vartheta_n \mu[n]\right) d\mu[n],$$

$$E_n^*(\eta[n], \vartheta_n) = \mathbb{E}_n^{-1}(\vartheta_n) \exp\left(-\eta[n]\, \vartheta_n\right),$$

$$\mathbb{E}_n(\vartheta_n) = \int_{\mathcal{E}_n} \exp\left(-\eta[n]\, \vartheta_n\right) d\eta[n], \tag{11.73}$$

where the Lagrange multipliers ϑ satisfy the system of equations

$$\mathcal{L}^{-1}(\vartheta) \int_{\mathcal{B}} \exp\left(-\langle \vartheta, T_{(n-p)}^{(r)} \mathbf{b} \rangle\right) T_{(n-p)}^{(r)} \mathbf{b}\, d\mathbf{b}+$$

$$+\mathbb{M}_n^{-1}(\vartheta_n) \int_{\mathcal{M}_n} M_n^*(\mu[n], \vartheta_n)\, \mu[n] d\mu[n]+$$

$$+\mathbb{E}_n^{-1}(\vartheta_n) \int_{\mathcal{E}_n} E_n^*(\eta[n], \vartheta_n)\, \eta[n] d\eta[n] = T^{(r)}[n], \tag{11.74}$$

$$n = \overline{p, 24}.$$

For *LDRR-P*, the RML algorithm has a form similar to (11.4.5.2, 11.4.5.2):

$$F^*(\mathbf{c}, \zeta, \lambda) = V^*(\mathbf{c}, \lambda)\, Z^*(\zeta, \lambda), \quad Z^*(\zeta, \lambda) = \prod_{n=p}^{24} Z_n^*(\zeta[n], \lambda_n),$$

$$V^*(\mathbf{c}, \lambda) = \frac{\exp\left(-\langle \lambda, R_{(n-p)}^{(r)}\, \mathbf{c}\rangle\right)}{\mathcal{V}(\lambda)},$$

$$\mathcal{V}(\lambda) = \int_C \exp\left(-\langle \lambda, R_{(n-p)}^{(r)}\, \mathbf{c}\rangle\right)\, d\mathbf{c},$$

$$Z_n^*(\zeta[n], \lambda_n) = \mathbb{Z}_n^{-1}(\lambda_n) \exp\left(-\lambda_n\, \zeta[n]\right),$$

$$\mathbb{Z}_n(\lambda_n) = \int_{\mathcal{Z}_n} \exp\left(-\lambda_n\, \zeta[n]\right) d\zeta[n],$$

$$G_n^*(\chi[n], \lambda_n) = \mathbb{G}_n^{-1}(\lambda) \exp\left(-\chi[n]\lambda_n\right), \qquad (11.75)$$

$$\mathbb{G}_n(\lambda_n) = \int_{\mathcal{G}_n} \exp\left(-\chi[n]\, \lambda_n\right) d\chi[n], \qquad (11.76)$$

where the Lagrange multipliers $\lambda = \{\lambda_p, \ldots, \lambda_{24}\}$ are determined from the system of equations

$$\mathcal{V}^{-1}(\lambda) \int_C \exp\left(-\langle \lambda, R_{(n-p)}^{(r)}\, \mathbf{c}\rangle\right)\, R_{(n-p)}^{(r)}\, \mathbf{c}\, d\mathbf{c} +$$

$$\mathbb{Z}_n^{-1}(\lambda_n) \int_{\mathcal{Z}_n} Z_n^*(\zeta[n], \lambda_n)\, \zeta[n] d\zeta[n] +$$

$$+\mathbb{G}_n^{-1}(\lambda_n) \int_{\mathcal{G}_n} G_n^*(\chi[n], \lambda_n)\, \chi[n]\, d\chi[n] = R^{(r)}[n], \qquad (11.77)$$

$$n = \overline{p, 24}.$$

According to (11.4.5.2, 11.4.5.2, 11.4.5.2, 11.4.5.2), the entropy-optimal PDFs belong to the exponential class parametrized by the corresponding Lagrange multipliers, which are given by the balance equations (11.4.5.2, 11.4.5.2, 11.4.5.2).

Algorithms 2 and 3 are oriented to calculation residual for balance equations and restoration of entropy-optimal PDFs.

11.4.6 Testing procedure, data and accuracy estimates

The testing procedure is applied to the following combination of the trained models:

- *LDRR-LA*, with the testing collection $\mathfrak{C}_{ts}[n], \mathfrak{D}_{ts}[n], \mathfrak{I}_{ts}[n]$;

- *LDRR-T*, with $T^{(r)}[n]$ from the testing collection;

Algorithm 2 Residual for balance equations.

---

    function BALANCE_L1(q, X, Y, I_b, I_e)
2:    $T = size(Y, 1)$ ▷ Sample size
      $K = size(X, 2)$ ▷ Dimension of data
4:    $\Delta = zeros(1, T)$ ▷ Model's residual vector
      for $t = 1 : 1 : T$ **do**
6:        $res_1 = 0$ ▷ Model's mean output
        for $k = 1 : 1 : K$ **do**
8:          $\Omega_b = \int_{I_b} \exp\left(b * \langle X(:,k), q \rangle\right) db$ ▷ $X(:,k)$ as column of X
          $B_k = \frac{1}{\Omega_b} \int_{I_b} b * \exp\left(b * \langle X_k, q \rangle\right)$
10:        $res_1 = res_1 + X(t, k) * B_k$
        $\Omega_e = \int_{I_e} \exp\left(e * q(t)\right) de$
12:        $res_2 = \frac{1}{\Omega_e} \int_{I_e} e * \exp\left(e * q(t)\right) de$ ▷ Mean value of noise
        $\Delta(t) = res_1 + res_2 - Y(t)$
      return Δ

Algorithm 3 Restoration of entropy-optimal PDFs.

---

    procedure ME_ESTIMATION(X, Y, I_b, I_e) ▷ Training procedure
      $T = size(Y, 1)$
3:    $K = size(X, 2)$
      $q_0(1 : T) = 0.25$ ▷ Initial approximation
      $Q_{me} = fsolve(F, q_0)$ ▷ Matlab's function $fsolve$
6:    $PDF_b = cell(1, K)$ ▷ Cells for PDFs of model parameters
      for $k = 1 : 1 : K$ **do**
        $\Omega_b = \int_{I_b} \exp\left(b * \langle x(:,k), Q_{me} \rangle\right) db$
9:      $PDF_b\{k\} = \frac{1}{\Omega_b} \exp\left(a * \langle X(:,k), Q_{me} \rangle\right)$
      $PDF_e = cell(1, T)$ ▷ Cells for PDFs of measurement noise
      for $t = 1 : 1 : T$ **do**
12:      $\Omega_e = \int_{I_e} \exp\left(e * Q_{me}(t)\right) de$
        $PDF_e\{t\} = \frac{1}{\Omega_e} \exp\left(e * Q_{me}(t)\right)$

- $LDRR\text{-}P$, with $R^{(r)}[n]$ from the testing collection.

When testing $LDRR\text{-}LA$, all components of the testing collections are used, and the model generates an output ensemble (the values of the lake areas) through sampling the entropy-optimal PDFs of its parameters and measurement noises.

Algorithm 4 Sampling algorithm for PDFs.

---

   procedure PDF_GENERATION_1D$(P, Xlim, N)$ ▷ P as objective function, $Xlim$ as constraints on x, N as ensemble size

     $dx = (Xlim(2) - Xlim(1))/100$

     $w = \max_{Xlim} P$ ▷ Maximum of function P

4:    $k = 1$ ▷ Counter of samples

     $ENS = zeros(1, N)$ ▷ Vector of samples

     while $k \leq N$ **do**

        $Rx = rand()$ ▷ Random number from 0 to 1

8:      $x = Xlim(1) + (Xlim(2) - Xlim(1)) * Rx$

        $Ry = rand()$

        if $w * Ry \leq P(x)$ **then**

           $ENS(k) = x$

12:     $k = k + 1$

When testing the trained $LDRR\text{-}T$ and $LDRR\text{-}P$ models, only the temperature and precipitation data from the testing collections are used. These models generate ensembles of the corresponding outputs for temperature and precipitation through sampling the entropy-optimal PDFs of their parameters and input sequences.

The accuracy of testing is characterized by the following indicators:

—the absolute mean squared error (in natural units)

$$AbsErr = \Delta = \sqrt{\frac{1}{10} \sum_{n=25}^{34} \left(\bullet^{(r)}[n] - \hat{\bullet}[n] \right)^2}; \qquad (11.78)$$

—the relative mean squared error

$$RelErr = \frac{\Delta}{\Delta_1 + \Delta_2},$$

$$\Delta_1 = \sqrt{\frac{1}{10} \sum_{n=25}^{34} (\bullet^{(r)}[n])^2}, \quad \Delta_2 = \sqrt{\frac{1}{10} \sum_{n=25}^{34} (\hat{\bullet}[n])^2}. \qquad (11.79)$$

11.4.7   Randomized forecasting of thermokarst lakes area evolution

The temporal evolution of the thermokarst lakes area is forecasted on time intervals exceeding the time interval for which the necessary data on the area and climatic parameters are available. In the case under consideration, the beginning of each forecasting interval refers to year 2008. We consider three time intervals as follows:

—short-term one, between years 2008 and 2013 ($n = \overline{35, 40}$);
—mid-term one, between years 2008 and 2018 ($n = \overline{35, 45}$);
—long-term one, between years 2008 and 2023 ($n = \overline{35, 50}$).

The temporal evolution of the thermokarst lakes area is forecasted using a composition of the trained randomized models (see Fig. 11.13) on the forecasting intervals specified above. The models from this composition are characterized by the entropy-optimal PDFs of their parameters, auxiliary random sequences, and measurement noises. The PDFs of the measurement noises are different for the time instants on the training interval: they depend on the time instant and the corresponding Lagrange multiplier. For example, the measurement noises for $LDRR\text{-}LA$ have the PDF $Q_n^*(\xi[n], \theta_n)$, $n = \overline{0, 34}$. For the forecasting procedure, we apply the PDFs averaged over the Lagrange multipliers.

Well, the blocks of the composite model are characterized by the following PDFs:

- for $LDRR\text{-}LA$, $P^*(\mathbf{a})$ and $Q^*(\xi, \bar{\theta})$, $\bar{\theta} = \frac{1}{34-p} \sum_{n=p}^{34} \theta_n$;

- for $LDRR\text{-}T$, $L^*(\mathbf{b})$, $M^*(\mu, \bar{\vartheta})$, and $E^*(\eta, \bar{\vartheta})$, $\bar{\vartheta} = \frac{1}{34-p} \sum_{n=p}^{34} \vartheta_n$;

- for $LDRR\text{-}P$, $V^*(\mathbf{c})$, $Z^*(\zeta, \bar{\lambda})$, and $G^*(\chi, \bar{\lambda})$, $\bar{\lambda} = \frac{1}{34-p} \sum_{n=p}^{34} \lambda_n$.

These PDFs are sampled on the corresponding forecasting intervals, and the predictive ensembles of trajectories (lakes area variations) are generated via Monte Carlo simulations.

11.4.8   Results of training, testing and forecasting the temporal evolution of thermokarst lakes area in Western Siberia

11.4.8.1   Randomized training (1973–1997)

Using ERML, we found the entropy-optimal PDFs of the model parameters, auxiliary random sequences, and measurement noises. The latter PDFs are parametrized by Lagrange multipliers.

Table 11.13: *LDRR-LA*: values of PDF parameters.

Zones	q_0	q_1	q_2	q_3
C	0.1174	0.1331	0.0214	−0.0723
D	−0.0969	0.1077	0.1712	0.0296
I	−0.0176	−0.1560	0.0892	−0.0678
Zones	q_4	q_5	q_6	q_7
C	−0.0034	−0.0100	−0.0761	−
D	0.0847	−0.1910	−0.0671	−
I	−0.0716	0.0191	−0.0002	0.0120

LDRR-LA The analytical expressions for the corresponding parametrized PDFs are given by (11.4.5.2, 11.4.5.2). Since *LDRR-LA* is linear, these PDFs belong to the exponential class:

$$P^*(\mathbf{a}, \theta) = \prod_{k=0}^{(p+2)} P_k^*(a_k), \quad P_k^*(a_k) = \frac{\exp(-q_k\, a_k)}{\mathcal{P}_k},$$

$$\mathcal{P}_k = \int_{A_k} \exp(-q_k\, a_k)\, da_k,$$

$$q_0 = \sum_{n=p}^{24} \theta_n, \quad q_k = \sum_{n=p}^{24} \theta_n\, S^{(r)}[n-k], \quad k = \overline{1, p},$$

$$q_{p+1} = \sum_{n=p}^{24} \theta_n\, T^{(r)}[n], \quad q_{p+2} = \sum_{n=p}^{24} \theta_n\, R^{(r)}[n],$$

$$Q^*(\xi, \bar{\theta}) = \frac{\exp(-\bar{\theta}\, \xi)}{\mathcal{Q}}, \quad \mathcal{Q} = \int_{\Xi} \exp(-\bar{\theta}\, \xi)\, d\xi, \quad \bar{\theta} = \frac{q_0}{24 - p}. \quad (11.80)$$

Table 11.13 contains the values of the PDF parameters for *LDRR-LA* and zones *C, D* and *I*. These values can be used in the expressions for the entropy-optimal PDFs to obtain the same results as presented below.

Figure 11.14 demonstrates some examples of the entropy-optimal PDFs $P^*(a_2, a_4)$ (11.14a) (the model parameters) and $Q^*(\xi)$ (11.14b) (the measurement noise), which were restored through learning for zone *C*.

A similar learning procedure is performed for the temperature and precipitation models for all zones *C, D* and *I*.

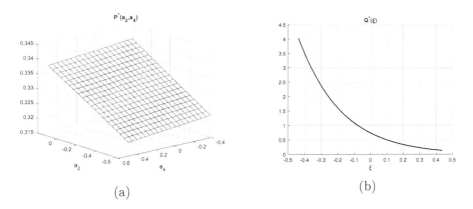

Figure 11.14: Examples of restored PDFs.

LDRR-T The analytical expressions for the corresponding PDFs are given by (11.4.5.2, 11.4.5.2). Due to the linearity of the model, we have:

$$W^*(\mathbf{b},\mu,\vartheta) = L^*(\mathbf{b},\vartheta)\,M^*(\mu,\vartheta),\ \ M^*(\mu,\vartheta) = \prod_{n=p}^{(24)} M_n^*(\mu[n],\vartheta_n),$$

$$L^*(\mathbf{b},\vartheta) = \prod_{k=0}^{p} L_k^*(b_k)\,,\ \ L_k^*(b_k) = \frac{\exp(-w_k\,b_k)}{\mathcal{L}_k},$$

$$\mathcal{L}_k = \int_{\mathcal{B}_k} \exp(-w_k\,b_k)\,db_k,$$

$$w_0 = \sum_{n=p}^{24} \vartheta_n,\ \ w_k = \sum_{n=p}^{24} \vartheta_n\,T^{(r)}[n-k],\ \ \ k = \overline{1,p},$$

$$M_n^*(\mu[n],\vartheta_n) = \frac{\exp(-\vartheta_n\,\mu[n])}{\mathfrak{M}_n},\ \ \ \mathfrak{M}_n = \int_{\mathcal{M}_n} \exp(-\vartheta_n\,\mu[n])\,d\mu[n],$$

$$E^*(\eta,\bar{\vartheta}) = \frac{\exp(-\bar{\vartheta}\,\eta)}{\mathfrak{E}},$$

$$\mathfrak{E} = \int_{\mathcal{E}} \exp(-\bar{\vartheta}\,\eta)\,d\,\eta,\ \ \bar{\vartheta} = \frac{w_0}{24-p}. \tag{11.81}$$

Table 11.14 shows the values of the PDF parameters for *LDRR-T* and zones *C, D* and *I*.

Table 11.14: *LDRR-T*: values of PDF parameters.

Zones	w_0	w_1	w_2	w_3	w_4
C	0.0431	0.0379	0.1703	−0.0864	–
D	−0.0237	−0.0373	0.1839	−0.1296	0.1612
I	−0.0002	0.0040	0.0677	−0.0912	0.1426

Table 11.15: *LDRR-P*: values of PDF parameters.

Zones	s_0	s_1	s_2	s_3	s_4	s_5	s_6
C	−0.0287	0.0089	0.0264	0.0006	−0.0200	−0.0214	0.0431
D	0.0876	−0.1451	−0.0841	−0.0709	0.1307	–	–
I	0.0571	−0.0890	−0.0166	0.1140	−0.0794	–	–

LDRR-P The analytical expressions for the corresponding PDFs are given by (11.4.5.2, 11.4.5.2). Due to the linearity of the model, we have:

$$F^*(\mathbf{c}, \zeta, \chi) = V^*(\mathbf{c}, \lambda)\, Z^*(\zeta, \lambda), \quad Z^*(\zeta, \lambda) = \prod_{n=p}^{(24)} Z_n^*(\zeta[n], \lambda_n),$$

$$V^*(\mathbf{c}, \lambda) = \prod_{k=0}^{p} V_k^*(c_k), \quad V_k^*(b_k) = \frac{\exp(-s_k\, c_k)}{\mathcal{V}_k},$$

$$\mathcal{V}_k = \int_{\mathcal{C}_k} \exp(-s_k\, c_k)\, dc_k,$$

$$s_0 = \sum_{n=p}^{24} \lambda_n, \quad s_k = \sum_{n=p}^{24} \lambda_n\, R^{(r)}[n-k], \quad k = \overline{1, p},$$

$$Z_n^*(\zeta[n], \lambda_n) = \frac{\exp(-\lambda_n\, \zeta[n])}{\mathcal{Z}_n}, \quad \mathcal{Z}_n = \int_{\mathcal{Z}_n} \exp(-\lambda_n\, \zeta[n])\, d\zeta[n],$$

$$G^*(\chi, \bar{\lambda}) = \frac{\exp(-\bar{\lambda}\, \chi)}{\mathcal{G}},$$

$$\mathcal{G} = \int_{\mathcal{G}} \exp(-\bar{\lambda}\, \chi)\, d\chi, \quad \bar{\lambda} = \frac{w_0}{24 - p}. \tag{11.82}$$

Table 11.15 contains the values of the PDF parameters for *LDRR-P* and zones C, D and I.

11.4.8.2 Testing (1998–2007)

The testing procedure is applied to the combination of the trained models with the testing data collections.

Table 11.16: Absolute and relative testing errors.

LDRR-LA			
Zone	C	D	I
$AbsErr$	0.3446	0.5354	1.3534
$RelErr$	0.0089	0.0135	0.014
LDRR-T			
$AbsErr$	1.2863	1.1239	1.0284
$RelErr$	0.0801	0.1084	0.2215
LDRR-P			
$AbsErr$	107.64	118.97	86.66
$RelErr$	0.1675	0.137	0.0947

This procedure is based on sampling of the optimal PDFs and further generation of the ensembles of random trajectories on the testing interval.

For zone C, the following ensembles are shown below:

- the lakes area ensembles generated by $LDRR$-LA (Fig. 11.15);

- the temperature ensembles generated by $LDRR$-T (Fig. 11.16);

- the precipitation ensembles generated by $LDRR$-P (Fig. 11.17).

Table 11.16 presents the indicators of testing accuracy for all zones C, D and I.

11.4.8.3 *Randomized forecasting (2008–2023)*

All forecasts are constructed using the composite model (Fig. 11.13) consisting of $LDRR$-LA, $LDRR$-T and $LDRR$-P. The entropy-optimal PDFs generate the ensembles of trajectories describing the temporal evolution of the thermokarst lakes area on the forecasting interval. The difference between the forecasting and testing intervals is that in this case, all data points are used to train the models, and the structural characteristics of the models correspond to Tables 11.10–11.12.

For each point, the mean value (*mean*) and standard deviation (*std*) are calculated over the entire ensemble. The resulting forecasts for all zones are combined in Tables 11.17–11.19, starting from year 2007 for 5, 10 and 15 years, respectively.

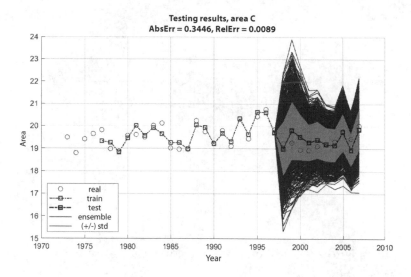

Figure 11.15: Simulation modeling of lakes area.

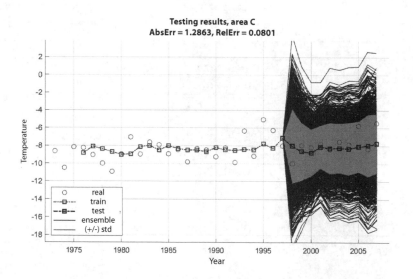

Figure 11.16: Simulation modeling of temperature.

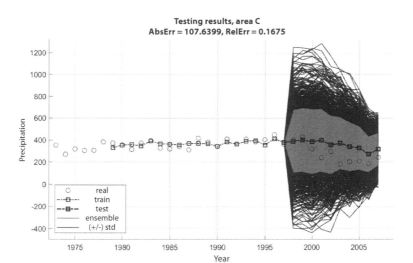

Figure 11.17: Simulation modeling of precipitation.

Table 11.17: Forecasted evolution of thermokarst lakes area up to year 2012.

Year	2007	2008	2009	2010	2011	2012
$C - mean$	19.9252	19.2820	19.4881	19.4444	19.7255	19.8765
$C - std$	-	0.5193	0.5509	0.5462	0.5623	0.5694
$D - mean$	21.3674	18.8605	20.7156	19.0790	20.1559	19.3715
$D - std$	-	1.2405	1.2599	1.1607	1.2085	1.0924
$I - mean$	53.2158	43.1694	50.1471	44.7717	48.9104	46.3077
$I - std$	-	2.3760	2.3276	2.2961	2.2025	2.1959

Figure 11.18 demonstrates the ensembles of forecasts for zones C, D and I. Here the trajectories of the mean values and standart deviation area are indicated by $(mean)$ and (std), respectively.

The evolution of the lakes area forecasted in this paper is characterized by mean values (Tables 11.17–11.19) and standard deviations from 2 to 4.5%. Also, the graphs for zones D and I (Fig. 11.17) have an oscillatory process, reaching a constant level after year 2018.

Table 11.18: Forecasted evolution of thermokarst lakes area up to year 2017.

Year	2013	2014	2015	2016	2017
$C - mean$	20.0226	19.9603	19.9349	19.8134	19.7374
$C - std$	0.6205	0.6542	0.6878	0.6878	0.6713
$D - mean$	19.922	19.5528	19.8171	19.6082	19.7359
$D - std$	1.1424	1.0787	1.1133	1.0887	1.1053
$I - mean$	48.6890	47.3009	48.2036	47.4963	47.8111
$I - std$	2.0383	2.0673	2.0253	2.0554	2.0402

Table 11.19: Forecasted evolution of thermokarst lakes area up to year 2022.

Year	2018	2019	2020	2021	2022
$C - mean$	19.6453	19.5491	19.4576	19.3742	19.3070
$C - std$	0.6367	0.5969	0.5569	0.5215	0.4901
$D - mean$	19.6215	19.6897	19.6372	19.6780	19.6546
$D - std$	1.0913	1.0979	1.0885	1.0911	1.0867
$I - mean$	47.5081	47.6778	47.6040	47.6912	47.6729
$I - std$	2.0584	2.0421	2.0484	2.0405	2.0443

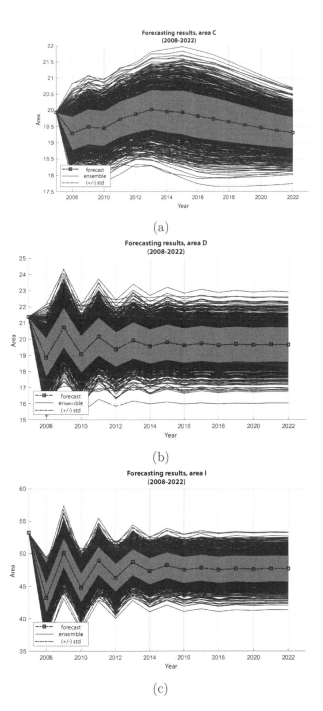

Figure 11.18: Forecast for 15 years.

MAXIMUM ENTROPY ESTIMATE (MEE) AND ITS ASYMPTOTIC EFFICIENCY

A.1 STATEMENT OF MAXIMUM ENTROPY ESTIMATION PROBLEM

Consider a scalar continuous function $\varphi(x, \theta)$ with parameters $\theta = \{\theta_1, \ldots, \theta_n\}$. Assume that this function is a characteristic of an object's model with an input x and an output y. Let $\mathbf{x}^{(r)} = \{x[1], \ldots, x[r]\}$ and $\mathbf{y}^{(r)} = \{y[1], \ldots, y[r]\}$ be given measurements. Note that the latter are obtained with random errors $\xi = \{\xi[1], \ldots, \xi[r]\}$, in the general case, different for different time instants.

Thus, after r measurements the model and observations are described by the equations

$$
\begin{aligned}
\hat{\mathbf{y}} &= \Gamma(\mathbf{x}^{(r)}, \theta), \\
\hat{\mathbf{v}} &= \hat{\mathbf{y}} + \xi,
\end{aligned}
\tag{A.1}
$$

where the vector function $\Gamma(\mathbf{x}^{(r)}, \theta)$ has the components $\varphi(x[t], \theta)$, $t = \overline{1, r}$; $\hat{\mathbf{v}}$ denotes the observed output of the model containing measurement noises of the object's output.

Assumption 1 *The random parameters are* $\theta \in \Theta \subset R^n$, $\Theta = [\theta^-, \theta^+]$.

DOI: 10.1201/9781003306566-A

Assumption 2 *The PDF* $P(\theta)$ *of the parameters is continuously differentiable on its support* Θ.

Assumption 3 *The function* φ *is bounded, i.e.,*

$$\varphi^- \leq \varphi(x[t], \theta) \leq \varphi^+, \tag{A.2}$$

for all $\theta \in \Theta$ *and* $x \in R^1$.

Assumption 4 *The random noise is* $\xi \in \Xi \subset R^s$, *where*

$$\Xi = \bigotimes_{t=1}^{r} \Xi_t, \qquad \Xi_t = [\xi^-[t], \xi^+[t]]. \tag{A.3}$$

Assumption 5 *The PDF* $Q(\xi)$ *of the measurement noises is continuously differentiable on the support* Ξ *and also has the multiplicative structure:*

$$Q(\xi) = \prod_{t=1}^{r} Q_t(\xi[t]). \tag{A.4}$$

The estimation problem is stated as follows: *find the maximum entropy estimates of the PDFs* $P^*(\theta)$ *and* $Q^*(\xi)$ *that maximize the generalized information entropy functional*

$$\mathcal{H}[P(\theta), Q(\xi)] = -\int_Q P(\theta) \ln P(\theta) d\theta -$$

$$-\sum_{t=1}^{r} \int_{\Xi_t} Q_t(\xi[t]) \ln Q_t(\xi[t]) \Rightarrow \max \tag{A.5}$$

subject to

—the normalization conditions of the PDFs given by

$$\int_{\Theta} P(\theta) d\theta = 1; \quad \int_{\Xi_t} Q_t(\xi[t]) d\xi[t] = 1, \, t = \overline{1, r}. \tag{A.6}$$

and

—the empirical balance conditions

$$\Phi[P(\theta), Q(\xi)] = \mathbf{y}^{(r)},$$

$$\Phi[P(\theta), Q(\xi)] = \{\Phi_1[P(\theta), Q(\xi)], \dots, \Phi_r[P(\theta), Q(\xi)]\}$$

$$\Phi_t[P(\theta), Q(\xi)] = \int_{\Theta} \varphi(x[t], \theta) P(\theta) d\theta + \int_{\Xi_t} Q_t(\xi[t]) \xi[t] d\xi[t], \tag{A.7}$$

$$t = \overline{1, r}.$$

The optimization problem (A.1–A.1) is of the Lyapunov type [77, 174]: such problems have an integral functional and also integral constraints. Hence, the entropy-optimal PDFs of the model parameters and measurement noises have the form

$$P^*(\theta \,|\, \mathbf{y}^{(r)}\,\mathbf{x}^{(r)}) = \frac{\exp\left(-\sum_{j=1}^{r}\lambda_j(\mathbf{y}^{(r)},\mathbf{x}^{(r)})\varphi(x[j],\theta)\right)}{\mathcal{P}(\lambda(\mathbf{y}^{(r)}),\mathbf{x}^{(r)})},$$

$$Q_t^*(\xi[t] \,|\, \mathbf{y}^{(r)}\,\mathbf{x}^{(r)}) = \frac{\exp\left(\lambda_t(\mathbf{y}^{(r)},\mathbf{x}^{(r)})\xi[t]\right)}{\mathcal{Q}_t(\lambda_t(\mathbf{y}^{(r)}),\mathbf{x}^{(r)})}, \quad t=\overline{1,r}. \qquad \text{(A.8)}$$

where

$$\mathcal{P}(\lambda(\mathbf{y}^{(r)},\mathbf{x}^{(r)})) = \int_{\Theta} \exp\left(-\sum_{j=1}^{r}\lambda_j(\mathbf{y}^{(r)},\mathbf{x}^{(r)})\varphi(x[j],\theta)\right)d\theta,$$

$$\mathcal{Q}_t(\lambda_t(\mathbf{y}^{(r)},\mathbf{x}^{(r)})) = \int_{\Xi_t} \exp\left(\lambda_t(\mathbf{y}^{(r)},\mathbf{x}^{(r)})\xi[t]\right)d\xi[t], \qquad \text{(A.9)}$$

$$t=\overline{1,r}.$$

Due to equalities (A.8) and (A.1), the entropy-optimal PDFs are parameterized by the Lagrange multipliers $\lambda_1,\ldots,\lambda_r$, which represent the solutions of the empirical balance equations

$$\frac{\mathcal{G}(\lambda(\mathbf{y}^{(r)},\mathbf{x}^{(r)}))}{\mathcal{P}(\lambda(\mathbf{y}^{(r)},\mathbf{x}^{(r)}))} + \frac{\mathcal{E}_t(\lambda_t(\mathbf{y}^{(r)},\mathbf{x}^{(r)}))}{\mathcal{Q}_t(\lambda_t(\mathbf{y}^{(r)},\mathbf{x}^{(r)}))} = y[t], \quad t=\overline{1,r}, \qquad \text{(A.10)}$$

where

$$\mathcal{G}(\lambda(\mathbf{y}^{(r)},\mathbf{x}^{(r)})) = \int_{\Theta} \varphi(x[t],\theta) \exp\left(-\sum_{j=1}^{r}\lambda_j(\mathbf{y}^{(r)},\mathbf{x}^{(r)})\varphi(x[j],\theta)\right)d\theta,$$

$$\mathcal{E}_t(\lambda_t(\mathbf{y}^{(r)},\mathbf{x}^{(r)})) = \int_{\Xi_t} \exp\left(-\lambda_t(\mathbf{y}^{(r)},\mathbf{x}^{(r)})\xi[t]\right)d\xi[t], \qquad \text{(A.11)}$$

$$t=\overline{1,r}.$$

The solution $\lambda^*(\mathbf{y}^{(r)}),\mathbf{x}^{(r)})$ of these equations depends on the sample $(\mathbf{y}^{(r)}),\mathbf{x}^{(r)}$ used for constructing the maximum entropy estimates of the PDFs.

A.2 EXISTENCE OF IMPLICIT FUNCTION $\theta(\tilde{\mathbf{Y}}^{(R)},\mathbf{X}^{(R)})$

According to the balance equations (A.10), (A.1), their second term is the mean value of the noise in each measurement t. The noises and also their characteristics are often assumed to be independent of the measurements, i.e., in the expression (A.3),

$$\Xi_t = \Xi = [\xi^-, \xi^+]. \tag{A.12}$$

Therefore, the mean value of the noise is given by

$$\bar{\xi} = \frac{\mathcal{E}_t(\lambda_t(\mathbf{y}^{(r)}, \mathbf{x}^{(r)}))}{\mathcal{Q}_t\left(\lambda_t(\mathbf{y}^{(r)}, \mathbf{x}^{(r)})\right)} \in \Xi. \tag{A.13}$$

The balance equations can be written as

$$W_t(\lambda \,|\, \tilde{y}[t], \mathbf{x}^{(r)}) = \int_{\Theta} (\varphi(x[t], \theta) - \tilde{y}[t]) \times$$

$$\times \exp\left(-\sum_{j=1}^{r} \lambda_j(\tilde{\mathbf{y}}^{(r)}, \mathbf{x}^{(r)})\varphi(x[j], \theta)\right) d\theta = 0, \tag{A.14}$$

$$t = \overline{1, r},$$

where

$$\tilde{y}[t] = y[t] - \bar{\xi}, \qquad \tilde{\mathbf{y}}^{(r)} = \{\tilde{y}[1], \dots, \tilde{y}[r]\}. \tag{A.15}$$

In the vector form, equations (A.2) are described by

$$\mathbf{W}(\lambda \,|\, \tilde{\mathbf{y}}^{(r)}, \mathbf{x}^{(r)}) = \mathbf{0}. \tag{A.16}$$

Equation (A.16) defines an implicit function $\lambda(\tilde{\mathbf{y}}^{(r)}, \mathbf{x}^{(r)})$. The existence and properties of this implicit function depend on the properties of the Jacobian matrix

$$J_\lambda(\lambda \,|\, \mathbf{y}^{(r)}, \mathbf{x}^{(r)}) = \left[\frac{\partial W_t}{\partial \lambda_i}, \,|\, (t, i) = \overline{1, r}\right], \tag{A.17}$$

which has the elements

$$\frac{\partial W_t}{\partial \lambda_i} = \int_Q (\varphi(x[t], \theta) - \tilde{y}[t]) \, \varphi(x[i], \theta) \times$$

$$\times \sum_{j=1}^{r} \exp\left(-\sum_{j=1}^{r} \lambda_j \varphi(x[j], \theta)\right) d\theta. \tag{A.18}$$

Theorem 1 *Assume that:*

a) the function $\varphi(\mathbf{x}^{(r)}, \theta)$ is continuous in all variables;

b) for any $(\mathbf{x}^{(r)}, \tilde{\mathbf{y}}^{(r)}) \in R^r \times R^r$

$$\det J_{(\lambda}(\lambda \,|\, \tilde{\mathbf{y}}^{(r)}, \mathbf{x}^{(r)})) \neq 0, \tag{A.19}$$

$$\lim_{\|\lambda\| \to \infty} \mathbf{W}(\lambda \,|\, \tilde{\mathbf{y}}^{(r)}, \mathbf{x}^{(r)}) = \pm\infty. \tag{A.20}$$

Then there exists a unique implicit function $\lambda(\tilde{\mathbf{y}}^{(r)}, \mathbf{x}^{(r)},)$ *with the domain of definition* $R^r \times R^r$.

Proof.

Due to assumption *a)*, the continuous function $\mathbf{W}(\lambda \,|\, \mathbf{y}^{(r)}, \mathbf{x}^{(r)})$ induces the vector field $\Phi_{(\tilde{\mathbf{y}}^{(r)}, \mathbf{x}^{(r)})}(\lambda) = \mathbf{W}(\lambda \,|\, \tilde{\mathbf{y}}^{(r)}, \mathbf{x}^{(r)})$ in the space $R^r \times R^r$.

Choose an arbitrary vector $\mathbf{u}$ in R^r and define the vector field

$$\Pi_{\mathbf{u}}(\lambda) = \Phi_{(\tilde{\mathbf{y}}^{(r)}, \mathbf{x}^{(r)})}(\lambda) - \mathbf{u}.$$

By condition (A.20) the field $\Pi_{\mathbf{u}}(\lambda)$ with a fixed vector $\mathbf{u}$ has no zeros on the spheres $\|\lambda\| = \varrho$ of a sufficiently large radius ϱ.

Hence, a *rotation* is well-defined on the spheres $\|\lambda\| = \varrho$ of a sufficiently large radius ϱ. For details, see [101].

Consider the two vector fields

$$\Pi_{\mathbf{u}^{(1)}}(\lambda) = \Phi_{(\tilde{\mathbf{y}}^{(r)}, \mathbf{x}^{(r)})}(\lambda) - \mathbf{u}^{(1)}, \quad \Pi_{\mathbf{u}^{(2)}}(\lambda) = \Phi_{(\tilde{\mathbf{y}}^{(r)}, \mathbf{x}^{(r)})}(\lambda) - \mathbf{u}^{(2)}.$$

hese vector fields are *homotopic* on the spheres of a sufficiently large radius, i.e., the field

$$\Omega(\lambda) = \alpha\Pi_{\mathbf{u}^{(1)}}(\lambda) + (1-\alpha)\Pi_{\mathbf{u}^{(2)}}(\lambda) = \Phi_{(\tilde{\mathbf{y}}^{(r)}, \mathbf{x}^{(r)})}(\lambda) - [\alpha\mathbf{u}^{(1)} + (1-\alpha)\mathbf{u}^{(2)}]$$

has no zeros on the spheres of a sufficiently large radius for any $\alpha \in [0, 1]$. Homotopic fields have identical rotations [101]:

$$\gamma(\Pi_{\mathbf{u}^{(1)}}(\lambda)) = \gamma(\Pi_{\mathbf{u}^{(2)}}(\lambda)).$$

The vector fields $\Pi_{\mathbf{u}^{(1)}}(\lambda)$ and $\Pi_{\mathbf{u}^{(2)}}(\lambda)$ are nondegenerate on the spheres of a sufficiently large radius; in the ball $\|\lambda\| \leq \varrho_1 < \varrho$, however, each of them may have a number of singular points. Denote by $\kappa(\mathbf{u}^{(1)})$ and $\kappa(\mathbf{u}^{(2)})$ the numbers of singular points of the vector fields $\Pi_{\mathbf{u}^{(1)}}(\lambda)$ and $\Pi_{\mathbf{u}^{(2)}}(\lambda)$, respectively. As the vector fields are homotopic,

$$\kappa(\mathbf{u}^{(1)}) = \kappa(\mathbf{u}^{(2)}) = \kappa$$

In view of (A.19), these singular points are isolated.

Now utilize *the index* of a singular point introduced in [101]:

$$\text{ind}(\lambda^0) = (-1)^{\beta(\lambda^0)},$$

where $\beta(\lambda^0)$ is the number of eigenvalues of the matrix $\Pi'_{\mathbf{u}}(\lambda^0) = J_\lambda(\lambda^0 \,|\, \tilde{\mathbf{y}}^{(r)}, \mathbf{x}^{(r)})$ with the negative real part. According to the definition, the value of this index depends not on the magnitude of $\beta(\lambda^0)$, but on its parity. Due to condition (A.19), all singular points have the same parity. Really, $J_\lambda(\lambda^0 \,|\, \tilde{\mathbf{y}}^{(r)}, \mathbf{x}^{(r)}) \neq 0$, and hence for any $\tilde{\mathbf{y}}^{(r)}, \mathbf{x}^{(r)} \in R^r \times R^r$ the eigenvalues of the matrix $J_\lambda(\lambda^0 \,|\, \tilde{\mathbf{y}}^{(r)}, \mathbf{x}^{(r)})$ may move from the left half-plane to the right one in pairs only: real eigenvalues are transformed into pairs of complex-conjugate ones, passing through the imaginary axis.

In view of this fact, the rotation of the homotopic fields (A.2) is given by

$$\gamma(\Pi_{\mathbf{u}}) = \kappa(-1)^\beta,$$

is the number of eigenvalues of the matrix $\Pi'_{\mathbf{u}}(\lambda)$ for some singular point.

It remains to demonstrate that the vector field $\Pi_{\mathbf{u}}(\lambda)$ has a unique singular point in the ball $\|\lambda\| \leq \varrho_1 < \varrho$. Consider the equation

$$\Pi_{\mathbf{u}}(\lambda) = \Phi_{(\tilde{\mathbf{y}}^{(r)}, \mathbf{x}^{(r)})}(\lambda) - \mathbf{u} = 0.$$

Assume that for each fixed pair $(\ddot{\mathbf{y}}^{(r)}, \mathbf{x}^{(r)})$, this equation has κ singular points, i.e., the functions $\lambda^{(1)}(\tilde{\mathbf{y}}^{(r)}, \mathbf{x}^{(r)}), \ldots, \lambda^{(\kappa)}(\tilde{\mathbf{y}}^{(r)}, \mathbf{x}^{(r)})$. Therefore, it defines a multivalued function $\lambda(\tilde{\mathbf{y}}^{(r)}, \mathbf{x}^{(r)})$, whose κ branches are isolated. (The latter property follows from the isolatedness of the singular points.) Due to condition (A.19), each of the branches $\lambda^{(i)}(\tilde{\mathbf{y}}^{(r)}, \mathbf{x}^{(r)})$ defines an open set in the space R^r, and

$$\bigcup_{i=1}^{\kappa} \lambda^{(i)}(\tilde{\mathbf{y}}^{(r)}, \mathbf{x}^{(r)}) = R^r.$$

This is possible if and only if $\kappa = 1$. Hence, for each pair $\tilde{\mathbf{y}}^{(r)}, \mathbf{x}^{(r)}$ from $R^r \times R^r$ there exists a unique function $\lambda^*(\tilde{\mathbf{y}}^{(r)}, \mathbf{x}^{(r)})$ for which the function $\mathbf{W}(\lambda \,|\, \tilde{\mathbf{y}}^{(r)}, \mathbf{x}^{(r)})$ will vanish. ■

Theorem 2 *Under the assumptions of Theorem 1, the function* $\lambda(\mathbf{y}^{(r)}, \mathbf{x}^{(r)})$ *is analytical in all variables.*

Proof

From (A.13) it follows that the function $\mathbf{W}(\lambda \,|\, \mathbf{y}^{(r)}, \mathbf{x}^{(r)})$ is analytical in all variables. Therefore, the left-hand side of equation (A.13) can be expanded into the generalized Taylor series [100], and the solution can be constructed in the form of the generalized Taylor series as well. The power elements of this series are determined using a recursive procedure.

■

A.3 ASYMPTOTIC EFFICIENCY OF MAXIMUM ENTROPY ESTIMATES

MEE yields the entropy-optimal PDFs (A.8) for the arrays of input and output data, each of size r. For the sake of convenience, consider the PDFs parameterized by the exponential Lagrange multipliers $z = \exp(-\lambda)$. Then equalities (A.8) take the form

$$
P^*\left(\theta, \mathbf{z}(\mathbf{y}^{(r)}, \mathbf{x}^{(r)})\right) = \frac{\prod_{j=1}^{r}[z_j(\mathbf{y}^{(r)}, \mathbf{x}^r)]^{\varphi(x[j],\theta)}}{\int_{\Theta}\prod_{j=1}^{r}[z_j(\mathbf{y}^{(r)}, \mathbf{x}^r)]^{\varphi(x[j],\theta)}\,d\theta},
$$

$$
Q_t^*(\xi[t], z_t(\mathbf{y}^{(r)}, \mathbf{x}^{(r)})) = \frac{[z_t(\mathbf{y}^{(r)}, \mathbf{x}^{(r)})]^{\xi[t]}}{\int_{\Xi_t}[z_t(\mathbf{y}^{(r)}, \mathbf{x}^{(r)})]^{\xi[t]}\,d\xi[t]}, \tag{A.21}
$$

$$
t = \overline{1, r}. \tag{A.22}
$$

Consequently, the structure of the PDF significantly depends on the values of the exponential Lagrange multipliers $\mathbf{z}$, which in turn depend on the data arrays $\mathbf{y}^{(r)}$ and $\mathbf{x}^{(r)}$.

Definition 1 *The estimates $P^*(\theta, \mathbf{z}^*)$ and $Q_t^*(\xi[t], z_t^*)$ are said to be asymptotically efficient if*

$$
\lim_{r\to\infty} P^*\left(\theta, \mathbf{z}(\mathbf{y}^{(r)}, \mathbf{x}^{(r)})\right) = P^*(\theta, \mathbf{z}^*),
$$

$$
\lim_{r\to\infty} Q_t^*(\xi[t], z_t(\mathbf{y}^{(r)}, \mathbf{x}^{(r)})) = Q_t^*(\xi[t], z_t^*), \quad t = \overline{1, r}; \tag{A.23}
$$

where

$$
\mathbf{z}^* = \lim_{r\to\infty} \mathbf{z}(\boldsymbol{y}^{(r)}, \boldsymbol{x}^{(r)}). \tag{A.24}
$$

Definition 2 *The conditions*

$$
\frac{\partial \mathbf{z}}{\partial \mathbf{y}^{(r)}} = 0, \qquad \frac{\partial \mathbf{z}}{\partial \mathbf{x}^{(r)}} = 0 \tag{A.25}
$$

are necessary and sufficient for

$$
\mathbf{z}(\mathbf{y}^{(r)}, \mathbf{x}^{(r)}) = \mathbf{z}^*. \tag{A.26}
$$

Consider the empirical balance equations (A.19) written in terms of the exponential Lagrange multipliers:

$$\Phi_t(\mathbf{z}, \tilde{\mathbf{y}}^{(r)}, \mathbf{x}^r)) = \int_\Theta \prod_{j=1}^r [z_j(\tilde{\mathbf{y}}^{(r)}, \mathbf{x}^r)]^{\varphi(x[j],\theta)} \times$$
$$(\varphi(x[t],\theta) - \tilde{y}[t])\, d\theta = 0, \qquad (A.27)$$
$$t = \overline{1, r}.$$

As it has been demonstrated above, equations (A.3) define an implicit analytical function $\mathbf{z} = \mathbf{z}(\tilde{\mathbf{y}}^r), \mathbf{x}^{(r)})$ for $(\tilde{\mathbf{y}}^r), \mathbf{x}^{(r)}) \in R^r \times R^r$.

Differentiating the left- and right-hand sides of these equations with respect to $\tilde{\mathbf{y}}^{(r)}$ and $\mathbf{x}^{(r)}$ yields

$$\frac{\partial \mathbf{z}}{\partial \tilde{\mathbf{y}}^{(r)}} = -\left[\frac{\partial \Phi}{\partial \mathbf{z}}\right]^{-1} \frac{\partial \Phi}{\partial \tilde{\mathbf{y}}^{(r)}}, \qquad (A.28)$$

$$\frac{\partial \mathbf{z}}{\partial \mathbf{x}^{(r)}} = -\left[\frac{\partial \Phi}{\partial \mathbf{z}}\right]^{-1} \frac{\partial \Phi}{\partial \mathbf{x}^{(r)}}.$$

In the resulting equations, all matrices are square and have dimensions $(r \times r)$.

Then, passing to the norms gives the equalities

$$0 \le \left\|\frac{\partial \mathbf{z}}{\partial \tilde{\mathbf{y}}^{(r)}}\right\| \le \left\|\left[\frac{\partial \Phi}{\partial \mathbf{z}}\right]^{-1}\right\| \left\|\frac{\partial \Phi}{\partial \tilde{\mathbf{y}}^{(r)}}\right\|, \qquad (A.29)$$

$$0 \le \left\|\frac{\partial \mathbf{z}}{\partial \mathbf{x}^{(r)}}\right\| \le \left\|\left[\frac{\partial \Phi}{\partial \mathbf{z}}\right]^{-1}\right\| \left\|\frac{\partial \Phi}{\partial \mathbf{x}^{(r)}}\right\|.$$

Lemma 1 *Assume that a square matrix A is nonsingular, i.e.,* $\det A \ne 0$. *Then there exists a constant $\alpha > 1$ such that*

$$\frac{1}{\|A\|} \le \|A^{-1}\| \le \frac{\alpha}{\|A\|}. \qquad (A.30)$$

Proof
Due to the hypothesis of the lemma, the elements of the inverse matrix,

$$a_{ik}^{(-1)} = \frac{A_{ki}}{\det A}, \qquad (k, i) = \overline{1, r}$$

are bounded:

$$a_{ik}^{(-1)} \le M < \infty, \qquad \|A^{-1}\| < \infty.$$

Hence, there exists a constant $\alpha > 1$ for which inequality (A.30) is satisfied. ■

Lemma 1 can be applied to the norm $\| \left[\frac{\partial \Phi}{\partial \mathbf{z}} \right]^{-1} \|$ of the inverse matrix. As a result,

$$\left(\| \frac{\partial \Phi}{\partial \mathbf{z}} \| \right)^{-1} \leq \| \left[\frac{\partial \Phi}{\partial \mathbf{z}} \right]^{-1} \| \leq \alpha \left(\| \frac{\partial \Phi}{\partial \mathbf{z}} \| \right)^{-1} . \tag{A.31}$$

where

$$\| \frac{\partial \Phi}{\partial \mathbf{z}} \| = r \max_{t,j} | \frac{\partial \Phi_t}{\partial z_j} |. \tag{A.32}$$

Lemma 2 *Assume that*

$$\| \frac{\partial \Phi}{\partial \mathbf{y}^{(r)}} \| \leq \varrho < \infty, \qquad \| \frac{\partial \Phi}{\partial \mathbf{x}^{(r)}} \| \leq \omega < \infty. \tag{A.33}$$

Then

$$\lim_{r \to \infty} \| \frac{\partial \mathbf{z}}{\partial \mathbf{y}^{(r)}} \| = \lim_{r \to \infty} \| \frac{\partial \mathbf{z}}{\partial \mathbf{x}^{(r)}} \| = 0. \tag{A.34}$$

Proof
Follows from (A.30–A.33) and Lemma 1.

APPROXIMATE ESTIMATION OF STRUCTURAL CHARACTERISTICS OF LINEAR DYNAMIC REGRESSION MODEL (LDR)

Consider a linear dynamic regression model (LDR) of the form

$$\hat{x}[n] = a_0 + \sum_{k=1}^{p} a_k \hat{x}[n-k] + a_{p+1}f[n] + \varepsilon[n], \quad n = \overline{p, N}, \quad \text{(B.1)}$$

where $\hat{x}[n]$ and $f[n]$ denote the model's output and input, respectively; p is the order of this model; $\varepsilon[n]$ gives a measurement noise; finally, $a_k, k = \overline{0, p+1}$, are some model parameters.

Let

$$f[0], f[1], \ldots, f[N], \qquad x[0], x[1], \ldots, x[N] \qquad \text{(B.2)}$$

be some available data on the input $f[n]$ and output $x[n]$ of the object under study.

DOI: 10.1201/9781003306566-B

Using these data and equations (B.1), we write the following regression equation in the vector-matrix form:

$$\mathbf{x} = X_{(p+2)}\mathbf{a} + \varepsilon, \tag{B.3}$$

where:

- the vectors $\mathbf{x} = \{x[p], \ldots, x[N]\}$, $\mathbf{a} = \{a_0, \ldots, a_{(p+1)}\}$, and $\varepsilon = \{\varepsilon[p], \ldots, \varepsilon[N]\}$;

- the matrix

$$X_{(p+2)} = \begin{pmatrix} 1 & x[p-1] & x[p-2] & \cdots & x[0] & f[p] \\ 1 & x[p] & x[p-1] & \cdots & x[1] & f[p+1] \\ \cdots & \cdots & \cdots & \cdots & \cdots & \cdots \\ 1 & x[N-1] & x[N-2] & \cdots & x[N-p] & f[N] \end{pmatrix}. \tag{B.4}$$

The vector $\mathbf{x}$ and matrix $X_{(p+2)}$ consist of the data (B.2).

B.1 ORDER P

The order p of this model is chosen by calculating the autocorrelation function $r(k)$ for the variable x based on the data (B.2):

$$r(k) = \frac{c_k}{\sigma^2}, \quad c_k = \frac{1}{N+1} \sum_{n=0}^{N+1-k} (x[n] - \bar{x})(x[n+k] - \bar{x}), \tag{B.5}$$

$$\sigma^2 = \frac{1}{N} \sum_{n=0}^{N} x^2[n].$$

The order p is the maximum value k for which $r(k) > \delta = 0.1$.

B.2 PARAMETERS RANGES

The approximate intervals for the PDFs of the LDR parameters are found by calculating the LS estimates and sample variance of the residual deviations. The LS estimates for (B.3) have the form

$$\hat{\mathbf{a}} = \left(X_{(p+2)}^{\mathsf{T}} X_{(p+2)}\right)^{-1} X^{\mathsf{T}} \mathbf{x}. \tag{B.6}$$

The residual deviations are described by the vector

$$\mathbf{e}_{ls} = \mathbf{x} - X_{(p+2)}\hat{\mathbf{a}}, \tag{B.7}$$

and the sample variance of the residual deviations is given by

$$\sigma^2 = \frac{1}{N-p} \sum_{n=p}^{N} e^2[n]. \tag{B.8}$$

The standard deviations for the model parameters are

$$s_{a_k} = \sqrt{\sigma^2 Q_{kk}}, \quad k = \overline{0, p+1}, \tag{B.9}$$

where Q_{kk} denote the diagonal elements of the covariance matrix $Q = \left(X_{(p+2)}^{\mathsf{T}} X_{(p+2)} \right)^{-1}$.

In view of the asymptotic normality of the LS estimates, the parameters ranges are selected by the rule

$$\{a_k^-, a_k^+\} = \{a_k - 3s_{a_k}, a_k + 3s_{a_k}\}, \quad k = \overline{0, p+1}. \tag{B.10}$$

Bibliography

[1] Dataset.

[2] KEEL Dataset repository. `https://sci2s.ugr.es/keel/datasets.php`. Accessed: 2019-07-03.

[3] ISO/IEC 2382:2015 information technology—vocabulary., 2015. URL: `https://www.iso.org/obp/ui/#iso:std:iso-iec:2382:ed-1:v1:en` (Accessed April 6, 2017).

[4] Test functions for optimization, 2015. (Accessed November 2, 2015).

[5] Virtual library of simulation experiments: Test functions and datasets, 2015. (Accessed November 2, 2015).

[6] About CUDA, 2017. URL: `https://developer.nvidia.com/about-cuda` (Accessed April 6, 2017).

[7] Intel many integrated core architecture, 2017. URL: `http://www.intel.com/content/www/us/en/architecture-and-technology/many-integrated-core/intel-many-integrated-core-architecture.html` (Accessed April 6, 2017).

[8] D. Achlioptas. Database-friendly random projections. In *PODS'01*, pages 274–281. Amer. Math. Soc., 2001.

[9] N. Ailon and B. Chazelle. Approximate nearest neighbors and the fast johnson–lindenstrauss transform. In *Proceedings of the 38th Annual ACM Symposium on Theory of Computing*, pages 557–563, 2006.

[10] S. A. Aivazyan and V. S. Mkhitaryan. *Applied statistics. Classification and dimension reduction.* Finansy i Statistika, Moscow, 1989.

[11] S. A. Aivazyan and V. S. Mkhitaryan. *Applied statistics and foundations of econometrics*. Yuniti, 1998.

[12] M. A. Aizerman, E. M. Braverman, and L. I. Rozonoer. *The method of potential functions in the theory of machine learning.* Nauka, Moscow, 1970.

[13] L. F. Amaral, R. C. Souza, and M. Stevenson. A smooth transition periodic autoregressive (stpar) model for short-term load forecasting. *International Journal of Forecasting*, 24(4):603–615, 2008.

[14] A. G. Arkadiev and E. M. Braverman. *Image recognition and machine learning.* Nauka, Moscow, 1964.

[15] M. Avellaneda. Minimum-relative-entropy calibration of asset-pricing models. *International Journal of Theoretical and Applied Finance*, 1(04):447–472, 1998.

[16] A. N. Averkin, M. G. Gaaze-Rapoport, and D. A. Pospelov. *Defining dictionary on artificial intelligence.* Radio i Svyaz, 1992.

[17] N. S. Bakhvalov, N. P. Zhidkov, and G. M. Kobel'kov. *Numerical methods.* BINOM, 2003.

[18] A. S. Bernstein, I. V. Pyshkin, Yu. S. Popkov, and R. G. Faradzhev. Analytic description of binary nonlinear sequential machines. *Automation and Remote Control*, 32(12):1920–1927, 1971.

[19] K. Beyer, J. Goldstein, R. Ramakrishnan, and U. Shaft. When is "nearest neighbor" meaningful? In *International conference on database theory*, pages 217–235. Springer, 1999.

[20] E. Bingham and H. Mannila. Random projection in dimensionality reduction: applications to image and text data. In *Proceedings of the 7th ACM SIGKDD International Conference on Knowledge Discovery and Data Mining*, pages 245–250. ACM, 2001.

[21] C. Bishop. *Pattern recognition and machine learning (Information Science and Statistics), 1st ed. 2006. corr. 2nd ed.* Springer, New York, 2007.

[22] L. Breiman. Bagging predictors. *Machine Learning*, 24(2):123–140, 1996.

[23] L. Breiman, J. H. Friedman, R. A. Olshen, and C. J. Stone. *Classification and regression trees. 1993*. Chapman Hall, New York, 1984.

[24] NA Bryksina and Yu M Polishchuk. Analysis of changes in the number of thermokarst lakes in permafrost of western siberia on the basis of satellite images. *Earth Cryosphere*, 19:100–105, 2015.

[25] P. Bühlmann and T. Hothorn. Boosting algorithms: Regularization, prediction and model fitting. *Statistical Science*, pages 477–505, 2007.

[26] L. Buitinck, G. Louppe, M. Blondel, F. Pedregosa, A. Mueller, O. Grisel, V. Niculae, P. Prettenhofer, A. Gramfort, R. Grobler, J. Layton, J. VanderPlas, A. Joly, B. Holt, and G. Varoquaux. API design for machine learning software: experiences from the scikit-learn project. In *ECML PKDD Workshop: Languages for Data Mining and Machine Learning*, pages 108–122, 2013.

[27] R. Calderbank, S. Jafarpour, and R. Schapire. Compressed learning: Universal sparse dimensionality reduction and learning in the measurement domain. Technical report, 2009.

[28] P. Campbell. Editorial on special issue on big data: Community cleverness required. *Nature*, 455(7209):1, 2008.

[29] E. J. Candes and T. Tao. Near-optimal signal recovery from random projections: Universal encoding strategies? *IEEE Transactions on Information Theory*, 52(12):5406–5425, 2006.

[30] M. S. Charikar. Similarity estimation techniques from rounding algorithms. In *Proceedings of the 34th Annual ACM Symposium on Theory of Computing*, pages 380–388, 2002.

[31] S. Chib and E. Greenberg. Understanding the metropolis–hastings algorithm. *The American Statistician*, 49(4):327–335, 1995.

[32] S. Chib and I. Jeliazkov. Marginal likelihood from the metropolis–hastings output. *Journal of the American Statistical Association*, 96(453):270–281, 2001.

[33] T. Cover and P. Hart. Nearest neighbor pattern classification. *IEEE Transactions on Information Theory*, 13(1):21–27, 1967.

[34] H. Cramer. *Mathematical methods of statistics*. Princeton University Press, 1962.

[35] B. S. Darkhovskii, A. Yu. Popkov, and Yu. S. Popkov. Monte carlo method of batch iterations: Probabilistic characteristics. *Automation and Remote Control*, 76(5):776–785, 2015.

[36] B. S. Darkhovsky, Yu. S. Popkov, A. Yu. Popkov, and A. S. Aliev. Method of generating random vectors with a given probability density function. *Automation and Remote Control*, 79(9):1569–1581, 2018.

[37] B. V. Dasarathy. Nearest neighbor ({NN}) norms:{NN} pattern classification techniques. 1991.

[38] A. Dasgupta, R. Kumar, and T. Sarlos. A sparse johnson–lindenstrauss transform. In *Proceedings of the 42nd ACM Symposium on Theory of Computing*, pages 341–350, 2010.

[39] W. Davis and D. Yen. *The information system consultant's handbook. Systems analysis and design*. CRC Press, 1998.

[40] J. De Beer and L. van Wissen. Europe: one continent, different worlds. population scenarios for the 21st century. Technical report, ISBN 0-7923-5840-6, 1999.

[41] A. P. Dempster, N. M. Laird, and D. B. Rubin. Maximum likelihood from incomplete data via the em algorithm. *Journal of the Royal Statistical Society. Series B (Methodological)*, pages 1–38, 1977.

[42] V. A. Ditkin and A. P. Prudnikov. *Operational calculus in two variables*. Fizmatgiz, Moscow, 1953.

[43] V. A. Ditkin and A. P. Prudnikov. *Integral transformations and operational calculus*. Fizmatgiz, Moscow, 1961.

[44] P. Domingos and M. Pazzani. On the optimality of the simple bayesian classifier under zero-one loss. *Machine Learning*, 29(2):103–130, 1997.

[45] J. J. Dongarra. *Sourcebook of parallel computing*. Morgan Kaufmann Publishers, San Francisco, 2003.

[46] D. L. Donoho. Compressed sensing. *IEEE Transactions on Information Theory*, 52(4):1289–1306, 2006.

[47] E. B. Dynkin. *Theory of Markov processes*. Prentice-Hall, 1961.

[48] W. Feller. *An introduction to probability theory and its applications. Vol. 2*. John Wiley & Sons, 2nd ed. edition, 1971.

[49] G. Fiedner. Hierarchical forecasting: Issues and use guidelines. *Industrial Management and Data Systems*, 101(1):5–12, 2001.

[50] A. F. Filippov. *Introduction to theory of differential equations*. URSS, 2010.

[51] R. A. Fisher. Two new properties of mathematical likelihood. *Proceedings of the Royal Society of London. Series A, Containing Papers of a Mathematical and Physical Character*, 144(852):285–307, 1934.

[52] R. A. Fisher. The use of multiple measurements in taxonomic problems. *Annals of Human Genetics*, 7(2):179–188, 1936.

[53] P. Flach. *Machine learning: the art and science of algorithms that make sense of data*. Cambridge University Press, 2012.

[54] J. W. Forrester et al. *World dynamics*. Waltham, MA, 2nd ed. edition, 1973.

[55] W. J. Frawley, G. Piatetsky-Shapiro, and C. J. Matheus. Knowledge discovery in databases: An overview. *AI magazine*, 13(3):57, 1992.

[56] Y. Freund, R. Schapire, and N. Abe. A short introduction to boosting. *Journal of Japanese Society For Artificial Intelligence*, 14(771-780):1612, 1999.

[57] J. Friedman, T. Hastie, and R. Tibshirani. *The elements of statistical learning: Springer Series in Statistics*, volume 1. Springer series in statistics Springer, Berlin, 2001.

[58] M. C. Fu, J. Hu, and S. I. Marcus. Model-based randomized methods for global optimization. In *Proceedings of the 17th international symposium on mathematical theory of networks and systems*, pages 355–363, 2006.

[59] E. Gamma, R. Helm, R. Johnson, and J. Vlissides. *Design patterns: Elements of reusable object-oriented software.* Addison-Wesley, USA, 1994.

[60] F. R. Gantmakher. Matrix theory, 1966.

[61] I. M. Gelfand, R. A. Silverman, et al. *Calculus of variations.* Courier Corporation, 2000.

[62] V. P. Gergel and R. G. Strongin. *Foundations of parallel computing for multiprocessor computer systems.* Lobachevsky State University press, Nizhny Novgorod, 2003.

[63] W. Gerstner and W. Kistler. Spiking neuron models, 2002.

[64] J. Geweke and H. Tanizaki. Note on the sampling distribution for the metropolis–hastings algorithm. *Communications in Statistics—Theory and Methods,* 32(4):775–789, 2003.

[65] I. I. Gikhman and A. V. Skorokhod. *Introduction to theory of random processes.* Nauka, 1977.

[66] A. Golan et al. Information and entropy econometrics—a review and synthesis. *Foundations and Trends® in Econometrics,* 2(1–2):1–145, 2008.

[67] A. Golan, G. Judge, and D. Miller. *Maximum entropy econometrics: robust estimation with limited data.* John Wiley & Sons, New York, 1996.

[68] A. V. Gorokhov and V. A. Putilov. *System dynamics of regional development.* Pazori, 2002.

[69] A. Grama, A. Gupta, and G. Karypis. *Introduction to parallel computing.* Addison-Wesley, Harlow, England, 2nd ed. edition, 2003.

[70] T. Hastie, R. Tibshirani, and J. Friedman. *The Elements of statistical learning: data mining, inference, and prediction.* Springer New York, 2009.

[71] T. Hong, P. Pinson, S. Fan, H. Zareipour, A. Troccoli, and R. J. Hyndman. Probabilistic energy forecasting: Global energy forecasting competition 2014 and beyond. *International Journal of Forecasting,* 32(3):896–913, 2016.

[72] J. Hopcroft, R. Motwani, and J. Ullman. *Introduction to automata theory, languages, and computation.* Addison-Wesley, 3rd ed. edition, 2007.

[73] David W. Hosmer J., S. Lemeshow, and R. X. Sturdivant. *Applied logistic regression*, volume 398. John Wiley & Sons, 2013.

[74] P. J. Huber. *Robust statistics.* Springer, 2011.

[75] L. Hyafil and R. L. Rivest. Constructing optimal binary decision trees is np-complete. *Information Processing Letters*, 5(1):15–17, 1976.

[76] P. Indyk and R. Motwani. Approximate nearest neighbors: towards removing the curse of dimensionality. In *Proceedings of the 30th Annual ACM Symposium on Theory of Computing*, pages 604–613, 1998.

[77] A. D. Ioffe and V. M. Tikhomirov. Theory of extremum problems. 1974.

[78] I. Jacobson, G. Booch, and J. Rumbaugh. *The unified software development process.* Addison-Wesley Longman, USA, 1999.

[79] A. K. Jain, R. P. W. Duin, and J. Mao. Statistical pattern recognition: A review. *IEEE Transactions on Pattern Analysis and Machine Intelligence*, 22(1):4–37, 2000.

[80] E. T. Jaynes. Information theory and statistical mechanics. *Physical Review*, 106(4):620–630, 1957.

[81] E. T. Jaynes. *Probability theory: the logic of science.* Cambridge University Press, 2003.

[82] T. S. Jayram and D. P. Woodruff. Optimal bounds for johnson–lindenstrauss transforms and streaming problems with subconstant error. *ACM Transactions on Algorithms (TALG)*, 9(3):1–17, 2013.

[83] W. B. Johnson and J. Lindenstrauss. Extensions of lipshitz mapping into hilbert space. In *Modern Analysis and Probability*, volume 26, pages 189–206. Amer. Math. Soc., 1984.

[84] I. T. Jolliffe. Principal component analysis and factor analysis. In *Springer Series of Statistics.* Springer, New York, 2002.

[85] M. A. Kaashoek, S. Seatzu, and C. van der Mee. *Recent advances in operator theory and its applications: The Israel Gohberg anniversary volume*, volume 160. Springer Science & Business Media, 2006.

[86] D. Kane, R. Meka, and J. Nelson. Almost optimal explicit johnson–lindenstrauss families. In *Approximation, randomization, and combinatorial optimization. Algorithms and techniques*, pages 628–639. Springer, 2011.

[87] D. M. Kane and J. Nelson. Sparser johnson–lindenstrauss transforms. *Journal of the ACM (JACM)*, 61(1):1–23, 2014.

[88] J. N. Kapur. *Maximum-entropy models in science and engineering*. John Wiley & Sons, 1989.

[89] Johanna Mård Karlsson, Steve W Lyon, and Georgia Destouni. Temporal behavior of lake size-distribution in a thawing permafrost landscape in northwestern siberia. *Remote sensing*, 6(1):621–636, 2014.

[90] M. Kendall and A. Stuart. *The advanced theory of statistics. Vol. 2: Inference and relationship*. Griffin, London, 1977.

[91] N. P. Kirillov. Conceptual models of technical systems with controllable states (a survey and analysis). *Artificial Intelligence and Decision Making*, 4:81–91, 2011.

[92] S.N. Kirpotin, Yu Polishchuk, and N. Bryksina. Abrupt changes of thermokarst lakes in western siberia: impacts of climatic warming on permafrost melting. *International Journal of Environmental Studies*, 66(4):423–431, 2009.

[93] M. V. Kiselev, V. S. Pivovarov, and M. M. Shmulevich. A text clustering method considering the co-occurrence of key terms, with application to the dynamical analysis of thematic structure of news flow. 2005.

[94] A. N. Kolmogorov. On representation of continuous multivariable functions using superpositions of continuous multivariable functions of smaller dimension. 114(5):953–956, 1957.

[95] A.N. Kolmogorov and S. V. Fomin. *Elements of theory of functions and functional analysis*, volume 1. Fizmatlit, Moscow, 1999.

[96] L. G. Komartsova and A. V. Maksimov. *Neurocomputers.* Bauman Moscow State Technical University Press, 2004.

[97] D. Kraft. A software package for sequential quadratic programming. Technical Report DFVLR-FB 88-28, DLR German Aerospace Center—Institute for Flight Mechanics, Koln, Germany, 1988.

[98] S. G. Krantz. *A handbook of real variables: with applications to differential equations and Fourier analysis.* Springer Science & Business Media, 2011.

[99] G. E. Krasner, S. T. Pope, et al. A description of the model-view-controller user interface paradigm in the smalltalk-80 system. *Journal of Object Oriented Programming*, 1(3):26–49, 1988.

[100] M. A. Krasnoselskii, G. M. Vainikko, P. P. Zabreiko, Ya. B. Rutitskii, and V. Ya. Stetsenko. *Approximate solution of operator equations.* Nauka, Moscow, 1969.

[101] M. A. Krasnosel'skii and P. P. Zabreiko. *Geometrical methods of nonlinear analysis*, volume 263. Springer, 1984.

[102] S. Kullback and R. A. Leibler. On information and sufficiency. *The Annals of Mathematical Statistics*, 22(1):79–86, 1951.

[103] S. D. Kuznetsov. *Great Russian Encyclodepia*, volume 11, chapter Information technology, page 493. 2008.

[104] M. B. Lagutin. Visual mathematical statistics. 2013.

[105] J. Langford. Tutorial on practical prediction theory for classification. *Journal of Machine Learning Research*, 6(Mar):273–306, 2005.

[106] A. V. Lapko, S. V. Chentsov, S. I. Krokhov, and L. A. Fel'dman. *Learning systems for data processing and decision making.* Nauka, Novosibirsk, 1996.

[107] M. V. Lawson. *Finite automata.* CRC Press, 2003.

[108] M. Ledoux. *The concentration of measure phenomenon.* Number 89. American Mathematical Soc., 2001.

[109] R. D. Levine and M. Tribus. Maximum entropy formalism. In *Maximum Entropy Formalism Conference (1978: Massachusetts Institute of Technology)*. MIT Press, 1979.

[110] A. V. Levitin. *Algorithms: Introduction to design and analysis*. Vil'yams, 2006.

[111] P. Lévy. Specific problems of functional analysis. 1967.

[112] Qianqian Liu, Mark D. Rowe, Eric J. Anderson, Craig A. Stow, Richard P. Stumpf, and Thomas H. Johengen. Probabilistic forecast of microcystin toxin using satellite remote sensing, in situ observations and numerical modeling. *Environmental Modelling & Software*, 128:104705, 2020.

[113] Miska Luoto and Matti Seppala. Thermokarst ponds as indicators of the former distribution of palsas in finnish lapland. *Permafrost and periglacial processes*, 14(1):19–27, 2003.

[114] L. A. Lusternik and V. I. Sobolev. Elements of functional analysis. 1965.

[115] J. R. Magnus and H. Neudecker. Matrix differential calculus with applications in statistics and econometrics. *Wiley series in probability and mathematical statistics*, 1988.

[116] O. Maillard and R. Munos. Compressed least-squares regression. In *Advances in neural information processing systems*, pages 1213–1221, 2009.

[117] J. K. Martin. An exact probability metric for decision tree splitting and stopping. *Machine Learning*, 28(2):257–291, 1997.

[118] A. B. Merkov. *Image recognition. Introduction to statistical learning methods*. URSS, Moscow, 2010.

[119] M. D. Mesarovic, D. Macko, and Y. Takahara. *Theory of hierarchical, multilevel, systems*. Academic Press, 1970.

[120] N. Metropolis and S. Ulam. The monte carlo method. *Journal of the American Statistical Association*, 44(247):335–341, 1949.

[121] D. D. L. Minh and Do Le Minh. Understanding the hastings algorithm. *Communications in Statistics—Simulation and Computation*, 44(2):332–349, 2015.

[122] M. Minsky and S. Papert. *Perceptrons: An introduction to computational geometry.* MIT Press, Cambridge, 1969.

[123] J. Neyman. Outline of a theory of statistical estimation based on the classical theory of probability. *Philosophical Transactions of the Royal Society of London. Series A, Mathematical and Physical Sciences*, 236(767):333–380, 1937.

[124] A. V. Oppenheim. *Discrete-time signal processing.* Pearson Education India, 1999.

[125] J. de D. Ortuzar and L.G. Willumsen. *Modelling Transport.* Wiley, 1995.

[126] G. S. Osipov. *Artificial intelligence methods.* Fizmatlit, 2011.

[127] K. Pearson. On lines and planes of closest fit to systems of points in space. *Philosophical Magazine*, 2(6):559–572, 1901.

[128] F. Pedregosa, G. Varoquaux, A. Gramfort, V. Michel, B. Thirion, O. Grisel, M. Blondel, P. Prettenhofer, R. Weiss, V. Dubourg, J. Vanderplas, A. Passos, D. Cournapeau, M. Brucher, M. Perrot, and E. Duchesnay. Scikit-learn: Machine learning in Python. *Journal of Machine Learning Research*, 12:2825–2830, 2011.

[129] Michiel Pezij, Denie CM Augustijn, Dimmie MD Hendriks, and Suzanne JMH Hulscher. Applying transfer function-noise modelling to characterize soil moisture dynamics: a data-driven approach using remote sensing data. *Environmental Modelling & Software*, 131:104756, 2020.

[130] O.S. Pokrovsky, L.S. Shirokova, S.N. Kirpotin, S. Audry, J. Viers, and B. Dupre. Effect of permafrost thawing on organic carbon and trace element colloidal speciation in the thermokarst lakes of western siberia. *Biogeosciences*, 8(3):565–583, 2011.

[131] V Polishchuk and Yu Polishchuk. Modeling and prediction of dynamics of thermokarst lake fields using satellite images. In *Permafrost: Distribution, Composition and Impacts on Infrastructure and Ecosystems*, pages 205–234. Nova Science Publisher, NY, 2014.

[132] YuM Polishchuk, AN Bogdanov, IN Muratov, VYu Polishchuk, A Lim, RM Manasypov, LS Shirokova, and OS Pokrovsky. Minor contribution of small thaw ponds to the pools of carbon and

methane in the inland waters of the permafrost-affected part of the western siberian lowland. *Environmental Research Letters*, 13(4):045002, 2018.

[133] Yury M. Polishchuk, Alexander N. Bogdanov, Vladimir Yu Polishchuk, Rinat M. Manasypov, Liudmila S. Shirokova, Sergey N. Kirpotin, and Oleg S. Pokrovsky. Size distribution, surface coverage, water, carbon, and metal storage of thermokarst lakes in the permafrost zone of the western siberia lowland. *Water*, 9(3):228, 2017.

[134] B. T. Polyak. *Introduction to optimization*. Nauka, 1983.

[135] B. T. Polyak and E. N. Gryazina. Hit-and-run: new design technique for stabilization, robustness and optimization of linear systems. *IFAC Proceedings Volumes*, 41(2):376–380, 2008.

[136] B. T. Polyak and M. V. Khlebnikov. Principle component analysis: Robust versions. *Automation and Remote Control*, 78(3):490–506, 2017.

[137] A. D. Polyanin and A. V. Manzhirov. *Handbook of integral equations*. CRC press, 2008.

[138] Yu. Popkov and A. Popkov. New methods of entropy-robust estimation for randomized models under limited data. *Entropy*, 16(2):675–698, 2014.

[139] Yu. S. Popkov. *Macrosystems theory and its applications (Lecture Notes in Control and Information Sciences, vol. 203)*. Springer, 1995.

[140] Yu. S. Popkov. On convergence of direct multiplicative methods in constrained entropy maximization problems. *Automation and Remote Control*, 59(4):534–544, 1998.

[141] Yu. S. Popkov. *Theory of macrosystems. Equilibrium models*. URSS, Moscow, 1999.

[142] Yu. S. Popkov. New class of multiplicative algorithms for solving of entropy-linear programs. *European Journal of Operational Research*, 174(3):1368–1379, 2006.

[143] Yu. S. Popkov. *Theory of macrosystems. Equilibrium models.* URSS, Moscow, 2012.

[144] Yu. S. Popkov, Yu. A. Dubnov, and A. Yu. Popkov. New method of randomized forecasting using entropy-robust estimation: Application to the world population prediction. *Mathematics*, 4(1):1–16, 2016.

[145] Yu. S. Popkov, Yu. A. Dubnov, and A. Yu. Popkov. Randomized machine learning: Statement, solution, applications. In *Intelligent Systems (IS), 2016 IEEE 8th International Conference on*, pages 27–39. IEEE, 2016.

[146] Yu. S. Popkov, O. N. Kiselev, N. P. Petrov, and B. L. Shmulyan. *Identification and optimization of nonlinear stochastic systems.* Energiya, 1976.

[147] Yu. S. Popkov, A. Yu. Popkov, and B. S. Darkhovsky. Parallel monte carlo for entropy robust estimation. *Mathematical Models and Computer Simulations*, 8(1):27–39, 2016.

[148] Yu. S. Popkov, A. Yu. Popkov, and Yu. A. Dubnov. *Randomized machine learning: from empirical probability to entropy randomization.* LENAND, Moscow, 2019.

[149] Yu. S. Popkov, A. Yu. Popkov, and Yu. N. Lysak. Estimation of characteristics of randomized static models of data (entropy-robust approach). *Automation and Remote Control*, 74(11):1863–1877, 2013.

[150] Yu. S. Popkov, Z. Volkovich, Yu. A. Dubnov, R. Avros, and E. Ravve. Entropy 2-soft classification of objects. *Entropy*, 19(4):178, 2017.

[151] Yuri S. Popkov, Vladimir Y. Polishchuk, Evgeny S. Sokol, Yury M. Polishchuk, and Andrey V. Melnikov. A randomized algorithm for restoring missing data in the time series of lake areas using information on climatic parameters. In *8th Scientific Conference on Information Technologies for Intelligent Decision Making Support (ITIDS 2020)*, pages 186–190. Atlantis Press, 2020.

[152] K. A. Pupkov, V. I. Kapalin, and A. S. Yushchenko. *Functional series in theory of nonlinear systems.* Nauka, 1976.

[153] J. S. Racine and E. Maasoumi. A versatile and robust metric entropy test of time-reversibility, and other hypotheses. *Journal of Econometrics*, 138(2):547–567, 2007.

[154] Brian Riordan, David Verbyla, and A. David McGuire. Shrinking ponds in subarctic alaska based on 1950–2002 remotely sensed images. *Journal of Geophysical Research: Biogeosciences*, 111(G4), 2006.

[155] I. Rish. An empirical study of the naive bayes classifier. In *IJCAI 2001 Workshop on Empirical Methods in Artificial Intelligence*, volume 3, pages 41–46. IBM, 2001.

[156] F. Rosenblatt. *The perceptron, a perceiving and recognizing automaton Project Para*. Cornell Aeronautical Laboratory, 1957.

[157] R. D. Rosenkrantz. *E.T. Jaynes: Papers on probability, statistics, and statistical physics*. Kluwer Academic Publishers, 1989.

[158] R. Y. Rubinstein and D. P. Kroese. The cross-entropy method: a unified approach to combinatorial optimization, monte-carlo simulation and machine learning, 2004.

[159] R. Y. Rubinstein and D. P. Kroese. *Simulation and the Monte Carlo method*, volume 707. John Wiley & Sons, 2007.

[160] G. Salton and M. J. McGill. Introduction to modern information retrieval. 1986.

[161] A. A. Samarskii and A. P. Mikhailov. Mathematical modeling. 1997.

[162] L. K. Saul and S. T. Roweis. Think globally, fit locally: unsupervised learning of low dimensional manifolds. *Journal of Machine Learning Research*, 4:119–155, 2003.

[163] M. Sewell. Ensemble learning. *RN*, 11(02), 2008.

[164] L. F. Shampine. Matlab program for quadrature in 2d. *Applied Mathematics and Computation*, 202(1):266–274, 2008.

[165] C. E. Shannon. Communication theory of secrecy systems. *Bell Labs Technical Journal*, 28(4):656–715, 1949.

[166] A. N. Shiryaev. *Stochastic financial mathematics*. Nauka, Moscow, 2002.

[167] A. S. Sigachev. Model of text as a set of numerical attributes. *Intellectual Technologies and Systems*, 16:127–135, 2006.

[168] I. M. Sobol. *Numerical Monte Carlo methods*. Nauka, Moscow, 1973.

[169] V. D. Soloviev, B. V. Dobrov, V. V. Ivanov, and N. V. Lukashevich. Ontologies and thesauri: A tutorial, 2006.

[170] G. Solton. Dynamic library information systems, 1979.

[171] A. S. Strekalovskii. *Elements of nonconvex optimization*. Nauka, Novosibirsk, 2003.

[172] R. G. Strongin, V. P. Gergel, V. A. Grishagin, and K. A. Barkalov. Parallel computing in global optimization problems. *Moscow State University Press*, 2013.

[173] R. G. Strongin and Ya. D. Sergeyev. *Global optimization with non-convex constraints: Sequential and parallel algorithms*. Springer Science & Business Media, 2000.

[174] V. M. Tikhomirov, V. N. Alekseev, and S. V. Fomin. *Optimal control*. Nauka, Moscow, 1979.

[175] I.G. Tsyganov. Neural network methods for automated analysis of information flows in real time. *Proceedings of International Youth Scientific and Technical Conference "Science Intensive Technologies and Intelligent Systems"*, pages 19–24, 2002.

[176] Ya. Z. Tsypkin. *Foundations of the theory of learned systems*. Nauka, Moscow, 1970.

[177] Ya. Z. Tsypkin and Yu. S. Popkov. Theory of nonlinear impulsive systems. 1973.

[178] H. L. van Trees. *Synthesis of optimum nonlinear control systems*. MIT press, 1962.

[179] V. N. Vapnik and A. Ya. Chervonenkis. *Theory of image recognition*. Nauka, Moscow, 1974.

[180] V. N. Vapnik and A. Ya. Chervonenkis. *Restoration of relations using empirical data*. Nauka, Moscow, 1979.

[181] S. S. Vempala. *The random projection method*, volume 65. American Mathematical Soc., 2005.

[182] S. Vine. *Options: A complete course for professionals*. Alpina Publishers, 2015.

[183] V. V. Voevodin and Vl. V. Voevodin. *Parallel computing*. BKhV-Peterburg, St. Petersburg, 2004.

[184] V. Volterra. *Theory of functionals and integrals and integro-differential equations*. Blackie, 1931.

[185] V. Volterra, J. Pérès. *Théorie générale des fonctionnelles* Tome 1. Généralités sur les fonctionnelles. Théorie des équations intégrales. Gauthier-Villars, 1936.

[186] J. von Neumann. Various techniques used in connection with random digits. *Appl. Math Ser.*, 12:36–38, 1951.

[187] K. V. Vorontsov. Machine learning (a course of lectures), 2013. URL=http://www.machinelearning.ru.

[188] K. V. Vorontsov. Mathematical methods of learning by precedents. a course of lectures (faculty of control and applied mathematics, moscow institute of physics and technology), 2013.

[189] W. Weidlich and G. Haag. *An integrated model of transport and urban evolution: with an application to a metropole of an emerging nation*. Springer, 1999.

[190] K. Q. Weinberger and L. K. Saul. Unsupervised learning of image manifolds by semidefinite programming. *International Journal of Computer Vision*, 70(1):77–90, 2006.

[191] I. H. Witten and E. Frank. *Data Mining: Practical machine learning tools and techniques*. Morgan Kaufmann, 2005.

[192] D. H. Wolpert. Stacked generalization. *Neural Networks*, 5(2):241–259, 1992.

[193] X. Wu, V. Kumar, J. R. Quinlan, J. Ghosh, Q. Yang, H. Motoda, G. J. McLachlan, A. Ng, B. Liu, S. Yu. Philip, et al. Top 10 algorithms in data mining. *Knowledge and Information Systems*, 14(1):1–37, 2008.

[194] N. G. Zagoruiko. *Applied methods of data and knowledge analysis.* Sobolev Institute of Mathematics, Novosibirsk, 1999.

[195] M.N. Zheleznyak. *Geotemperature field and permafrost in the southeast of the Siberian platform.* Nauka, Novosibirsk, 2005.

[196] A. Zhigljavsky and A. Zilinskas. *Stochastic global optimization*, volume 9. Springer Science & Business Media, 2007.

[197] E. G. Zhilyakov. Quatization errors in estimation of autocorrelation functions of gaussian processes. *Problems of Information Transmission*, 18(3):90–94, 1982.

[198] Yu. I. Zhuravlev, V. V. Ryazanov, and O. V. Sen'ko. *Image recognition: mathematical methods, program system, applications.* FAZIS, 2006.

[199] A. V. Zorin and M. A. Fedotkin. *Monte Carlo methods for parallel computing.* Moscow State University Press, 2013.

[200] Frieda S Zuidhoff and Else Kolstrup. Changes in palsa distribution in relation to climate change in laivadalen, northern sweden, especially 1960–1997. *Permafrost and periglacial processes*, 11(1):55–69, 2000.

Index

Note: Locators in *italics* represent figures and **bold** indicate tables in the text.